The Greater Genius?

Sir Marc Isambard Brunel, 1813
by James Northcote RA
National Portrait Gallery, London

The Greater Genius?

A biography of
Marc Isambard Brunel

by

HAROLD BAGUST

Ian Allan
PUBLISHING

First published 2006

ISBN (10) 0 7110 3175 4
ISBN (13) 978 0 7110 3175 3

Published by Ian Allan Publishing

an imprint of Ian Allan Publishing Ltd, Hersham, Surrey KT12 4RG.
Printed in England by Ian Allan Printing Ltd, Hersham, Surrey KT12 4RG.

Code: 0604/B

Visit the Ian Allan Publishing website at www.ianallanpublishing.com

Contents

Acknowledgements 5
Foreword by Professor Emeritus Heinz Wolff 7
Introduction 9

1. French Origins 11
2. America 17
3. England 22
4. Blockmaking at Portsmouth 32
5. Battersea and Financial Problems .36
6. Woolwich 40
7. The Boot Factory at Battersea 44
8. Chatham Dockyard 47
9. Disasters Leading to the Debtors' Prison 51
10. Back in Harness 58
11. The Tunnel — the Rotherhithe Shaft and the Great Shield 62
12. Tunnelling — the First Phase 71
13. More Problems 77
14. The Long Pause 81
15. Alternative Interests 84
16. Tunnelling Resumed 87
17. Epilogue 97

Appendices

1. Certificate of Citizenship — United States of America 101
2. Letter to the Navy Board, 2 May 1817 102
3. The Footway Bridge at Liverpool 103
4. Letter to the Thames Tunnel Directors dated 22 January 1828 104
5. Letter to Mr Coxson on the Advantages of Steam when Crossing to Ireland . . .105
6. Letter Concerning Postal Services and Roads 106
7. Robespierre 107
8. Taylors of Southampton 108
9. The Dockyard and Town of Portsmouth 111
10. Maudslay and Bramah 113
11. 1814 Report on Prisons 114
12. Patents Filed by Marc Brunel 122
13. *The Works at the Head of the Tunnel now in progress under the River Thames* 148
14. *The Banquet in the Thames Tunnel, 1827* 149
15. The Brunel Watercolour of 1835 150

Marc Brunel (1769-1849) — Comparative Chronology 151
Bibliography 155
Useful Addresses 157
Index 158

The Greater Genius?

This is a biography of Sir Marc Brunel, the father of Isambard Kingdom Brunel, whose genius is considered by many to be the direct result of his father's influence and earlier experiments.

It was Sir Marc who first used the 7ft railway gauge. It was he who designed the first bascule bridge, decades before Tower Bridge was constructed. He was responsible for mass-produced pulley blocks for Nelson's Navy and boots for Wellington's Army. Most importantly it was he who designed and built the first tunnel under a river in the world.

He filed 18 patents and was responsible for many of the designs which his famous son developed and has been credited with.

This book provides the facts — it is left to the reader to decide who was the greater genius.

'. . . I met the head of the house . . . I had liked the son [Isambard], but at our very first meeting I could not help feeling that his father far excelled him in originality, unworldliness, genius, and taste.'

Charles Macfarlane, 1829

Acknowledgements

More than six years of research have gone into this book, during which time I have received assistance from so many sources that it is possible I may have omitted some names in the following list. If this is so, I hope they will accept my sincere apologies for any unintentional omissions. I have made no attempt to put them in either alphabetical or chronological order, neither are they in order of importance to the narrative — I am sincerely grateful for *all* the help I have received from whatever source.

I am particularly indebted to Judi and Alan Pike of the Patent Office, Newport, for obtaining Marc Brunel's English patents. Thanks to their efforts, detailed specifications of these are published in book form for the first time in this volume.

The owner of *Ferme Brunel* in Hacqueville was extremely helpful, and the Mayor of Hacqueville, M. Jean-Jacques Pilinski, opened the town's archives and spent several hours guiding us through records of the Brunels. The Barandou family devoted almost an entire week with us in Agen working upon translations of some of the Brunel letters, together with my son Colin and his partner Caroline, without whose persistent and enthusiastic research much of the important details would not have been uncovered. Ron Eatwell, a former chief librarian at Surrey University, suggested several contacts in America who could assist, as did Professor Mark Greenaugh of Richmond University in Virginia, USA.

Pencil and chalk sketch of Marc Brunel by William Brokedon, c1834. *National Portrait Gallery, London*

Daguerreotype of Marc Brunel by an unknown photographer, *c*1845.
National Portrait Gallery, London

The help and encouragement I have received from Professor Heinz Wolff of Brunel University, Uxbridge, is invaluable and I am extremely grateful to him for allocating so much of his valuable time to penning the Foreword.

The officials of the Brunel Tunnel Exhibition at Rotherhithe were particularly helpful, as were the staff of the Bristol University Library, and the Maritime Archives and Library at Liverpool; the Woolwich & District Antiquarian Society also provided several leads. The Institution of Mechanical Engineers proved to be a mine of information as did the Chatham Dockyard Historical Society. The latter put me in touch with Dr Philip MacDougall to whom I am indebted for permission to use part of his paper *The Saw Mills of Chatham Dockyard*. The Royal Institute of British Architects Library holds several letters and papers concerning the Thames Tunnel which they made available to me, as did the Plymouth Library Service and the National Maritime Museum. Mrs Lenore Symons, the Archivist at the Royal Institution, searched its Quarterly Journals for details of Michael Faraday's lectures and demonstrations of Brunel's experiments. Carol Morgan put the facilities of the Institution of Civil Engineers Library at my disposal and guided my researches through its collection of Brunel archives.

Portsmouth Museums and Records Service introduced me to the Portsmouth Naval Base Property Trust which allowed me to inspect the Brunel Blockmaking Machinery held in the Apprentices Museum and the Blockmaking factory buildings in the Dockyard. Wandsworth Council's Local History Library provided copies of the Poor Rates and Road Rates entries for the years of the operation of the Battersea sawmill.

My thanks are also due to my researchers Mrs M. Holland, Mrs Alison Mowat and Mr Graham Davies. Professor Malcolm Andrews of Kent University, Editor of *The Dickensian*, provided a great deal of information on London Prisons, and Miss Marylynn Rouse's knowledge of the Taylor family of Southampton was responsible for correcting errors in the original manuscript.

Southampton's Archives Department offered assistance with the location and details of the Taylor Workshops. Martin Woodward, the proprietor of the Bembridge Maritime Museum on the Isle of Wight, spent a lot of time helping me to search for identifiable Taylor's blocks among his many exhibits. Reg Silk supplied details of relevant articles in several engineering journals from his collection, and through these I became aware of many of the errors contained in earlier publications. I am also obliged to Cornell University and to the International Committee for the Conservation of the Industrial Heritage. The New York State Historical Association provided confirmation of Marc Brunel's association with the Champlain Canal and several engineering projects in and around New York, but was unable to confirm his official appointment as the city's Chief Engineer although he is mentioned in that capacity in several reports.

I am grateful to Sue Constable, the Shoe Heritage Officer of Northampton Museum & Art Gallery, for her assistance in tracing Marc Brunel's involvement in the boot and shoe industry and in introducing me to other avenues of research. St Bride Printing Library provided details of the Stanhope Press, and Koenig & Bauer of Würzburg were most co-operative over the original *Times* installation.

Henry Vivian-Neal of the Friends of Kensal Green Cemetery provided information correcting statements in earlier publications regarding Brunel's involvement (or lack of) in the design of the cemetery, and Pam Cole of the Pugin Society corrected errors regarding A.W.N. Pugin's involvement in the same project.

The line drawings are almost entirely the work of Julie Williams, who has been responsible for illustrating my books and articles for the past 25 years, and most of the unattributed photographs were supplied by family members. I am particularly obliged to another of my sons, Dr Jeff Bagust, who spent many hours enhancing illustrations taken from antique books and photographs, and for the Brunel family tree.

Most of my family have been involved in one way or another, and I am extremely grateful to all of them for their support and patience.

Harold Bagust
January 2006

Foreword

by Professor Emeritus Heinz Wolff

For the past 22 years I have worked as a research professor at Brunel University. I therefore have a reasonable familiarity with the life and achievements of the better-known Brunel, that is Isambard Kingdom Brunel, after whom the University is named. There is no direct historical connection, except that the University wanted to adopt the name of a great engineer, and that it started life in Great Western Railway territory in Acton. By coincidence its Uxbridge campus includes the remains of a cutting on a line running from West Drayton to Uxbridge via Cowley, which until 1881 was broad gauge. Nostalgic railway enthusiasts have now reinstated a section of broad-gauge rail which is illustrated in the book.

This biography will serve to focus attention on Isambard's father Marc whose originality of thought may well have been greater than that of his offspring. However, the Thames Tunnel aside, the visibility of his work was much less, being largely concerned with the improvement of manufacturing processes.

I am fascinated by 19th century technology and the vigour and cruelty of the society in which it flourished. In expansive and unwise moments I have been heard to say that 'everything worth inventing was invented in Victorian times or just before'. This is of course not true in a world dominated by electronics, but the basis of our civilisation is anchored in the beautiful transparency of Brunel-type technology, where a passably intelligent person can understand how a machine works by no more than close inspection and reasoning. As an aside, the disenchantment of modern young people with engineering must in part be due to the opacity of most contemporary technology, where the inability to see the 'wheels go round' is a great discouragement.

When I first saw the manuscript of this book, I recognised it as a stoutly scholarly work with a significant part of the text being in the form of appendices, bibliographies and acknowledgements. Whilst I expected to learn a great deal about Marc Brunel (1769-1849), who in my consciousness had been much eclipsed by his son, I anticipated it would be rather hard going, so I knuckled down to it, but was most agreeably surprised! For me it became an absolute page-turner, in spite of the technical difficulty in treating a typed manuscript in that manner.

What the author has managed to do superbly is not only to do justice to the qualities of a man he clearly much admires, but also to do so by setting him firmly into the context of contemporary society. The gift of making the background come alive is worthy of Charles Dickens (1812-1870) whose active life overlapped that of Marc Brunel. He was even born in the same part of Portsmouth where the family of Marc Brunel lived for a time.

There is a real Dickensian flavour in the manner in which the subject has been treated; at times I felt as if I was a spectator and could feel the frustration which beset the inventor when faced with the obtuseness of officialdom and company directors, the corruption endemic in the administration of the Navy and Dockyards and the not infrequent financial embarrassments.

Where else could I have read about the mechanisation of the dockyards, and in passing discovered that until 1790 the vast majority of shoes and boots were 'straights', not left and right footed as one would expect? The reason was that to manufacture left and right versions of the same size and design would have required two different lasts. It was not until the invention of the pantograph lathe, when it became possible to produce mirror image lasts economically, that lefts and rights could be contemplated commercially. I turn a few pages and get a history of the vicissitudes of London Bridge. The appropriate appendix provides me with a working knowledge of the operation of the King's Bench debtor's prison in which Brunel spent some 88 days, without availing himself of one of the many arcane practices under which, against a fee, prisoners were allowed to live outside and even at home, if within a given radius of the prison. Having been reduced to penury by the failure of the government agencies to pay him his just dues, he ingeniously and effectively blackmailed the same government into paying his debts, and thereby setting him free, by threatening to emigrate to Russia.

What must impress any reader is the sheer acuteness, versatility and courage of Marc Brunel's mind. Whilst capable of planning and executing civil engineering projects, he was totally at home

with moving parts; to my mind a powerful distinction between types of engineers. His aptitude in the use of tools and in making devices with his own hands, apparent when he was still a child, was I am sure, one of the factors in training a receptive mind to greatness. Most of his inventions are concerned with machines. In this respect he differs markedly from his famous son who was essentially a civil engineer who sculpted the landscape, and either came to grief (the Atmospheric Railway), or had to seek advice when his concept involved moving components. Even the tunnelling shield, which made the Thames Tunnel possible, was essentially a machine with 9,000 individual parts.

Professor Heinz Wolff demonstrates the 7ft gauge at Brunel University. Note the longitudinal sleepers which were a feature of the broad gauge. *C. Bagust*

I am astonished how quickly challenges thrown up in the course of a major project such as the Thames Tunnel, itself over 15 years under construction, could be met by an almost ad hoc construction industry. If it became necessary, almost overnight, to procure hundreds or even thousands of bags filled with clay, with hazel rods run through them, and to anchor them on the bed of the Thames so as to block a hole in the roof of the tunnel workings, this could be achieved without mechanisation, but with real talent in rapid improvisation and procurement.

Marc Brunel lived at a time when engineering was changing from a craft into an industry. The companies which could build machines to his designs were themselves led by substantial pioneering intellects. They served as the training ground for the next generation of engineers, who introduced the necessary precision into manufacture to make interchangeable and standardised parts possible. Joseph Clement, whose life straddles this conversion, worked for both Bramah and Maudslay (the reader will find them in an appendix), built Babbage's Difference Engine and trained Joseph Whitworth, who gave his name to a whole system of standardised threads for nuts and bolts.

An air of probity, goodness, consideration and trust emanates from the persona of Marc Brunel as portrayed by the author. Harsh business practice was not his forte, his trust was frequently betrayed and at times betrayal served as a stimulus for more invention just so as to have another business to provide him with a living. He truly lived by his wits in the nicest possible way! I would dearly have liked to have made his acquaintance in person, but reading this book is a very, very good substitute.

Brunel University
August 2005

Introduction

In 1957, L. T. C. Rolt published his book, *Isambard Kingdom Brunel*. I have been an admirer of Brunel ever since childhood when I often spent a Saturday morning on Platform One of Paddington Station gazing at the magnificent engines which pulled the 'Cornishman', the 'Riviera Express' and the 'Bristolian' to far off places in the West of England. I was compelled then to purchase a copy of Rolt's book at the outlandish price of seven shillings and sixpence (37½p). From that moment Rolt, in my estimation, was second only to the Almighty and his book was regarded as the final word on I. K. Brunel.

In every respect I accepted his word and overlooked the many errors and false statements which his manuscript presented as fact and it was several years before I had cause to question the words of 'the master'. This came about through my interest in later years in the inventions of Isambard's father, Marc Brunel. Naturally I initially turned to Rolt for information, only to find that out of the 340 pages devoted to the exploits of Isambard, the entire career of his father is covered in just 19 pages at the beginning of the book. I started to delve a little deeper into Marc's life and, as I did so, began to realise that Rolt viewed Isambard through strongly rose-tinted spectacles, glossing over his many faults and emphasising his good points, almost to the extent of hero-worship.

Any writer attempting to unravel the life of Marc Brunel is immediately beset with the problem of separating fact from fiction. This is primarily because much of his correspondence and diary entries from his early life were destroyed after his death, presumably by his widow who wished to prevent their publication for various reasons. In later years this inevitably led various authors to fill in the blanks with imagined incidents. Rolt is not alone in repeating mistakes which have occurred in earlier and later books on the subject of the Brunels. Adrian Vaughan (1991) restated the oft-repeated error that Sophie Kingdom (Marc's wife) was the granddaughter of Thomas Mudge, the famous English clockmaker. This is not so and a possible explanation for the original error is explained in Chapter One. Rolt claimed that there were no strikes and no fatalities during the construction of the Thames Tunnel and that throughout the entire project, nothing but harmonious relations existed between Marc and his workers. We now know that this too is untrue; a number of workers died through accidents and working in the filthy conditions underground. There were also fairly frequent strikes, mainly caused by wages being paid weeks after they were due, but sometimes in protest about working conditions. Adrian Vaughan also includes a suggestion of rivalry between father and son for which I have found no documentary evidence.

Professor Angus Buchanan's *Brunel: The Life and Times of Isambard Kingdom Brunel* is much better researched but repeats the claim that Marc Brunel designed Kensal Green Cemetery in collaboration with Augustus Welby Northmore Pugin. This is not borne out by the Company's records nor by Professor Curl's book *Kensal Green Cemetery* (Phillimore, 2001) which lists the details of all 48 architects who entered the competition for a design for the cemetery. This error is repeated in Sally Dugan's *Men of Iron* which was published in 2003 to accompany the Channel 4 series of the same name, and is also reiterated under the entry for Marc Brunel in the latest *Dictionary of National Biography* (2004).

In 1970 Paul Clements published his book *Marc Isambard Brunel* in which some but not all of the earlier errors were eliminated, and a more balanced view of Marc appears for the first time. However, that was over 35 years ago, and our knowledge of the Brunels has advanced somewhat since then as research has become easier through the computerisation of public records. Nonetheless, it is still very difficult to separate the truth from fiction in many instances, often through the carelessness on the part of those responsible for keeping public records in the past.

Two instances will suffice to illustrate the point. The first example states that work commenced on the construction of a wooden bridge between Chelsea and Battersea in 1771; of this there is no doubt, but dates for the completion and opening of the bridge to the public vary between 1773 and 1777 according to which account one accepts. I have taken the official records in the archives of the Wandsworth Council as being most probably accurate at 1777, although this may refer to the time when it was officially opened to wheeled traffic. The second example refers to the Highway Rate books and the Poor Rate books of the same Council for the period 1804-27. They hold several entries which defy positive explanation. The property which we know was part of the Battersea sawmill occurs time and again under a file number but a different name appears as the owner on several occasions. It may be that Marc Brunel registered it in the name of his manager who was replaced at different periods but at least two entries show the ratepayer as a *Mr Burnet*. Could this simply be a misspelling of Brunel? Later, during the time when Brunel was incarcerated in the King's Bench prison, the sawmill site is shown as being under the control of a Mr Mudge. Had Marc's brother-in-law been brought in to hold the property during the imprisonment, or is it simply a coincidence that a Mudge appeared on the scene at this time?

Even at Hacqueville, Marc's birthplace in France, problems arise. The name there is spelt 'Marc Isambart Brunel' and appears so on both the memorial in the town and on the plaque mounted at the entrance to the farm. But most of the correspondence from Marc bears the clear signature 'Marc Isambard Brunel'. The issue of his name becomes more complicated as it appears that he was never known to family and friends by the name Marc during his lifetime, being referred to instead as Isambart in France and Isambard in England. This leads to obvious difficulties once Marc's

Oil painting of Marc Brunel by Samuel Drummond, *c*1835. *National Portrait Gallery, London*

granddaughter Lady Noble was deliberately misleading her readers when she agreed with Beamish. It must be remembered that the so-called 'Draft Riots', which occurred in 1863, resulted in most of the City of New York's public buildings being burned to the ground and large swathes of public records were consequently destroyed. Nevertheless, the institutions I have contacted have no records of a Chief Engineer named Marc Brunel. These include the Avery Architectural & Fine Arts Library, New York City Municipal Archives, the New York Historical Society, the New York State Historical Association, the Smithsonian Institution and the State Education Department of the New York University at Albany.

Throughout the book, various dates are subject to argument and I have taken those generally accepted by professional historians to be correct. The disagreements are usually caused by faulty entries in parish registers by semi-literate clerks, the fact that there was no standard spelling or universal dictionaries until after the 18th century, and the fact that many male children were given the same name as their father. There were, for example, at least three Walter Taylors and three Thomas Mudges. What is more, the father who registered the birth of his child was often unsure of the spelling of his own name and so allowed the clerk to influence him, with the result that it becomes very difficult to trace a family tree. I have checked and double checked all the dates included in the book as far as I have been able, but future research may uncover errors, for which I apologise.

Some would say that it is pointless to argue whether father or son was the greater, as both were geniuses and each gained from association with the other. To many, Isambard Kingdom Brunel was supreme in that his efforts produced many more lasting and visible monuments to his skills. It must be understood, however, that much of his knowledge originally came from, or was initiated by, his father. Marc was always supportive and encouraging and assisted to a great extent in guiding his son's efforts into the right channels. Moreover, Marc derived nothing from *his* father, who consistently opposed all his plans and aspirations, even forcing him to study for a position in the clergy against his will. The fact that he refused to submit to his father's demands and insisted upon following his own bent in spite of the constant opposition, confirms in my view that his genius was the greater since his progress resulted entirely from his own efforts.

In the following chapters I have felt the need to emphasise the achievements of Marc Brunel. He has lived too long in the shadow of his famous son. I hope therefore that this book will make an important contribution in bringing Marc's achievements before a wider public.

son had been born. For the purposes of this book, the father is referred to as Marc throughout and any reference to Isambard is referring to his son, Isambard Kingdom Brunel. Similarly, Marc's wife is Sophie in the narrative, and his eldest daughter Sophia, to avoid confusion.

One of the more frustrating problems has been that encountered with the New York authorities. Here it is continually stated by previous authors that Marc was Chief Engineer for several years before coming to England. Nowhere does Marc's name appear in the New York records in that capacity. In fact none of the edifices he is said to have designed, such as the cannon factory and harbour improvements, are positively attributed to him. The sole exception is the Park Theater (or Bowery), the design of which one writer claims was the work of Marc Brunel, but even this is disputed by others, one of whom attributes it to another French émigré.

This is not to say that Brunel was lying when he described his American exploits to Beamish his biographer, or that his

1.

French Origins

Marc Isambard Brunel was born on 25 April 1769 at Hacqueville in Normandy, on the family farm *Ferme Brunel* which his ancestors had held as tenants and part landowners for over 330 years. The first recorded tenant was Jean Brunel who was born on the farm in 1490, but other Brunels occupied the property prior to that date. By the time of Marc's birth, the Brunels were not peasant farmers, but neither were they aristocracy. Being wealthy farmers, they controlled several acres of rich farmland with a considerable number of employees, had little or no interest in politics except as regards maintaining the status quo, and were therefore supporters in general of the monarchy and the establishment. Their landlords, the noble Le Coulteux family, held them in high regard: the rent was paid regularly and the property kept in good heart. There were probably minor disagreements from time to time, but these did not damage the relationship. The Brunels' prosperity steadily increased, especially when they gained control of the local posting arrangements and the farm became a staging post for coaches travelling between Rouen and Paris, Hacqueville lying about halfway between the two cities.

When, on 25 April, Marie Victoire Lefèbvre, the wife of Jean Charles Brunel, bore her third child and second surviving son, the parents were overjoyed. (It is interesting to note that on Marc's baptism certificate, his mother's name was written as Angelique Victoire Lefèbvre. However, the majority of sources on Marc Brunel refer to her as Marie.) It was planned that, according to tradition, Marie and Jean Charles' first son, Charles Ange, would eventually take over the farm whilst the church would accept the younger. (Daughters were a liability until successfully married off, preferably to a wealthy suitor or one able to bring more land into the family holdings.) The child was christened Marc Isambart, apparently on the wishes of his godfather and great uncle, the Abbé Lefèbvre, who insisted on the child being christened after St Mark on whose feast day he was born. Marc Isambard was destined to become one of the major influences of the industrial age and the father of Isambard Kingdom Brunel whose fame as the great railway engineer has, to this day, overshadowed the genius of his father in so many history texts.

Brunel family tree.

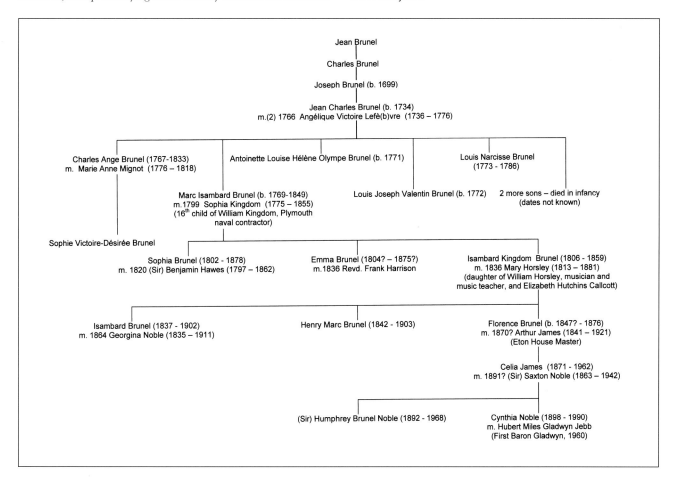

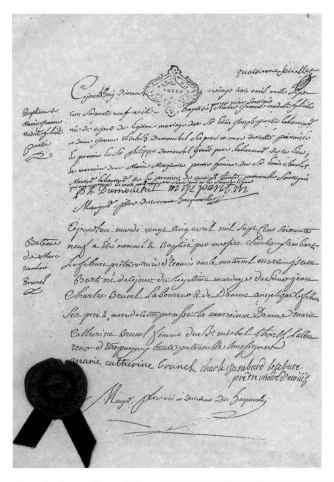

Although Hacqueville itself has now been bypassed by a major road, the small town remains little changed from the time when Marc left it. Many of the original buildings still stand and *Ferme Brunel,* although in parts rebuilt and refurbished, still retains the same layout and several of the original buildings that Marc would have known in his childhood. The entrance is through an archway almost opposite the church, into a large farmyard with barns and stabling for more than a dozen horses and a substantial pond. It is thought that both a smithy and a wheelwright's workshop were included among the buildings.

By the middle of the 18th century France was still an agrarian nation. Although Nicolas Cugnot, a military engineer, had demonstrated a three-wheeled steam-driven gun carriage with a speed of 3mph in 1770, it was never developed through lack of support from the government and the public. After the French Revolution nearly all projects had to be submitted to official departments for approval before proceeding, and this discouraged all but the most persistent inventors. Even before the Revolution, the unrest among the population made investors reluctant to consider schemes which they regarded as risky. The banks knew the country was close to bankruptcy and refused to assist and as a result, many ideas and inventions in France never left the drawing board through lack of finance.

By contrast the industrialisation of Britain was gaining pace. Newcomen had installed his first steam engine at Dudley Castle in 1712. Watt had patented his separate condenser in 1769. Murdoch had displayed a model locomotive at Redruth in 1784, and in 1797 Trevithick produced models of both stationary and locomotive engines. In 1774 James Watt and Matthew Boulton had founded a company near Birmingham for the manufacture of pumping engines. They patented the sun-and-planet motion, the expansion principle, the double engine, the parallel motion, a smokeless furnace and the governor.

It was against this backdrop that young Marc was growing up. His mother died when he was seven and his father started him on a classical education which was not at all to the boy's liking. It often resulted in his father frequently punishing him by shutting him up in a gloomy room where portraits of numerous forebears glowered down at him. The gaze of one stern abbé disturbed the lad so much that one day he dragged a heavy table across the room, clambered onto it and cut out the eyes of the canvas with his pocket-knife. This was the last straw as far as his father was concerned. Marc, now aged eight, was despatched to the College at Gisors, a nearby town where boys were schooled to become army officers, dressed in military coats with cocked hats, swords and powdered wigs. He quickly showed proficiency in drawing and mathematics but his Latin and Greek fell far below the required standards. Holidays were spent sketching a variety of buildings including the Brunels' farmhouse and the Château Gaillard of King Richard the Lionheart, and watching the local wheelwright and blacksmith at their craft.

All his life Marc was deeply attracted to music. He learned to play the flute and at the age of 10, after studying the details of the

Top: One of the surviving original buildings that Marc would have known. This one appears to have had living quarters above the barn or stables. *C. Bagust*

Top right: One of the buildings of *Ferme Brunel* thought to have been a granary originally. *C. Bagust*

Above: The entrance to *Ferme Brunel* from Rue Marc-Brunel. It would originally have been the width of the beamed attic above. Note the plaque on the pillar to the right of the doorway which details the Brunel connection. *Author*

Above right: Inside the entrance arch. The open doorway in the centre of the picture is said to be the original entrance to the farmhouse; Marc's bedroom was either in the attic or over the gateway. *Author*

time he was 12 Marc could carve and perform with the precision and elegance of a skilled cabinet maker.

After lessons he would run down to the quayside and make drawings of the great ships lying at anchor in the river or being unloaded at the dockside. He sketched the various craft and made detailed drawings of their rigging, and of the types of tackle used to transfer the cargoes from ship to shore. He examined steering gear, windlasses, masts and navigation equipment, watched the cordage being repaired and spliced and saw for himself the skills of the shipwrights and sail makers.

* * *

harpsichord, he made a working model of a machine which combined the sounds of both instruments. No details of this model have been discovered but his father was not impressed when shown this miniature masterpiece and was convinced that his son's efforts would be better channelled in other directions. Marc was therefore transferred at the age of 11 to the Seminary of Saint Niçaise at Rouen. There he reluctantly donned the sombre vestments required by the clergy whilst his father bemoaned his son's stupidity. The Seminary's Superior however was far from being the oppressive disciplinarian hoped for by Jean Charles. He quickly recognised the boy's artistic talents and arranged for him to have special instruction. He also had him tutored in carpentry. By the

Marc was brought up in a normal family within the Roman Catholic faith, but had no great leanings towards taking the cloth as was the custom for second and subsequent sons. The power of the church was such that there were always vacancies for young men to enter the priesthood. Most French sons after the first-born automatically donned the cloth, although a few chose the armed forces or politics, but the priesthood was usually the preferred choice.

The Superior at the Seminary saw immediately that Marc was endowed with exceptional talents but had no particular leanings towards the church. When the boy sold his hat in order to purchase a special tool he had seen in a shop window, he relayed these

Above: Rue Marc-Isambart Brunel runs past the church into the town centre. The entrance to *Ferme Brunel* is just out of shot in the lower right-hand corner of the photograph. *Author*

Below right: The nameplate on the street corner in Hacqueville. *C. Bagust*

feelings to Jean Charles who decided to send Marc away where he could cause fewer problems. He was sent to stay with a cousin, François Carpentier, and his English wife in Rouen. François had left the sea and was now the American Consul in that city: by lodging with them the lad could be tutored by their friend Monsieur Dulague, an expert in naval matters, and thus be in a more favourable position to apply for a commission in the navy as an officer cadet.

Vincent François Jean Noel Dulague was Professor of Hydrography, an astronomer and hydrographer of the first rank, and author of the text books on navigation used in all French naval institutions. His agreement to accept Marc as a pupil could only have been on the recommendation of the Superior at the Seminary, for Dulague would not normally have accepted a country boy. It was a tremendous opportunity, and to Marc's delight, his father's initial objections were finally overcome and an agreement reached.

There is a divergence of opinion as to whether Jean Charles or his tutor was responsible for getting Marc into the naval service. It would have been almost impossible for a common farmer (no matter how wealthy) to have obtained an audience with Louis XVI's Minister of Marine without the backing of someone in high authority, and the Brunels were not that well connected. On the other hand, Dulague was known and respected by all in naval circles and well able to bring his pupil's qualifications before those able to recognise them. The fact is that Marc was accepted as a cadet on a new frigate named in honour of the Maréchal de Castries, and at the age of 16 he found himself making several voyages to the West Indies in the service of his country. It was during this time that he designed and constructed a quadrant of brass and ebony with the help of money reluctantly given by his father. It was made with such accuracy that he required no other throughout his time spent in the navy. Few other details are available about his achievements during this period, other than that

it is known that he designed a coffee-husking machine at Guadeloupe in 1790, and that, during the ship's occasional calls at American ports, he acquired more than a working knowledge of the English language. Details of the coffee-husking machine have never been discovered nor has it been possible to trace the movements of the vessel on which Marc served. The *Maréchal de Castries* returned to Rouen and paid off her crew in January 1792 and Marc now faced an uncertain future in a land where the Revolution was in its third year.

At that time the city of Rouen, the ancient capital of Normandy and divided by the River Seine, was served by two bridges, the higher one built of stone and the lower of iron with a central section which could be raised to allow the passage of tall ships. Rouen was an important seat of learning and commerce. It was a major port, although it was 78 miles from the sea, and it was fourth in importance after Marseilles, Le Havre and Bordeaux. In addition to the naval establishments it was also the headquarters of the third *Corps d'Armée*.

Products from all over the world came into France via Rouen. The fact that France had been at war with England on and off for many centuries had never had a significant effect on trade. The businessmen of both countries regarded war as a sideline to their enterprise, and the two carried on side by side without interfering with each other. So it happened that even at the beginning of the French Revolution, salesmen from the English woollen industry continued to travel in France to contact their customers, and French traders brought their wines and wares to England without much difficulty. However, the situation changed after the execution of Louis XVI in 1793 when it became common knowledge that asylum for the royals had been offered by England, and it was widely assumed that the English were about to invade in an attempt to restore the monarchy. Forthwith, all English nationals discovered on French soil were classed as spies. By 1793 many of the senior officers in the French army and navy had been arrested and subsequently guillotined by the revolutionaries as possible royalists, and the services had been further weakened by the promotion of underlings to positions quite beyond their capacity, simply because they were supporters of the Commune.

* * *

After his discharge from the navy Marc was invited to take up residence again with his kinsfolk, the Carpentiers in Rouen, where he returned to his old room and rejoined their household. François Carpentier, together with most of the inhabitants of Normandy, shared Marc's royalist sympathies and when the royal family was imprisoned following the sacking of the Tuileries on 10 August 1792, the men of Rouen formed themselves into a National Guard to defend themselves against the sans-culottes (the breech-less ones). Marc had a very narrow escape on 12 August, only two days after the sacking of the Tuileries, when he attempted to intervene on behalf of the Swiss Guard who had escaped on the 10th, probably from Courbevoie. His diary entry reads: 'On entering the gates of the barracks we were surrounded by thousands of the mob, so that we could not have escaped had they closed up on us, but they were intimidated by the sight of a few of the National Guard coming down the Cours Dauphin.'

In January 1793 Marc and François visited Paris. The capital was full of rumours concerning the king's trial. On the evening of the 16th, just four days before the sentence of death was pronounced, Marc publicly predicted the eventual demise of Robespierre to a crowd assembled in the Café de l'Echelle near the Palais Royale. This was extremely foolhardy, but Marc felt strongly that the masses were being misled by the revolutionaries and abhorred the atrocities which were being reported daily in the press. Only a diversion created by Louis Taillefer, a future member of the Chamber of Deputies, enabled Marc and François to escape. They gained the shelter of a nearby inn where they hid until nightfall before returning to Rouen. The following day barricades were erected around Paris.

The memorial to Marc Brunel outside the church in Hacqueville. *C. Bagust*

* * *

On 4 December 1792 a 16-year-old girl named Sophie Kingdom had arrived in Le Havre from Portsmouth. As an orphan, she had been sent to France by her elder brother (and guardian) to learn French, the knowledge of which was considered essential for any young lady of quality. She was accompanied by some friends, Monsieur and Madame Longuemar. However, they did not stay long once they heard of the murder of two acquaintances who were accused of being royalists. Sophie would have gladly accompanied them back to England, but a severe illness meant she had to stay behind, so she was left in the care of a good friend of the Longuemars, Monsieur Carpentier. On 17 January 1793, at the Carpentier's house, Marc Brunel met Sophie for the first time. The debonair young Frenchman and the beautiful young English girl were immediately attracted to each other. However, they ran into considerable opposition from Madame Carpentier who had taken on the role of unofficial chaperone. Nevertheless the romance rapidly blossomed, so much so that, according to Beamish, while Sophie was admiring Marc's first attempt at an oil painting, pointing out which parts pleased her most, Marc turned to Madame Carpentier and exclaimed, '*Ah! ma cousine quelle belle main!*'('Ah, my cousin, what a beautiful hand') to which Madame Carpentier replied, '*Oui, mais elle n'est pas pour toi.*' ('Yes, but she is not for you.') It is thought that the pages of Marc's diaries which were destroyed after his death may have included more details of his courtship with Sophie, but as it is, hardly any information exists concerning the early months of their relationship.

Most of the books published in the past 50 years insist that Sophie Kingdom was either the granddaughter of Thomas Mudge or his niece. She was neither. Thomas Mudge (1717-94) was born in Exeter, the second son of Zachariah Mudge, a clergyman and schoolmaster who moved to Bideford when Thomas was very young and became master of a grammar school there. Showing no desire to follow in his father's footsteps, Thomas was sent at the age of 14 to London to be apprenticed to George Graham, one of the most eminent of English clockmakers who had learned his trade at the hands of Thomas Tompion to whom he was chief assistant for many years.

The Mudges were well connected, and family friends included Dr Samuel Johnson and the Reynolds family of Plympton whose son Joshua became famous as a portrait painter. After Graham's death, Thomas Mudge started in business on his own account at the age of 34, at the *Sign of the Dial and Three Crownes* in Fleet Street, London, where he rapidly acquired a reputation as one of England's outstanding clockmakers. He invented the lever escapement *c*1757, the greatest single improvement ever applied to pocket watches which remained a major feature of every pocket timekeeper for the next 200 years. From 1771 he worked upon the development of the chronometer and sent his first effort for trial in 1774; he was eventually awarded £3,000 (an enormous amount in those days) by a Committee of the House of Commons. In the same year he left London for Plymouth where he remained for the rest of his life. He refused to patent any of his designs or inventions and in 1776 was appointed Watchmaker to George III.

Sophie Kingdom was the 16th child of William Kingdom and Joan Spry. William was a Plymouth naval contractor who died soon after Sophie was born. Sophie's sister Elizabeth married Thomas Mudge's son (also named Thomas) and they produced 11 children of whom three died in infancy. One of these three was named Mary Sophie and it was this Sophie who was the granddaughter of Mudge Senior. Sophie Kingdom was Sophie Mudge's aunt.

* * *

Unfortunately Marc had no choice but to leave Sophie behind in France following the scuffle in Paris during which he had predicted the demise of Robespierre and the crowd had turned against him. With the help of the American vice-consul at Le Havre who had been contacted on Marc's behalf by François Carpentier, he obtained a passport on the pretext of travelling to America to purchase grain for the navy, and left Rouen on horseback bound for

Above: Thomas Mudge (1717-94) by Luigi Schiavonetti. *National Portrait Gallery, London*

Right: Mudge's escapement. *J. Williams* (Illustration taken from *The Marine Chronometer.*)

Le Havre, some 80 kilometres (50 miles) distant. Unfortunately his horse stumbled and threw its rider, and Marc was concussed and lost consciousness for a short while. By the time he came round the horse had disappeared and it seemed that his cause was lost, for without transport he had no chance of eluding his pursuers and reaching the ship before it sailed.

Fortunately a passing coach stopped in the *Cours de Daville* and the occupant asked if he could be of assistance. Marc immediately recognised the Revolution's Navy Minister who was travelling to Le Havre to interview a group of cadets. Marc was invited into his cabriolet and completed his journey in style seated beside Gaspard Monge. A mathematician, Monge was the inventor of mechanical drawing, and during the Revolution was appointed Navy Minister with responsibilities that included the design and construction of gun and gunpowder factories. He was well known throughout France, and his theories and methods had been studied and followed by Marc in the normal course of his education. The two therefore must have had much in common to discuss during the coach journey. Monge however was never a tutor of Marc as claimed by some authors. Marc was eventually delivered to the quayside where he boarded the American ship *Liberty*. He set sail for New York on 7 July 1793.

It is said that Marc lost his passport when thrown by his horse and boarded the ship without it. On 19 August, whilst at sea, Marc's ship was challenged by a French frigate. If discovered, Marc was clearly at risk of being taken straight back to France. Showing

great ingenuity, Marc designed and produced in a matter of two hours a forged passport which the boarding party accepted as genuine. He even rubbed it under his feet to give it an air of age. Some think the forgery story highly unlikely on the grounds that he would have needed to have access to paper of similar quality, to inks of the required colours and the official stamps and seals which a genuine passport would have carried. Furthermore, it would have been necessary to borrow a genuine document to copy from another passenger (with a risk of being betrayed), and then to produce the finished article in two hours. However, the story is confirmed by Marc's own hand in a journal entry of 1837 where he recounts how back in the summer of 1793, he had had a narrow escape and to save himself had made a false passport.

Eighteenth-century passports were usually no more than letters of recommendation from an official in the country of origin to a person of authority in another country, for the specific purpose of assuring the recipient of the integrity of the traveller. They were never intended during this era to be used as a permit to leave a country or to enter another. The passport carried by Marc was probably little more than a letter of introduction to an American supplier of grain; it may not even have been written on an official note heading, and so it would have been fairly easy to forge a replacement as no seals or official stamps were involved.

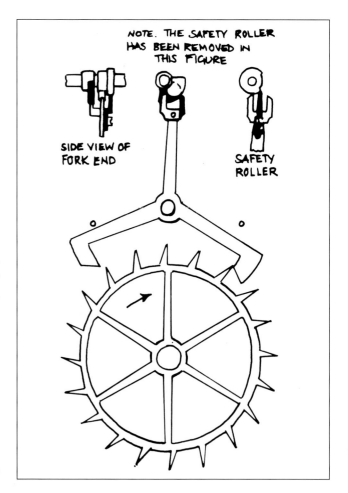

2. **America**

On 6 September 1793 the *Liberty* anchored off Sandy Hook and the next day Marc reached New York. Unfortunately a small French squadron, which had come from St Domingo, was sheltering in the harbour. He referred to the crew as being nothing more than a set of banditry, and several of them threatened to hang him and his colleagues for being a cargo of Royalists. Marc and a few others made a hasty exit and managed to get to a lodging house in Hanover Place. As he had no wish to risk being taken back to France, he decided to travel to Philadelphia where there was a fledgling French community.

At this period, Philadelphia was far more developed than the majority of cities in America, including New York. It was the model metropolis, with one newspaper describing it as 'the great seat of American affluence, of individual riches and distinguished philanthrophy'. It was commonly referred to as the Mother City, and its layout and infrastructure were frequently copied in the planning of other cities throughout the country. Marc stayed in Philadelphia in idleness for a short time, spending his evenings in the back room of a bookshop owned by Moreau de St Mery, in lengthy discussions with compatriots including the crafty Talleyrand (Charles Maurice de Talleyrand-Périgord) who was barred from England but eventually became very influential in France, especially during Napoleon's heyday, although at the time he and Marc met he was opposed to the Revolution. After relaxing for a short period in Philadelphia, Marc decided to contact two fellow émigrés, Pharoux and Desjardins, who had been passengers with him on the *Liberty*. After a long and painful journey, 'in a gig without springs over almost impassible roads', he joined them in Albany.

The pair were apparently well endowed financially and planned to survey a large tract of land lying between the 44th parallel, the course of the Black River and Lake Ontario, with the idea of parcelling it out for sale to migrants. Marc, being well qualified for the project and also an agreeable companion, was welcomed into the partnership. On 18 September the trio, accompanied by three Red Indian guides and equipped with guns, axes, supplies and tents, set out from Schenectady to paddle up the Mohawk River.

Little is known of the outcome of this venture. Communications were sparse and the distances involved enormous. Writing to his old friend Monsieur Allard in 1831, Marc recalled how he felt a wonderful sensation of safety in America, having escaped from the government of Robespierre, and how, on one adventure up country, he met with bears who he thought were less ferocious than the agents of that despot. It is not known how long the party was away or the success of the expedition, but early in 1794 they returned to Albany. Here the group boarded a river boat bound for New York which grounded on a sandbank in the Hudson River and was stranded for two tides.

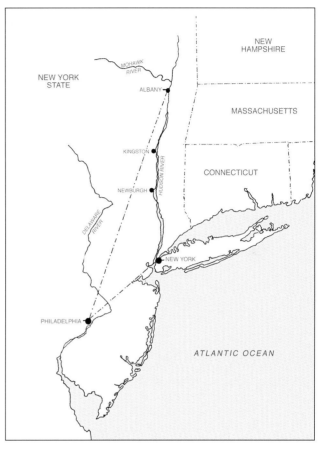

Route taken by Marc Brunel after reaching New York, according to Lady Celia Noble. *J. Williams*

During this period of enforced rest a traveller named Thurman, a wealthy New York merchant, came aboard. Discussions followed in which the three Frenchmen expounded their plans for cutting a canal to connect the Mohawk to Lake Oneida and opening a waterway whereby barges could ply between New York and the Great Lakes. Thurman was initially impressed, but soon realised that the immediate beneficiaries of the scheme would be the French trio, since the proposed canal would be flanked by their land. His enthusiasm therefore cooled. They then put forward a new plan to link the Hudson River via a canal to Lake Champlain which would eventually provide a thoroughfare between New York and the St Lawrence. This time Thurman agreed to finance a survey. Work commenced in May and the trio were appointed to supervise the project. According to Marc's journal of 1840, he recalled that on 17 May 1794 he had killed a rattlesnake on an island in Lake Champlain known as Rattlesnake Island. This event must have remained strong in his memory for him to recall it so vividly 46 years later.

Thurman soon discovered that Marc was a marvel of resource and invention who cleared blocked watercourses and skirted round swamps like a natural-born engineer. He was also adept at bypassing obstacles and obstructions which were encountered from time to time. It was during this project that Marc finally abandoned all ideas of a naval career and decided to concentrate on civil engineering.

It is agreed by most authorities, including the International Committee for the Conservation of the Industrial Heritage, that Marc Brunel was responsible for the surveying of the Champlain Canal and was assisted by Benjamin Wright. However, work on the project was shelved when financial backing was not forthcoming and it was not until *c*1815 that Wright was appointed to continue operations assisted by James Geddes, but by that time Marc was a British citizen with eight patents to his name.

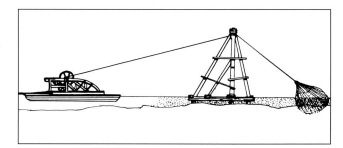

* * *

For most of his life Brunel was toying with more than one idea at a time, and this period was no exception. Even while the canal was under consideration, he entered a competition to design a new Capitol building for Washington which was intended to reflect the importance of Congress. The design which Marc submitted was inspired by the Palace of Versailles. The judges pronounced it outstanding and the intelligentsia of New York and Washington were delighted. There are several conflicting accounts as to why it was finally passed over: one claimed that the cost was too high for the authorities to accept, whilst another stated that it was rejected because it was considered too elaborate and ornate for a Republican building. This was at a time when the USA was strongly anti-monarchist and very touchy about the influence of the British Empire on American life.

Even so, the design was not entirely lost: it was apparently modified and used for the construction of the Palace Theater in the Bowery (sometimes called the 'Park' or the 'Bowery' Theater). This is however disputed by two books; the first entitled *Iconography of Manhattan Island* and the second, the *Biographical Dictionary of American Architects* (1956); both claim that another Frenchman,

Joseph Mangin, was responsible for the design. However, *A History of New York City to 1898* described the playhouse as having been designed by 'émigré engineer Marc Isambard Brunel' and went on to describe how the Park Theater 'cost a princely $130,000 and was specifically calculated for the comfort and convenience of the city's respectable classes'.

More confusion lies in the fact that some books state that the Bowery and the Park were separate institutions. In 1985, Professor John W. Frick published a paper entitled *New York's First Theatrical Center* which included the following lines: 'The Park and Burton's remained bastions of the genteel theatergoer, but the Bowery, the National and the Chatham became recognised by the working classes as their own special places of entertainment.' From this it would appear that the Park Theater and the Bowery were separate places. At this time there was an area of New York which was occupied by several theatres, similar to London's West End today. This is confirmed by *Reminiscences of an Octogenarian of the City of New York* written by Chas Haswell and published in 1897. In it there is an illustration of Chatham Street (Park Row) showing the Park Theater among several similar establishments. Further confusion is caused by a reference on page 496 where the Park Theater was also called Old Drury. It is interesting to note that this author, writing at the very end of the 19th century, used the English spelling of the word 'theatre'. Whoever was responsible for the design of the Park or Bowery Theater would not see his creation survive for long. The theatre burned down in 1821.

Brunel received no payment for this design but his reputation as an engineer and architect had been established. In the autumn of 1796, after taking American citizenship (see Appendix 1), he was

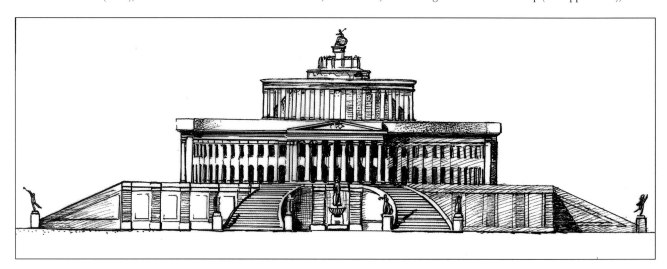

Left: A drawing of Brunel's 'Plan for Raising Obstructions in the Rivers'.
R. Beamish

Bottom left: Design by Marc Brunel for the US Capitol building.
R. Beamish

Right: Exterior of the Park Theater designed by Marc Brunel.
Author's collection

Below: Interior of the Park Theater designed by Marc Brunel.
Author's collection

Below right: A design by Marc Brunel for a bank in Wall Street.
R. Beamish

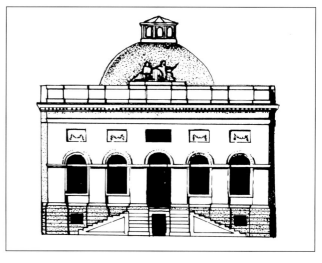

appointed Chief Engineer to the city of New York. Here he designed houses, docks, commercial buildings, an arsenal and a cannon factory. The latter was so advanced technically that its fame spread rapidly throughout America and Europe. He also completely redesigned the New York waterfront to improve its efficiency. It is a great pity that no records exist to confirm his involvement in these American projects. It is very probable that the information was destroyed during the 'Draft Riots' which began in New York in the summer of 1863.

During the Civil War, as a result of the Northern troops suffering severe losses, Abraham Lincoln introduced a law that all males between the ages of 16 and 50 would be conscripted into the Northern army. However, those able to pay $300 could be excused from call-up and could send another person to replace them. This was at a time when the average labourer's annual pay was less than

$500. It quickly became obvious that the new recruits would be drafted from the poor and immigrant classes. Minor riots occurred in several cities, but when the drawing of names began in New York on 11 July 1863, mobs took to the streets en masse, assaulting residents, taunting the police, attacking Draft Offices and burning most of the government buildings in the city. Damage to property was estimated at over $1.5 million and vast quantities of official records were lost at the hands of looters and arsonists.

* * *

In 1798 Brunel was invited to dine with Alexander Hamilton, an honest and intelligent man who had been George Washington's First Secretary to the Treasury. They were united by temperament and by an instinctive distrust of Republican France, which at that time was waging undeclared naval war against the United States. Hamilton was continually warning George Washington about French schemes and plots. Although no longer in office, he remained a confidant of Washington and soon afterwards was made a Major-General, second in command of the United States Army. Another émigré, Monsieur Delabigarre, was also present at the dinner. He had recently returned from England and soon the conversation turned to ships, navies and battles, and in particular to the British victories at Camperdown and St Vincent which had aroused the interest of the American public. Noting Brunel's detailed understanding of seamanship and warship design, Delabigarre mentioned the problems faced by Pitt's government as they attempted to expand the Royal Navy. Among other difficulties, ships' pulley blocks were a constant supply problem and the manufacturers were under continual pressure to increase production. The rough sketches which Brunel produced and showed to Hamilton a few days later delighted him; Hamilton would have approved any scheme which promised to check the French or put them at a disadvantage. Marc himself recognised in his journal many years later that it was at Hamilton's house that he first came up with the elements that would lead to the invention of block machinery. Hamilton encouraged Marc in his intention to sail for England and offered him a letter of introduction to Earl Spencer, the Navy Minister in Pitt's government. (He was an ancestor of Diana, Princess of Wales.) This introduction was to have far-reaching effects for Brunel later in life.

* * *

Sophie Kingdom, meanwhile, remained happily in Rouen, supporting herself by teaching English to girls whose families could afford her services, and paying little attention to the political situation in the country. All this changed when Louis XVI was executed in 1793 and England joined the alliance against France. Following the execution of the king, a committee of public safety

Right: Thought to be the only known portrait of Sophie Brunel in existence. It is by an unknown artist (possibly Marc Brunel himself) and it was previously owned by Lady Celia Noble. It was auctioned after her death and sold to an unidentified purchaser. *J. Bagust*

Below: Miniature self-portrait of Marc Brunel believed to have been painted during his courtship of Sophie and sent to her when he was in the United States. *National Portrait Gallery, London*

was convened which included members of all the opposing factions, many of whom were atheists and enemies of Robespierre (see Appendix 7). Soon afterwards, the 'Terror' was instigated. It is not clear how far Robespierre was implicated in its enforcement although it certainly met with his approval. Here now was an opportunity for him to wipe out many of his enemies and any others opposed to his plans. In October 1793 the Council of the Revolution decreed that all English nationals were to be arrested as spies and their papers and possessions confiscated.

The Terror was at its height. Sophie was arrested, put in an open cart and taken to Gravelines, a small port about midway between Calais and Dunkirk. She must have thought she was about to be executed. Eventually, after an uncomfortable journey lasting several days, the nuns at a convent in Gravelines received her and offered such comfort as they were able, although they were themselves prisoners in their own convent. The building was being used as a prison because all the normal prisons were already filled with 'enemies of the people' awaiting 'Madame Guillotine's' attention. Sophie had no option but to share the hardships imposed on the nuns. They slept together on boards and existed on a diet composed almost entirely of bread mixed with straw. Through a little grated window an old servant of the Carpentiers managed occasionally, at great personal risk, to pass up a few luxuries such as a tiny jug of milk or a piece of cheese. Soon afterwards Carpentier himself was arrested for having harboured a spy, but the charges were dropped although it was some time before he was released from prison.

Many good republicans testified to Sophie's excellent character and confirmed that she had taught their daughters. The Carpentiers never relented in their efforts to secure her release, but to no avail. Every day she was taunted by her jailers, 'Your turn tomorrow', as they dragged that day's victims into the courtyard. This was a most terrifying experience for an 18 year old to witness on her own in a strange country, deprived of any means of communicating with the outside world. When almost all hope of a reprieve had been lost, unexpected news came; Robespierre had fallen from power. On 24 July 1794 the gates of the prisons were flung open and the prisoners, including Sophie Kingdom, were free at last.

Details of Sophie's incarceration are difficult to come by. It seems it was not a part of her life she liked to recall and it was subconsciously erased from her memory. However, extensive research has revealed a few more details of the convent. It was originally called 'Nazareth' of the Order of St Clare and a 19th century manuscript contains an account of the years 1793-95 with a few details of the suffering endured by the nuns. On 12 October 1793 the convent was surrounded by guards, and the papers and property of the nuns seized. Five days later the communities of Benedictine and Poor Clares from Dunkirk were brought to Gravelines as prisoners, consisting of 43 persons, making a total of 77 prisoners at the convent. A few days later 'the Commissioners' arrived and effaced all tokens of royalty and nobility both within and without the convent and removed all the sacred vessels, vestments and ornaments. They also closed and sealed the church and sacristy, and the nuns were informed that they were to be removed to Compiegne. They were constantly harassed and alarmed by visits from committee members and minor officials making fresh inventories, carrying away sacred ornaments and defacing whatever crosses, holy names and religious tokens remained in any part of the convent.

In one of these visits, however, they experienced so much humanity from the Procurator Syndic of Bergnes as to cause him to declare that the proposed removal of the nuns to Compiegne was impracticable, and also to write to the Republican Minister of the Interior to that effect. For 18 months the three communities were confined together and suffered intense privations and numerous afflictions, particularly the shortage of fuel in a very severe winter. They were reduced to cutting up the cupboards and wainscoting of the house and the trees in the garden to survive. They were allowed only the English equivalent of twopence a day each to cover all their requirements — food, heating, clothing and medicines — and during their period of confinement their Chaplain and four Sisters died.

When Robespierre fell from power they secured their freedom, but seeing no prospect of an end to their miseries in France, they decided to return to England. They left Gravelines on 29 April 1795, sailed from Calais on the 30th and reached London on 3 May. The Duchess of Buckingham provided for them during their stay in London and eventually made a retirement house available for them at Gosfield in Essex.

* * *

Following the storming of the Bastille on 14 July 1789 most English observers believed that an official organisation was set up to take over the running of affairs in France, and this opinion is still held by many in England today. Nothing could be further from the truth;

the Bastille had been stormed by a group consisting mainly of ignorant peasants led by a few hotheads, a disorganised, starving rabble with little in mind other than the immediate object of releasing the prisoners and thereby forcing their problems to the notice of the authorities and the king in particular. But immediately several other factions joined in, and the Commune was set up consisting of a dozen or more opposed and opposing groups, each determined to advance its own ideas; consequently the Commune was at war with itself right from the beginning.

Robespierre was the leader of one of these groups and the most fanatical of the leaders of the French Revolution, and when the Terror was instigated it gave him the opportunity he needed to eliminate many of his enemies and others opposed to his plans. On 19 March 1794 he had Hébert (one of his main antagonists) and his associates arrested and five days later they were executed. On the 30 March another group, Danton, Camille Desmoulins and their confederates, were arrested, and on 5 April they too were guillotined. Robespierre was now the virtual dictator of France and able to put all his schemes into operation, especially those emanating from Rousseau. He was determined to increase the pressure of the Terror, especially in Paris and on 10 June brought in a new law under which even the appearance of justice was abandoned; since no witnesses were to be called or examined, courts became simply places of condemnation and sentencing. As a direct result of this law, between 12 June and 28 July 1794, a total of 1,285 victims perished on the guillotine in Paris alone. Mobile guillotines were transported on wagons to outlying areas of the country and set up in town and village squares where residents were encouraged to inform on suspected royalist sympathisers who were dragged out and executed on the spot without trial. Many innocent people who had offended someone in the community or who had become unpopular for some reason were denounced as 'Enemies of the People' — it was an excellent way of removing an annoying neighbour or terminating a pending lawsuit.

The total number of people guillotined or otherwise murdered during the Revolution has always been disputed; according to the source consulted it can vary between 10,000 and 60,000, but D. Greer's book *The Incidence of The Terror in the French Revolution* published in 1935, suggested a figure of between 35,000 and 40,000 and this is now accepted by most authorities.

* * *

The Carpentiers were still in the same house at Rouen when Robespierre died, and welcomed Sophie with open arms. They carefully nursed her back to health before arranging for her return to England the following year. She had been unable to acquaint Marc of her predicament during her incarceration at Gravelines and it is probable that he was unaware of her situation until after her release from captivity. On 7 February 1799 Marc Brunel left America for England aboard the *Halifax*. His first priority was to contact Sophie and then to contact the Navy immediately afterwards.

England

3.

During the period that Marc spent in America he converted to Protestantism. At that time much of the population of the eastern seaboard of the United States was staunchly Puritan in origin. Although New York itself was populated by a variety of nationalities holding different religious beliefs, the majority of senior posts in the city were reserved for members of the Protestant communities. Perhaps it was necessary therefore for Marc to abandon his old faith before he could take up any position of importance in the city. He might also have been advised to convert before travelling to England, because at that time Roman Catholics were viewed with suspicion by those in authority and still subjected to a considerable amount of prejudice. This, coupled with his French nationality, would make serious consideration of his plans by any English government agency unlikely. It is also worth considering that his involvement with Sophie Kingdom might have been a factor in his conversion. Her family would have been expected to oppose any suggestion of marriage to a Catholic at a time when England was being threatened by members of the Holy Roman Empire, including France, Holland and Spain. It is however beyond doubt that Marc arrived in New York a Roman Catholic and departed from America a few years later, a convert to Protestantism.

* * *

After a month at sea, the *Halifax* put into Falmouth on 7 March. On disembarking, Marc travelled to London where he met Sophie's brother at Somerset House. He took rooms in the parish of St Mary Newington and immediately contacted Sophie who was then lodging with her brother in the capital.

Sophie was now 23 years old. She had grown into a beautiful woman but had lost none of her girlish charms and happy disposition. Marc's appearance was described by Beamish as being of below average height, with a conspicuously large head, which nevertheless did not detract from the symmetry of his body. His forehead was apparently so striking that a friend of Beamish, on meeting Brunel for the first time, was tempted to exclaim, 'Why, my dear fellow, that man's face is all head!' An impressive forehead was said to be a distinctive feature of the Brunels. Sophie had been besieged by suitors since her return from France but had remained constant to Marc and according to Lady Gladwyn, the great-granddaughter of Isambard Kingdom Brunel, one of Sophie's suitors was a certain Mr Walter Taylor from Southampton, of blockmaking fame. Sophie turned his proposal down, leading Marc to recount triumphantly in his later years how he had managed to supplant the blockmaker in two ways. Marc was the only man Sophie had seriously considered as a partner in marriage, and in spite of the fact that the Kingdoms were suspicious of all Frenchmen (as were most Englishmen at that time), the couple announced their engagement.

Brunel moved into lodgings in Canterbury Place, Lambeth, and within a month of coming to England he had filed his first patent application. This described a 'duplicate writing and drawing machine', which he called a 'Polygraph' (see Appendix 12A). Although this was his first patent application in England, it was far from being his first invention; when only 17 he had designed and constructed the ebony and brass sextant which accompanied him throughout his naval career. He had also invented an ingenious little machine for winding cotton-thread into balls and a system of making the small boxes used by druggists, which up until then had been predominantly manufactured in Holland. He had also designed a nail-making machine, a machine for hemming and stitching and a device for shuffling playing cards. At this time Marc

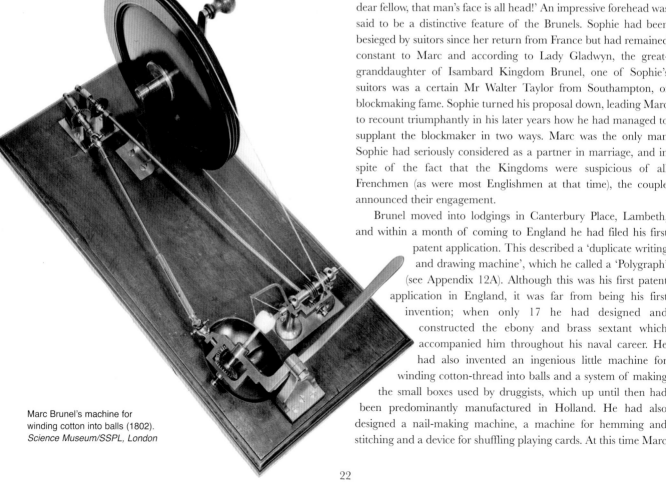

Marc Brunel's machine for winding cotton into balls (1802).
Science Museum/SSPL, London

spent most of his waking hours making drawings of the machines he was designing for the mass production of ships' pulley blocks, but the drawings were intelligible only to engineers. He knew he would need working models to get his ideas accepted by the Admiralty.

During the early summer of 1799, Marc made the acquaintance of a fellow émigré named de Bacquancourt. He introduced him to Henry Maudslay who owned a small engineering business at Wells Street in the West End of London. Maudslay had been apprenticed to and later became manager of Joseph Bramah's company, where he learned his skills under the influence of this outstanding craftsman (see Appendix 10). Bramah was a superior lock-maker who also manufactured the first fountain pen, an ingenious banknote-numbering machine, a beer engine and the first practical water closet many years before Crapper. In 1795 he had patented the first hydraulic press and was famous for his 'unpickable lock' which he displayed in his shop window, offering a reward of 200 guineas to anyone who could pick it. (This reward remained unclaimed for 67 years.) He was also one of the first people to investigate the use of screw-propulsion in ships.

Maudslay left Bramah's employment after a dispute over wages and set up his own workshop in Wells Street. He was easily the greatest mechanic of his generation. He invented the screw-cutting lathe (which for the first time enabled standard screw-threads to be interchangeable), the slide-rest lathe, the planing machine and the micrometer. He also invented a process for printing calico. It is fair to say that without Maudslay's expertise and advice, the careers of both Marc and Isambard would have been very different, for Maudslay (and later his company) became involved in most of their projects. Many famous engineers obtained their initial instruction and knowledge from an apprenticeship under Maudslay's eagle eye. These included

Top: 'Men of Science Living in 1807-8' (detail) by Sir John Gilbert. Henry Maudslay is standing centre with hand half raised. *National Portrait Gallery, London*

Above: Joseph Bramah (1748-1814), inventor of the hydraulic press, water closet, banknote numbering machine etc, and instigator of screw propulsion in ships. *National Portrait Gallery, London*

Bramah's water closet produced years before Crapper. *Science Museum/SSPL, London*

married on 1 November 1799 at the church of St Andrew in Holborn and moved into rooms in Bedford Street, just a few minutes' walk from Maudslay's workshop.

According to Dickinson who wrote an article on the Taylors of Southampton for the Newcomen Society in 1955, the vast improvements in the machinery for making ships' pulley blocks was due to the combined efforts of Sir Samuel Bentham of the Navy Board in initiation, of Marc Brunel in design and of Henry Maudslay in mechanical precision, and obscured the improvements made more than a generation earlier by the Taylor family.

The use of block and tackle in ships can be traced as far back as Egyptian models and wall paintings dating from 2000BC. By the use of a pair of blocks, each with one sheave, the amount of purchase is doubled; with two pairs of sheaves, quadrupled and so on. A block consisted of a shell, normally of hard wood, mortised to take a shiver or sheave turning on a pin through the centre of the shell. A serving of rope in a groove round the shell had an eye for attachment to the point of support. Originally the shell was fashioned with an axe and the mortise cut with chisel and mallet. According to Dickinson, the sheaves of lignum vitae (the hardest of commercial woods derived from a tropical American tree and resistant to salt-spray) were sawn off the end of a log by hand or cross-cut saw, and consequently were not absolutely equal in thickness. The sheave was turned on a wood lathe, the hole was bored with an auger and the pin, also of lignum vitae, was similarly turned. The finished article was rough, and instances are on record of blocks catching fire due to the friction developed in use. A ship's safety could also be endangered by the sailors being unable to furl the sails quickly enough due to the ropes sticking or even seizing in the blocks. The only remedy in these cases was for the sailors to climb the rigging, remove the blocks and scrape the sheaves and pins with their knives — an extremely dangerous operation, especially in a storm or battle.

* * *

J. P. M. Pannell, in his article portraying the Taylor family as pioneers in mechanical engineering, some of which has been reproduced below, wrote that during the mid-18th century in St Michael's parish, Southampton, two men working in a dark damp cellar developed an idea that was to be of immediate value to the Navy and which became the foundation of a business which supported their descendants for nearly 100 years. They were Walter Taylor (1734-1803) and his father Walter Taylor, the second and third of a generation of three Walter Taylors, all of whom were highly skilled craftsmen carpenters (see Appendix 8).

The first Walter Taylor was a prosperous Southampton master builder and carpenter who had an ambition to build a house for himself at Newport on the Isle of Wight. To this end he purchased the materials of the ruined Netley Abbey on the outskirts of Southampton and planned to ship the stone to his site in Newport. However, during the demolition, a falling stone fractured his skull, and a blunder by the surgeon whilst attempting to remove the fragments of broken bone resulted in his death.

Joseph Whitworth, famous for his compressed steel gun and the 'Whitworth screw thread'; Richard Roberts, who was responsible for designing and manufacturing machinery for the cotton industry and railway locomotives; Joseph Clement, the builder of Babbage's calculating machine, now acknowledged to be the pioneer of modern computers; and James Nasmyth, the inventor of the steam hammer, the steam pile driver and a hydraulic punching machine. By the time Maudslay was contacted by Marc Brunel, he was already producing blockmaking machinery for Taylors of Southampton, one of the main suppliers to the Navy. By selecting Maudslay to work with him, Brunel had, according to *The Times*, shown true discrimination, 'and thus, possibly, was laid the foundation of one of the most extensive engineering establishments in the kingdom, and in which, perhaps, a degree of science and skill has been combined and applied to mechanical invention and improvement scarcely exceeded by any other in the world'.

Work at Wells Street was concentrated initially on two models for the blockmaking enterprise — a sheave-making machine and one for mortising. Marc was a regular visitor to Maudslay's workshop for most of 1799, and there was able to modify his designs and ideas as problems in production were encountered. Although he was rapidly whittling down the savings he had brought with him from America (he was earning nothing), he and Sophie remained confident that his machines would be readily accepted and so they faced the future with complete confidence. They

His son, the second Walter, was engaged as a carpenter on a ship trading to the Levant. After several uneventful voyages he was taken prisoner by the French and imprisoned in Rouen, where he claimed to have constructed an electrical machine using an ostrich egg in place of a glass cylinder. During his time at sea he was concerned by the poor-quality lifting tackle and pumps with which all ships were fitted in this period and on his return to England he began to consider ways of raising the standards of the equipment. During his absence his son, the third Walter, had become apprenticed to a blockmaker named Messer who was in business in Westgate Street, Southampton, making pulley blocks under contract to the Navy. Whilst the younger Walter was completing his apprenticeship, his father, now freed by the French, visited all the blockmaking workshops he was able to reach (and it was a major industry at that time) in order to obtain an overall view of the methods in use. Soon after completing his investigations, Messer died and the Taylors were able to acquire the business and equipment on favourable terms. They were now in a position to put their own ideas into effect.

Since theirs was a new process, they needed privacy to develop ideas and designs for new machinery. To this end, a workshop was established in a cellar which had no windows and little ventilation. According to Pannell, the Bugle Street area of Southampton was soon filled with rumours that the Taylors were in league with the devil, producing artefacts on his behalf. It was under these conditions that the prototypes of machines were constructed which were to become part of the first plant for the mass production of blocks to a much higher standard than had previously been possible. The Taylors constructed a slide with a hand-operated saw to cut the sheaves to an exact thickness, and a boring machine to drill holes perfectly true. They also constructed a treadle-operated wood lathe on which to turn the sheaves and pins. The prototypes were driven by manpower, but later they used horses and later still a steam engine manufactured for them by Boulton & Watt.

Having progressed thus far, in 1759 they submitted specimen pattern blocks to the Board of Ordnance which after close inspection and comparison with the older productions agreed that 'Messrs. Taylor shall supply all the gun-tackle blocks for the Navy and continue to do so as long as required'. Gun-tackle blocks were simple products compared with blocks for the rigging, chiefly because they were used at ground level and any imperfections could be remedied in reasonably safe conditions. Encouraged by Hans Stanley, one of the Lords of the Admiralty and MP for Southampton, a complete set of topsail yard blocks for a 74-gun ship was produced as an experiment and Walter Taylor II took them to London for the inspection of the Lords of the Admiralty after a lengthy delay. They directed the Navy Board by letter dated 21 April 1761 to institute experiments with them and to submit a report. The resulting order was for blocks for 'main topsail Halyards and Topsail Tyes for ten ships of the line', intending them to be tried and tested on the Mediterranean station.

This very large order involved the Taylors in great expense, leading to financial difficulties. The officers of Plymouth Yard reported to the Admiralty on 28 April 1762 that they had 'provided Messrs Taylor & Son with the dimensions and every information they required to make their proposed New Invented Blocks by, but upon enquiry they find there is a Commission of Bankruptcy taken out against them and they are absconded, so that they have had no blocks for them to make the trial with'. This was an exaggeration; the Taylors were hard pushed for funds to enable them to complete this massive order, but they were certainly not bankrupt nor had they absconded.

Walter Taylor II had died in 1759, leaving his son at the age of 25 to succeed to the business and continue negotiations with the Navy Board. He managed to get the experiments transferred to Deptford, and when they proved successful the Board decided to equip the entire fleet with the new blocks as quickly as possible. However, Sir Thomas Slade, Surveyor to the Navy, expressed doubts as to whether the Taylor organisation was capable of fulfilling such a large order, whereupon Walter invited him to inspect the premises at Southampton to see the plant. After a visit, Slade was satisfied and the contract awarded. How Walter Taylor managed to persuade Sir Thomas that such a tiny workshop could produce blocks at the required rate remains a mystery. The blocks were accepted by the Board and paid for at once (a rare occurrence

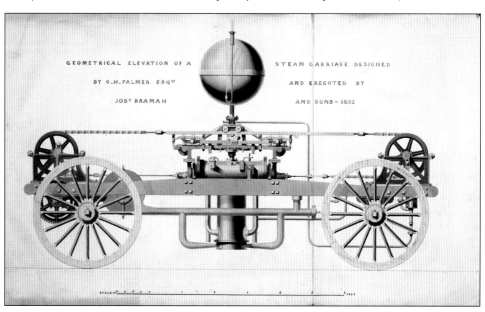

The steam carriage designed by G. H. Palmer and manufactured by Joseph Bramah *c*1832. *Science Museum/SSPL, London*

Above: Walter Taylor II holding his circular saw.
Southampton City Heritage Services

Right: Diagram of a Taylor pump with pendulum valves. *Institution of Civil Engineers, London*

Below: Map showing the location of the Taylor workshops in Southampton (not to scale). *J. Williams*

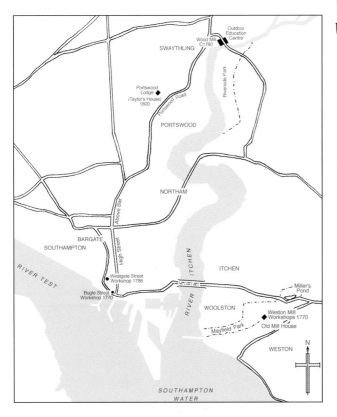

indeed) thus enabling Taylor to discharge the debts which had been incurred by the firm. The Taylors' connection with the Admiralty lasted for 50 years, a period of critical build-up for the naval battles against the yet unborn Napoleon and years of growing reputation and prosperity for Walter Taylor III.

The production of high-quality blocks had an immediate effect on the efficiency of the fighting ships. Captain Bentinck, a naval officer of advanced outlook, suggested that a series of further tests be conducted with the Taylors' blocks. Up until this time it was believed that a larger sheave offered a greater purchase, and this had led to the use of unnecessarily large and cumbersome blocks. Captain Bentinck took the courageous decision to have his ship, *HMS Centaur*, rigged entirely with Taylor's blocks of half the usual size. The trials confirmed that these were fully effective, much cheaper, and reduced the weight on the masts alone by 26 hundredweight (over $1\frac{1}{4}$ tons).

In 1770, the Navy's entire stock of blocks at Portsmouth was destroyed by fire. Taylor was contracted to replace them as quickly as possible and suggested the total conversion to the new smaller sizes. Naturally there were objectors, but these were silenced after official tests. Together, Taylor and Bentinck drew up a table of standards and thereafter the Navy was equipped with blocks in accordance with the approved tables. An important feature of Taylor's blocks was the care taken to reduce friction. Although lignum vitae was used for the carefully bored sheaves which ran on a turned pin and was a great improvement on those of his predecessors, Taylor was not satisfied. He experimented and then patented several methods whereby the life of the wearing surfaces could be extended. One of these was the substitution of iron pins for the lignum vitae. However, his most important improvement was to fit a flanged bush or 'coak' of antifriction metal. This coak was riveted through the sheave to a loose flange and was renewable. The success of these bushed flanges enabled Taylor to guarantee his sheaves against failure for seven years — a phenomenal achievement in those times.

The expansion of the Taylor empire from the cellar at Westgate to the Weston Mill and finally to the Woodmill workshop is described in Appendix 8, but a contemporary guidebook of Southampton provides details of what was happening at Woodmill in 1802: 'Mr Taylor has erected a curious manufactory for ships blocks . . . from which His Majesty's Dockyards are supplied with all kinds of blocks, from the largest made use of in a first rate man-of-war to the smallest wanted for boats. Everything necessary for expediting the blocks are likewise manufactured here.' The guidebook goes on to explain that 'The dexterity with which these

works are carried on will at once astonish and delight the beholder. Mr Taylor has a patent from the Crown for the security of his invention, and we are happy to hear he is amply remunerated for his ingenuity.'

As well as the Taylors, credit must also be given to Sir Samuel Bentham for the improvement in machinery that made ships' pulley blocks. Samuel Bentham was born in London on 11 January 1757, educated at Westminster and apprenticed to the Master Shipwright at Woolwich Dockyard at the age of 14. At the termination of his apprenticeship he went to sea. At the age of 23 he disembarked at St Petersburg in Russia and entered the service of Catherine the Great. He travelled widely in Russia as a navy official and an Inspector of Mines and Metal Foundries and other engineering projects including the Fontanka Canal. In 1782, after only two years' service, he was made a Knight of the Russian Order of St George. He was instrumental in introducing some revolutionary heavy armament into the Russian Navy, which resulted in a much smaller Russian force defeating the Turks in 1788. He returned to England in 1791, and in 1793 patented several woodworking machines including one for mortising and a mechanical wood-planer. From 1795 he devoted the next 20 years to building up England's naval defences during the critical period of the Napoleonic wars, and he was an early advocate of explosive-shell weapons. He also developed the Arrow class of sloop which was widely used against the French Navy.

He made great advances towards eradicating the corrupt practices which threatened the lives of British seamen. Corruption and maladministration were endemic from top to bottom of the Navy. Ships often put to sea with sails that shredded in the first gale and ropes that disintegrated within weeks. Unscrupulous suppliers made huge profits by delivering rancid meat and stale, weevil-infested bread and biscuits. Even the contracts for the construction of the ships were awarded against a system of bribery by officials who were supposed to be 'servants of King George'. It was never proved how much the Admirals in London benefited from these practices, but there is no doubt that they were aware of them and encouraged them. Therefore when Bentham began his purge they did their best to discredit him.

In 1796 Bentham was appointed Inspector-General and began to overhaul the administration of the naval dockyards, during which he introduced several technical innovations including some of his own inventions. The supply of blocks to the navy was a continual problem, and at Portsmouth Dockyard he installed a 12-horsepower steam engine to drive woodworking machinery.

In 1801 Marc filed an application for a patent on his blockmaking machines (see Appendix 12B) and at about the time of the birth of his first child, a daughter who was named Sophia after her mother, they moved to Gerrard Street, Soho, and later to a house in Queen Square Place, just east of Southampton Row, still only a few minutes' walk from Maudslay's workshop.

Marc's next move was to contact Earl Spencer, the Navy Minister in Pitt's government, to present his letter of introduction explaining his ideas regarding the mass-production of ships' blocks. Spencer was impressed and agreed to approach the Admiralty with Brunel's proposals. Without Spencer's assistance, it is unlikely that Marc would ever have been able to contact the Admiralty with any hope of success. Their Lordships were (and probably still are) of the opinion that their knowledge of naval matters far exceeded that of any mere 'landlubber', especially one of French origin. Earl St Vincent was now the First Lord of the Admiralty, and after the usual delays and difficulties normal in such high places, but assisted by the powerful influence of Earl Spencer and the recommendations of Brigadier-General Sir Samuel Bentham, the system was accepted for initial trials.

Every British man-of-war was fitted with approximately 1,400 of these blocks (plus a quantity held in reserve in the hold) and the continuity of supply was a constant headache for the Navy Board who had placed the main contract with Taylors of Southampton. Marc's first instinct was to approach the Taylors to offer them his new machines and methods, but Samuel Taylor, managing the company for his ailing father, felt that the firm could not afford the outlay involved to replace their existing plant. Nor could they consider closing down the premises for several months whilst the new plant was being installed. He has been criticised for his lack of foresight but it should be appreciated that the designs submitted by Marc Brunel to Taylor were not those completed designs of blockmaking machinery as later constructed in Portsmouth. At that time Brunel had only designed the shaping machine which was a type of lathe for turning the block to its finished oval form.

Sophie's brother, who was then an under-secretary to the Navy Board, agreed to approach Taylor on Marc's behalf, since his connections with the Admiralty often brought him into contact with the company. Samuel Taylor replied to his initiative on 5 March in the following terms:

'I will just describe in a few words how we have made our blocks for upwards of twenty-five years — twenty years to my own knowledge. The tree of timber, from two to five loads' measurement, is drawn by the machine under the saw, where it is cut to its proper length. It is then removed to a round saw where the piece cut off is completely shaped, and only requiring to be turned

EXPLANATION of the CONSTRUCTION & USE of the Patent Block Mill.

1 The horizontal horse wheel, or great sweep, is from 18 to 20 feet diameter. The upper surface is 9 or 10 inches broad perfectly smooth and runs correctly horizontal.

2 The Vertical or Friction wheel. Of this there are two sorts double and single; the first runs with two leathers round its circumference, giving motion to two lathes, or to a lathe and a saw frame and is thus constructed. Its centre consists of a solid cylinder of about 20 or 24 Inches diameter and 8 or 9 Inches long, with a surface perfectly smooth; the extreme circumference of the wheel is from 3 to 4 feet diameter carries two leathers, — by being hollowed towards the middle and rising towards the edges, thus ◁ by which the leathers running on opposite sides do not interfere with each other the felley is about 5 Inches thick, and about 3 Inches deep. The single wheel differs from the other only in the shape of the felley which has one curvature only thus ◠ and is about 3 Inches square. The Axis of these wheels is of Steel, about one Inch diameter.

3 A Square Iron box inclosing patent rollers, on which the Axis of the friction wheel works. This box is fastened to

4 A frame of Oak, the sides and ends of which are 6 Inches broad and 4 Inches deep the length of the frame in the clear is 6 feet and its breadth about 15 Inches. This frame is fastened by hinges at one end to a piece of Oak fixed to the joist or floor and is occasionally lifted by the lever, removing thereby the friction wheel from being acted upon by the great sweep wheel.

5,5 Are the two levers used to lift the friction wheel from the great sweep by a cord from each lever to one lathe or to one saw frame. Thus when the workman wants to stop the machine he pulls upon the cord, and raising the friction wheel about one inch from the great sweep its motion ceases. And when he wishes to renew his work he lets the cord go, and immediately the two wheels coming into contact the motion is renewed.

One disadvantage attends the use of the double wheel, which is that its removal from the great sweep causes the two machines to rest.

6 Are the leather bands which pass round the friction wheel to the wheel of those Machines which are to be set in motion.

Its uses

From the previous explanation of the construction its uses may be readily conceived Five, six, or seven friction wheels may be adapted to the great sweep in proportion to its size, the power used to turn it and extent of the building containing it; and these wheels may be double or single. — The great ends to which this invention is applied is the cutting of sheaves by the saw frame; the turning of sheaves, pins &c by the common lathe; & the boring of sheaves blocks &c by a lathe with a sliding puppet. It may be necessary here to explain the nature of the lathe with the sliding puppet & also the construction of the saw frame. This lathe differs from the common lathe in this. The puppet, on which the work to be bored is fixed, is larger than ordinary, & slides on brass; it is pulled backward from the mandrel by a weight hanging through a sheave in the leg. There is likewise fixed in the sliding puppet two sheaves, and in the other leg two sheaves; through which a rope passes having its standing part fixed to the leg, and when its tail is pulled on by the workman it presses the sliding puppet which has on it the sheave &c to be bored towards the bit in the mandrel; and when the hole is made, the workman lets go the rope & the weight draws back the puppet. The saw frame is worked by the leather band going round the wheel; the parts of which marked 8 are boards 6 or 7 inches wide which slide in front being dovetailed. These are brought out to a little beyond the edge in order to keep the band from slipping off. The timber to be sawed is confined near the saw by blocks & a wedge & at the other end by means of a screw which pressing down on it prevents its rising

A double Friction wheel

Walter Taylor's patent blockmill. Originally powered by horses, it was later steam driven. *Southampton City Heritage Services*

under the saw. The one, two, or three, or four mortises are cut in by hand, which wholly completes the block, except with a broad chisel cutting out the roughness of the teeth of the saw, and the scores for the strapping of the rope. Every block we make (except more than four machines can make) is done in this way, and with great truth and exactness. The shivers are wholly done by the engines, very little labour is employed about our works, except the removing the things from one place to another.

'My father has spent many hundreds a year to get the best mode, and most accurate, of making the blocks, and he certainly succeeded; and so much so, that *I have no hope of anything ever better being discovered, and I am convinced there cannot.* At the present time, were we ever so inclined, we could not attempt any alteration. We are, as you know, so much pressed, and especially as the machine your brother-in-law has invented is wholly yet untried. Inventions of this kind are always so different in a model and in actual work.'

The receipt of such a letter was a great disappointment to the Brunels, who had convinced themselves that Taylor would be very keen to take up their proposals. Marc put the setback behind him, however, and with Sophie's encouragement sought another interview with Earl Spencer who, following Pitt's resignation, had been forced to vacate his office as Navy Minister.

Marc Brunel, unlike many of his contemporaries, was a cultured and literate gentleman who had travelled widely. The upper classes in England therefore found him socially agreeable and acceptable. His written English was perfect (as proved by his journals) but his speech still carried that slight French accent which so intrigues the English in general (Charles Boyer and Maurice Chevalier had similar appeal in later years), and as a former officer in the Royalist French Navy, Marc did not attract the class prejudice which would have stifled the ambitions of the son of an English farmer. He was warm, pleasant and considerate and deferred, sometimes excessively, to English customs. When Earl Spencer introduced him to Sir Samuel Bentham, the two took to each other immediately and remained friends until the latter's death. Through him, Marc was introduced to several significant personalities of the day, including Faraday and Babbage, and he became an intimate of the landed ruling class which he much admired. It is interesting to note that some years later, in March 1823, Marc recorded in his diary a visit to Charles Babbage. During his stay, Babbage showed Marc his machine for solving calculations, 'to a very extensive reach'. Marc thought it extremely ingenious. Little did he know that he had just witnessed the machine that would become the forerunner of today's modern computer. Brunel commented in his diary that he thought the drawbacks to producing such a machine would be great, involving much expense as well as a large amount of labour. He was to be proved right on both counts.

Bentham's resignation from his position of Inspector-General of Naval Works in the Pitt government had not affected his reputation among those of his friends and acquaintances (the honest ones) who knew the details of his work for the Navy, and the improvements he had already brought about in that institution which, until his appointment, had scarcely progressed since the time of Samuel Pepys. He arranged for Brunel's models to be demonstrated before the Lords Commissioners of the Admiralty, who agreed to a series of trials being conducted.

* * *

The battle of Trafalgar was only about four years into the future and the tonnage of the Navy had increased by more than 60 per cent during the previous decade. Taylors of Southampton were now producing about 100,000 blocks per year and they thought their products so superior to those of their rivals that each block was guaranteed for seven years and the sheave stamped with the initials WT followed by the year of manufacture. But the whole system of supplying to the Navy was corrupt and contaminated by nepotism

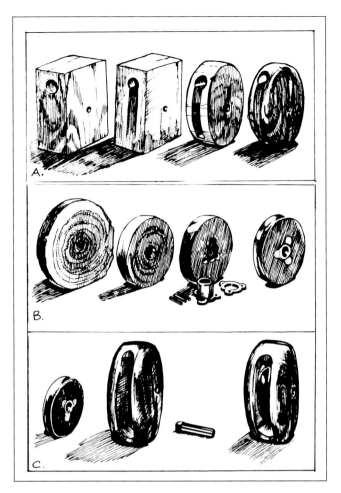

A.

B.

C.

Dockyard. His was the decision that counted and, after the normal delays and protracted discussions, the Navy Board finally had no option but to agree.

Towards the end of 1802 the Brunels moved from London to a modest house in Portsea, a short walk from the dockyard, which enabled Marc to be in a better position to supervise the installations. Writing in his journal in 1841, Marc recalls how 5 November 1802 was the day he first entered the office of the dockyard at Portsmouth — a memorable day, he notes, as it marked the first day of his public career. Bentham had been successful in persuading the Navy Board to reward Brunel with a sum equivalent to the savings from one year's *full-scale* operation of the new plant. He was also given an expense allowance of one guinea a day whilst engaged on His Majesty's service, plus an extra 10 shillings a day whilst in Portsmouth or when travelling, and coach fares at cost.

It had taken four years for the machinery to be designed, perfected and manufactured by Henry Maudslay, but skilled men were needed for its erection at the Portsmouth Dockyard. As such men were in great demand at a time when the Industrial Revolution was in full flow, they had first to be obtained and instructed. By October 1803, with the plant fully assembled, Marc began working with six labourers and two carpenters, but by the following May, he had only managed to train two actual blockmakers. His application to the Navy Board for extra labour received no answer, and this at a time when the war with Napoleon was in progress.

The Board then appointed a general manager named Burr. He was an inexperienced mechanic who took an instant dislike to Brunel and constantly allowed outsiders to crowd into the yards

and graft from top to bottom — a system acceptable (to some) in peacetime but quite intolerable under crisis conditions. Consequently the government had appointed Brigadier-General Sir Samuel Bentham with the object of bypassing the Navy Board, a department which, in the words of Beamish, 'seems to have been only calculated to enlarge patronage, decrease responsibility, and multiply the links in the official drag-chain of the naval service'. In 1802, Bentham sent his engineering manager, Simon Goodrich, on a tour of the works of several

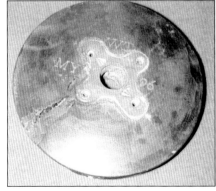

while work was in progress, charging them a shilling a time to enter. In the spring of 1805, Marc wrote to Bentham, 'This frequent admission of visitors is of great hindrance to the men at work: the place was, the whole morning, crowded with visitors much to the annoyance of the Service.' He ended by pleading with Bentham to surround the wood-mill with a fence to prevent intruders gaining access. When Burr objected, Marc locked the door to the mill and posted a man on the door with instructions to admit no persons other than officers in uniform.

Being a Frenchman by birth, Marc was

English blockmakers including those of Walter Taylor in order to compare the different methods of production. From Goodrich's report it appeared that only William Dunsterville of Plymouth could compete with Taylor's methods, and he was already using circular and reciprocating saws and a variety of other machinery similar to those in use at the Taylor works. So whilst Samuel Taylor reflected complacently on his father's 'perfect' methods, Bentham was exploring every avenue towards increased productivity.

On 15 April 1802, Bentham officially recommended the installation of Brunel's blockmaking machinery at Portsmouth

still regarded with suspicion. He was not allowed into the Portsmouth Dockyard without a special permit, which he was forced to carry with him at all times. The annoyance of responsibility without authority and the constant problems of admittance to the yard grated on Marc's nerves so much that he refused to continue and resigned from his post. He was only persuaded to reconsider his resignation after being 'waited upon by some important gentlemen' who claimed that 'nothing could go on at the wood-mills without his immediate attendance'. One of these gentlemen was undoubtedly Bentham.

4. Blockmaking at Portsmouth

It was fortunate for Marc Brunel that during the discussions and arguments that were in progress between Bentham and the Admiralty concerning improvements to the supply of blocks, a fire destroyed the Navy's entire stock in the Portsmouth warehouse. Thus their Lordships were forced to a decision which in normal circumstances would probably have been delayed for years. There had been smaller fires at the dockyards on several previous occasions when most of the buildings were constructed of wood, many with thatched roofs. The most memorable however was that caused by a Scottish arsonist called James Aitken ('John the Painter') whose attempt in 1777 to destroy the entire dockyard was thwarted only by a combination of bad luck and ignorance. He had calculated upon the ships, loaded with explosives and ammunition, being unable to move from their berths alongside the quays at low tide. When the conflagration began, however, the alarm was instantly raised and the ships were able to leave their berths because it was high tide. An investigation uncovered the lax security measures in place at the dockyard and resulted in the strict regulations in force during Marc Brunel's employment in the block factory, and led to the rebuilding of most of the damaged buildings in brick and stone.

Financial matters were of little concern to Marc Brunel so long as he and his family were adequately provided for; his interest lay in his inventions. Had he been more of a businessman he would have insisted upon some kind of written agreement, but he naively assumed that the Navy Board would compensate him fairly without resorting to such trivialities as legal documentation. In a letter addressed to the Lords of the Admiralty in June 1802, Marc pleaded for his past two years of intense experimentation and development to be taken into account when considering compensating him for his invention. This letter appears to have received no reply and it was not until Bentham interceded on his behalf that their Lordships finally instructed the Navy Board to accept Bentham's proposals. This they did with bad grace and thereafter placed many obstacles in the way of full production, thus delaying payment for several years.

The Navy Board's attitude towards Marc is obvious from the letter sent to him dated 5 May 1803:

'I am commanded by my Lords Commissioners of the Admiralty, to acquaint you, that they have given orders that an allowance be made to you of one guinea per diem during the time you may be employed in erecting your machine for the manufacture of blocks in His Majesty's dockyard at Portsmouth, to commence from the 16th of September last, and to continue until the same shall be completed; together with the amount of coach hire actually paid, or to be paid, by you in your journeys to and from that place; and also, an allowance of ten shillings per diem, for extra expenses, during your absence from town; and that their Lordships will consider what farther reward may be proper to be made to you for your invention, whenever the extent of the advantages likely to be derived by the public from it, shall be fully ascertained.'

It may be deduced from this that the commissioners of the Navy Board were going to do their utmost to delay or avoid payment without actually refusing to acknowledge the instructions from the Admiralty. Although payment eventually gave Brunel a boost to his financial position, had he been more astute he could have had a steady income for life and would have been able to avoid many of the problems which later beset him. Marc and Sophie were now living in a house in Britain Street, Portsea, less than a mile from the Naval Dockyard. The allowance of 10 shillings a day while in Portsmouth would have covered their rent and day-to-day expenses but little else.

* * *

The first set of machinery was produced by Maudslay and installed by the middle of 1803. Brunel paid for this and recouped his outlay from the Navy Board, but payment for the later installations in 1804 and 1805 (costing £12,000) was made by the Board direct to Maudslay. This substantial contract enabled Maudslay to move his workshop to larger premises at 75 Margaret Street, London, where he was able to devote his skills to even larger projects including marine engines. A total of 45 machines were finally installed at Portsmouth. Three of these are on display at the Dockyard Apprentices Museum in Portsmouth Dockyard; others are held by the Science Museum in South Kensington and by the Smithsonian Institution in Washington, USA.

The major part of the Portsmouth plant was installed and operating within three years of Bentham's letter of approval. This included the replacing of a single 12-horsepower steam engine with one of 30 horsepower which drove the overhead shafts by which the machinery was operated. This steam engine was purchased from Boulton & Watt and a second, even larger one, was ordered from Murray & Wood. It is interesting to note the intense rivalry which was endemic to British industry in those days of rapid expansion. Boulton & Watt secretly purchased the land adjoining the premises of Murray & Wood, thereby preventing the latter from expanding their factory on the same site.

Brunel's first great achievement was to produce a block which, although showing very little improvement over Taylor's, could nevertheless be manufactured by unskilled labour at 10 times the rate, each being identical. The blocks were unique in that they required from the operators neither much judgement nor skill. Within months of the installation, the plant was

Left: Portsmouth blockmills front elevation in 2005. Originally two separate three-storey buildings, the yard between was roofed over to give a large ground floor workshop with a bridge connecting the upper floors. The railings in front of the building enclose a dry dock. *J. Bagust*

Below: View of the overhead shafting in Brunel's blockmills at Portsmouth. *J. Bagust*

regularly producing blocks of previously unknown precision by men, the majority of whom were unable to read or write. Ten unskilled men could, in the same number of hours, achieve the equivalent output of 110 skilled workers under Taylor's system. There were however constant problems, many of them caused by the continual intervention or deliberate obstruction by the Navy Board.

It seems obvious that both the government and Brunel must have benefited from the rapid introduction of his machines and methods, but on the contrary, every possible impediment was devised to hinder production (see Appendix 9). Labour to be trained was in short supply and the authorities were in no hurry to supplement it. Every application to the Board for more workmen was either ignored or sidestepped. By May 1804, Brunel, by his own efforts, had found *two* more blockmakers and trained them; his application to the Board for a third received no answer. He had by now completed the third set of machines, but had no workmen to operate them. He advised the Board that in time of war the Navy would require roughly 70,000 blocks each year of a certain size, and his machines were capable of delivering that quantity by 1 August if all the materials could be provided.

By July 1804 we find Marc still complaining about the lack of operators to man the machines, but now another problem was troubling him. Brass coaks and shivers were hard to come by, as they could not be made by the present founder of the yard. In August of the same year, he wrote to

General Bentham that, without any authority from the Navy Board, he had been compelled to procure them from Maudslay. There were further problems in January 1805 when Brunel wrote to Bentham regarding the problems experienced with the coak-maker: 'The difficulties and delays which are experienced to procure anything wanted, make him lose as great a part of his time in applying for, as in getting them.'

Eventually after many ineffectual requests for additional men to work the machines and frustrated by the restrictions placed upon him, Brunel decided that unless he was given complete control he would resign from the project. He was however prepared to stay on for such time as would be necessary for the new larger engine to be installed. When Mr Burr was appointed by the Board to take over complete control at Portsmouth, thus demoting Brunel to second in command, he was outraged, particularly since Burr was illiterate and had no knowledge of machinery or manpower handling. Burr was a constant irritation to Brunel, especially concerning the employment of labour. Brunel's method was to take on one or two men at a time and to give them a thorough training before allowing them control of the machines. Burr had other ideas. He engaged 15 boys without consulting Brunel and this resulted in a fall rather than an increase in production. The age of the boys is not stated but it was normal at that time for children to be employed in factories from the age of nine upwards for a maximum 12-hour day. On 3 July Marc recorded that 'since these boys were entered for the wood-mill, the circular cutters of the old scoring engine were broken by the inexperience of these workmen'. The Navy Board hated the thought of a civilian having any kind of power in the dockyards and did their utmost to ensure that all authority would be lodged firmly in the hands of dockyard officers who were employed by and directly responsible to them.

In July 1805, Sir Samuel Bentham was suddenly withdrawn from his duties as Inspector-General of Naval Works. This was ostensibly to undertake a mission to Russia but was more probably as a result of intrigue by members of the Navy Board who were opposed to his efforts to reduce corruption in that body. Whatever the true reason, his arrival in Russia was unexpected even by the British Embassy there. His remit was to supervise the building of warships for the British Navy, which it was said could not be produced in England because of the shortage of suitable oak in the country. However the Surveyor-General reported in 1806 that each 74-gun ship required the timber from 2,000 trees, each one 100 years old. Since an acre of land could sustain about 40 oaks, 50 acres were needed to produce the timber required for building one such vessel. He further reported that as the quantity of waste land in the kingdom in 1806 amounted to over 20 million acres, it should have been easily possible for the ships to have been built in English dockyards. This adds credence to the claim that Bentham was removed from office before he could cause any more trouble.

Brunel continued complaining about the delay in supplies and the quality of the timber purchased which was often unsuitable for use in blockmaking. On 18 August 1808, he complained to the Navy Board that 'from want of proper supplies of *lignum vitae* of large dimensions, the mill is working to considerable disadvantage . . . the machines, not being properly supplied with materials, are not kept regularly employed'.

The war with Napoleon had now been under way for five years but there was still little sign of urgency in the hierarchy of the Navy Board. Furthermore they now attempted to cheat the inventor by comparing the prices inflated by the war against the peacetime production of blocks so as to reduce the apparent advantages of the new methods to the detriment of Brunel.

In 1803 the Taylor's contract was renewed for 21 years, with optional break clauses at seven and 14. However, in March 1805, six months before Trafalgar, the contract was cancelled by the

Admiralty. Brunel's plant in Portsmouth was now well able to cope with the Navy's requirements. Walter Taylor III began proceedings for breach of contract, which would probably have been successful, but he died before his appeal could be considered.

Henry Maudslay worked with Marc Brunel in the planning and construction of the machines, and he was soon joined by the draughtsman Joshua Field who had also acquired his skills under Bramah's instruction. Joshua Field was a well-educated man; he became a founder member and vice-president of the Institution of Civil Engineers and was made a Fellow of the Royal Society in 1836. The firm now traded as Maudslay, Sons & Field.

In spite of numerous applications by Brunel for a settlement, the Navy Board always discovered a reason to delay payment. In 1808 the plant produced over 130,000 blocks of every size and species. The value of these articles was estimated at £50,000. In January 1809 it was found that the quantity manufactured during the last three months of 1808 would give an average of 160,000 a year, the value of which was estimated at not less than £54,000. The machinery was complete. It was capable of supplying all the blocks previously produced by Portsmouth, Plymouth, Deptford, Woolwich, Chatham and Sheerness and at a fraction of the cost. Brunel had spent over £2,000 of his own money producing models of machines, and conducting various experiments using more than 100 species of tropical wood in an effort to find a substitute for the costly lignum vitae; he even went aboard a Portuguese warship at Portsmouth which had recently arrived from South America in order to collect specimens of various Brazilian woods. Yet still the Navy Board prevaricated.

The distress which Brunel suffered in consequence began to affect his health. He was attacked with 'nervous fever' (today's nervous exhaustion) and was for some weeks unable to give his attention to any business. Eventually, in August 1808, the Admiralty directed that a payment be made, not of the promised amount, but of £1,000 on account! Brunel was devastated. He had already expended more than double that amount over the past six years on drawings, models and experiments in the service of the government. He had also refused several offers to design projects elsewhere while his main efforts were concentrated on the blockmaking plant at Portsmouth. It took another two years before the Admiralty finally consented to a payment of just over £17,000. This followed reams of correspondence and lengthy discussions, often involving several persons of renown, including Lord Spencer

who had publicly stated that the minimum amount to which Brunel was entitled was £20,000.

It was only after Marc Brunel's death in 1849 that detractors began to emerge from the shadows. In 1852, over 20 years after the death of Sir Samuel Bentham, his widow credited Bentham with the invention of the block machines and discredited Brunel's role in the matter. James Nasmyth, one of Maudslay's apprentices, was another who went out of his way to attribute the plant's invention to the work of his master and completely ignored Brunel's innovations. It says a lot for the integrity of Joshua Field that in a letter to Richard Beamish he stated:

'The works in progress when Mr Brunel arrived were a new steam engine and some buildings *intended* for the reception of machinery, which General Bentham *had proposed* to erect but *had not erected*. The General had already introduced saws of various kinds, and machines for tongueing, grooving and rabbetting timber; but there was no *machinery whatever* especially applicable to blockmaking. That was altogether the invention of Mr Brunel. The character of the drawings was different from any we ever had before — the proportion of the parts — the whole thing, in short; and I never once heard, during all the time of my connection with the dockyard, with General Bentham, with Mr Goodrich, and with Mr Maudslay, that any one ventured to deny Mr Brunel's claims to be the *sole inventor of the block machinery*.'

For many years after its construction, the block machinery at Portsmouth was an accepted part of the itinerary of distinguished sightseers. Czar Alexander I was very impressed with the installation and did his utmost to persuade Brunel to travel back with him to St Petersburg (an offer which Marc had reason to be grateful for in later years). Sir Walter Scott was conducted round the machinery in 1816 and described it as a wonderful sight. Without doubt, British victory in numerous sea battles of the 19th century was made more certain by the introduction of Brunel's blocks, for the chances of the blocks jamming or catching fire had now been almost totally eliminated.

From the very beginning, the efficiency of the Portsmouth installation was never in doubt. Within three years it had recouped the initial outlay and was still in production after the conclusion of World War One. Brunel's blocks were even used during the Second World War in the landing craft which transported men to the Normandy beaches on D-Day. Parts of the plant are still used today when replacement blocks are needed for Nelson's ship *HMS Victory*.

5. Battersea and Financial Problems

During the time spent in the installation of the blockmaking machinery in Portsmouth Dockyard, Marc Brunel was experiencing considerable financial difficulties. He had already paid Maudslay £1,200 for the working models out of his own pocket, and although 100,000 blocks had been produced in 1805, the Navy was demanding even greater production to equip the fleet which was being rapidly expanded to meet the ever-present French threat. The agreement with the Navy Board was that they would award him the equivalent of the savings made in one year's full-scale production of the plant. The amount saved would therefore increase in direct ratio to the increase in production, so the longer he waited before seeking a settlement, the more he would eventually be awarded. But he was already the father of two little girls (his second daughter was called Emma) and Sophie was pregnant again. The need for extra income was becoming more urgent by the day.

Before they moved to Portsea in 1802, Brunel had filed a patent specification entitled 'Trimmings and Borders for Muslins, Lawns and Cambrics' (see Appendix 12C). However, because of his preoccupation with work on the block machinery, this project had been put aside. His growing dilemma now forced him to resurrect the idea but he realised that it would be a long-term investment, unlikely to yield any substantial income for several years, even if he was successful in finding a backer. The patent specification is not illustrated and is mainly interesting because of the vast industrial knowledge which it reveals.

In 1805 Marc suggested an improved apparatus for the bending of timber, which in those days was simply heated in a kiln, then moulded to the required shape, with the result that it often distorted or fractured as it cooled. He proposed the construction of an apparatus whereby the timber would be softened by the action of steam or boiling water, and then moulded by a slow process to any required form, without being removed from the equipment during the entire operation. In the same year he filed a patent for 'Saws and Machinery for Timber Sawing' (see Appendix 12D). It was an arrangement of circular saws through which the logs would be propelled on sliding carriages. His design called for a saw made in several segments in order to reach a diameter of 5ft or over when required, with each segment having a 'V' profile to enable it to be locked at the radius line into the notched trailing edge of the adjacent segment. He produced working drawings and ordered a complete saw from Maudslay: this reached the Portsmouth Dockyard towards the end of 1806, together with the second 30-horsepower Murray & Wood steam engine which Maudslay had been testing at his Margaret Street workshop.

If Brunel had known how disgracefully the Navy Board had treated Walter Taylor over his block contract, it is unlikely that he would have gone forward without a definite order from the Board. It appears, however, that he was ignorant of Taylor's problems at that time. He went ahead with no more than verbal encouragement

and was fortunate that, on this occasion, the Board stood by its commitments. He was well aware that, unless he could improve production of the blocks, his reward would be reduced. In any case he was convinced that his saw had much wider commercial possibilities and he still retained the patent.

About this time he also designed a saw to cut the curved staves for casks. Maudslay produced a model which was tested at Portsmouth and demonstrated at Deptford Victualling Yard. It promised to revolutionise the extremely skilled art of barrel-making by producing 60,000 dozen staves of various sizes for dry and light casks during a 12-hour day using only seven men and a saw whetter. Barrels were an important part of British industry, being used for transporting wine, spirits, gunpowder, salted meat, fruit and manufactured goods, including coinage and works of art such as statuary. They were essential to the victualling of ships, both for the Royal Navy and the Merchant fleet, and the quantity produced annually was enormous. It is all the more surprising, therefore, that no further reference to Brunel's design has been discovered, except for a statement in a letter dated March 1807, to the effect that the Master Cooper at Deptford was finding it satisfactory, so it must have gone into production.

It was in the spring of 1806 that Marc wrote in his Journal, 'On the 9th of April, and at five minutes before one o'clock in the morning, my dear Sophia was brought to bed of a boy'. The child was christened Isambard Kingdom. He was to become the apple of his father's eye and one of the world's greatest civil and railway engineers.

Later that same year, Marc filed another patent specification entitled, 'Cutting Veneers' (see Appendix 12E). It is described in detail in the specification, but, put simply, it involved locking a log to the bed of a machine which could be raised by jacks in microscopic degrees; a sharp blade then traversed the surface of the plank, shaving off a veneer of the desired thickness. By now Marc was beginning to realise that inventions, no matter how ingenious, do not necessarily make money; unless they can be put into production they must always remain little more than a brilliant idea, yielding nothing for the inventor. Marc was in sore need of an income sufficient to support his growing family until such time as the Navy Board honoured its commitments regarding the blockmaking plant at Portsmouth.

It was at this point that Brunel was joined by a Mr J. H. Farthing and entered into partnership with him in a sawmill business at Battersea on the south bank of the Thames. It occupied 476ft of river frontage and was described as having easy access to the bridge (Battersea) and Chelsea market. Little is known about Mr Farthing, but it was against Marc's character to work in harness with a partner, so the arrangement was probably forced upon him by the circumstances in which he now found himself. Farthing was to be the financier of the new business, possibly because he was a banker or someone with access to working capital. Marc was to control the practical side of the enterprise. Although the above details appear in several earlier works on Brunel, they have never been confirmed. The *London Directory* of 1829 gives a clue as to who Farthing was. The entry states that he was an 'Improved Copying and Writing Machine and Polygraphic manufacturer of 42 Cornhill'. It is possible that Marc either sold his 'Polygraphic' machine designs to Farthing or licensed their manufacture out to him. It is also likely that Brunel and Farthing were known to each other before the Battersea sawmill was contemplated.

From the beginning there were difficulties at Battersea. The executors of the late owner were loath to part with the river frontage to the partners; they felt it could be better developed to their advantage. Once the lawyers were brought into negotiations, it was inevitable that the transactions would be prolonged, tedious and costly. The longer the delays, the less enthusiastic Farthing grew, especially when some teething troubles were encountered with the design of some of the new machinery for the sawmill. It had been decided to use the mill as a private factory for supplying blocks to merchant ships, and the East India Company had shown interest after a group of its directors inspected the Navy's plant at Portsmouth in November 1806. Eventually the idea of a private blockmaking factory was abandoned, probably on the insistence of Farthing, but he was supportive of proceeding with the circular saw and the veneer-cutting machine, and in December Maudslay was asked by the Navy Board to quote for another circular saw of improved design.

In March 1807 Marc wrote to the Admiralty and to the Board of Revision for the Merchant Marine, drawing their attention to the advantages of his circular saws: 'I do not hesitate to say that the price actually paid for sawing will be reduced from three shillings per hundred feet to six pence.' The approbation of the Master Cooper of the Victualling Yard at Deptford had been obtained for the production of cask staves by means of Marc's circular saws, and 'should the Admiralty decide to establish circular saws and sawmills at dockyards it would be necessary for me to prepare drawings and estimates'.

Farthing appears to have been discouraged by the lack of immediate response to these approaches, but Brunel kept him informed about progress in other directions, including the veneer-cutting which 'exceeds my expectations', adding that he had cut veneers from one-eighth to one inch in thickness during trials at Portsmouth.

Still the Navy Board withheld payment for the work at Portsmouth, and in a letter of November 1808 to a Mr Poole, Marc pointed out that 10 months had elapsed since the blockmills had been pronounced capable of supplying all the Navy's requirements. The reply from the Board suggested that he should prepare an itemised estimate of the economies resulting from one year's working of the block machines, but they positively refused to provide him with details of the prices that had been paid in the past for the hand-made blocks.

Brunel now requested that his allowance be restored, but this was promptly refused. He pointed out that delays (often caused by the Board) and rising labour rates had inflated the cost of the machines, which in 1809 produced 150,000 blocks, at prices which equated to £54,000. He had spent over £2,000 of his own money

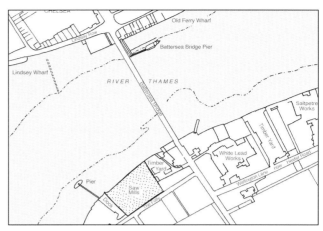

Approximate location of Brunel's Battersea sawmills *c*1814 (not to scale). *J. Williams*

to produce models of machines — whereas an officer employed by the Board in an engineering capacity would have been reimbursed from public funds. He enclosed a statement of the expenses incurred since the termination of his allowance, and requested that, if the Navy Board were unable to pay him, he would be obliged if they would forward it to the Lords Commissioners of the Admiralty, 'and kindly beg them at the same time to inform me if my services are in any way further required'.

The teething troubles from which the sawmill suffered were eventually overcome and the partners prospered due largely to Farthing's prudent management. Their main business was converting timber, cutting veneers and selling circular saws, but they also received commissions for the improvement of primitive sawmills up and down the country, including that of James Borthwick of Leith, near Edinburgh.

Borthwick was a sawmiller who planned to modernise the family sawmill at Leith with a view to developing the greater use of the available water supply. A trickle of letters to Brunel in 1806 had become a steady stream by 1811 and, together with his friend Moncrieff, Borthwick had visited the Brunels at Portsea before their move to Chelsea in 1807. Everything that Brunel developed, Borthwick considered, including the production of veneers. In March 1808, Marc mentioned in a letter to Borthwick that his machine had cut over 50,000sq ft of mahogany *without attendance*. 'I have never met with so much trouble in one single machine,' he states; 'all obstacles seem, however, to be removed, excepting those which proceed from want of trade, which appears to be severely felt here.' Borthwick's reply contained further questions and a firm decision to modernise, and it is in his response to this letter that Brunel warned of possible pitfalls awaiting an unwary application of the new technology: 'Now that it is your intention to set aside the whole of your old works to be replaced by others on an improved principle; whether my terms meet with your approbation or not, I cannot too strongly recommend you to be very cautious in the selection of the means you are about to adopt'. British-grown elm, oak and ash logs tended to be irregular in shape, and Brunel warned against installing saws that were claimed to cope with all types of timber:

'The apparatus is likely to be too heavy in some instances and too light in others. The best use must be made of the available water power: it is not enough that logs should be brought near to the carriage by the assistance of machinery. I have introduced the use of a windlass of a new design which saves all labour in the disposition of the logs on the carriages'.

Brunel then went on to say: 'The confidence you place in me makes me regret that the distance which separates us is so great, the more particularly so since I am at present engaged in the settlement of my affairs with the Government, and also, in extending my concern here in the way of cutting veneers, which I have accomplished by means of circular saws of nine feet in diameter in the most perfect manner and with very little waste of wood.'

Both Ainslie's plan of Edinburgh & Leith of 1804 and Thomson's plan of Leith dated 1825 show Borthwick's sawmill sited on the south side of the harbour at Leith. Kirkwood's plan of Edinburgh & Leith of 1817 also shows this. It is not possible from the above however to deduce the exact extent of the enterprise, but it occupied the best part of a quarter of a mile of the waterfront. It was flanked on one side by a mill (possibly a flour mill) owned by a Mr W. Boyd, and a builder's yard was on the other side, although a new road was proposed some time prior to 1804, which would have crossed the northern end of the Borthwick property. However, the 1825 map still shows the new road as 'Proposed' so it was probably not completed during the lifetime of the company.

James Borthwick was a man of many talents, and ran the Leith sawmills in partnership with his elder brother Patrick, an enterprise which had been passed down to them through several earlier generations. When James first approached Marc Brunel concerning the re-equipping of the sawmill, the business was well established and prospering. However, Britain was suffering from Napoleon's naval blockade which was gradually stifling trade with Europe. James Borthwick made several journeys to Norway, Finland and Russia to purchase supplies of timber, mainly softwoods from Norway and Finland and oak from Russia, the main supplier of hardwoods to Britain at that time. During these journeys he often indulged his passion for painting and produced several watercolours of the Norwegian coastline and villages. How many of his timber purchases actually reached Scotland is unknown, but it may be supposed that a proportion was confiscated by the French before reaching these shores.

From the time he first contacted Marc Brunel at Portsmouth, several years passed before James finally placed the order for the equipment in March 1810. Moreover, the brothers had lengthy discussions about the current trading conditions before finally agreeing to proceed. Even so, Brunel cautioned them before accepting the contract — he was well aware of the problems the French were causing, and at that time no early solution to the international situation could be foreseen. Nonetheless the new plant was installed and in operation by 1812, making the Borthwick sawmill easily the most efficient in the north. Unfortunately business continued to decline even after the demise of Napoleon, and in 1824,

at the age of 42, James left the sawmills and took a post as Manager of the North British Insurance Company in Edinburgh, a position he filled for the next 34 years. He still retained his share of the timber business but the day-to-day management was in the hands of Patrick. By 1826, with the sawmills rapidly declining, James felt compelled to return to assist his brother, and tendered his resignation to the directors of the Insurance Company. They refused to accept it and he stayed on until his retirement in 1858, eight years before his death. James and Patrick had obtained outside finance to fund the refurbishment of the sawmills under Brunel's plans, and when the steady decline in trade made it difficult to service the loan, the creditors closed in and the business was declared bankrupt in 1828.

* * *

It was during the summer of 1807 that the Brunels moved to number 4 Lindsey Row, Chelsea. Theirs was one of seven dwellings formed by the subdivision of Lindsey House originally built on the northern bank of the Thames by Sir Theodore Mayerne, the physician to James I. This stately terrace housed several eminent occupiers before and after the Brunels, including Joseph Bramah, the painters John Martin, William Dyce and J. M. Turner, and Jane and Thomas Carlyle. Martin and Turner had an arrangement with the local boatmen to awaken them if there was a particularly impressive cloud formation at daybreak, both being keen cloud painters. The Brunels' immediate neighbour was the novelist Anne Manning, with whom they formed a close and affectionate relationship.

In front of the house were steps going down to the river and it is often claimed that Isambard learned to swim here. This is highly unlikely. At that time the Thames was little more than an open sewer used by the entire population of London for the disposal of every type of rubbish. It was not quite so polluted at Chelsea as lower down, but it still stank with raw sewage, animal offal, dead dogs, cats, rats and other carcasses. Several small rivers flowed into the Thames including the Fleet, which was simply an outlet for the waste products from Smithfield Meat Market and Farringdon Vegetable Market, both of which lined its banks. The Fleet was not conduited until the mid-1880s, by which time it had become notorious as an overnight depository for the bodies of people murdered by the cut-throats and robbers who infested the area surrounding the Fleet and Newgate prisons. At the time of Isambard's birth and for nearly 50 years afterwards, no parent would have considered allowing their beloved child to risk entering the waters of the Thames — in fact even the Thames boatmen were renowned for being non-swimmers. If they were unfortunate enough to fall into the water, they knew their chances of surviving the pollution were very small.

Until 1750 London Bridge was the only bridge over the Thames below Kingston. When Westminster Bridge was originally proposed, it met with considerable opposition over 10 years from the ferrymen and the Church who owned the ferry rights. Similar opposition was encountered from the ferrymen working between Battersea and Chelsea when a wooden bridge was first proposed. Until then the Battersea area south of the Thames was still quite rural and

undeveloped, since most industry and population was located on the northern side of the river and crossing was a major obstacle. In 1776 Earl Spencer obtained leave by Act of Parliament to build a bridge at or near the ferry at his own expense, and formed a company to finance it. It cost between £15,000 and £20,000 and was constructed entirely of wood. (The estimates for a stone bridge were found to be beyond the company's resources.) It was 726ft long and 24ft wide, with spans varying between 15ft 6in to 32ft. It was first opened to foot passengers for a toll of a halfpenny, then in stages until it could take a full complement of wheeled vehicles. The bridge had a number of bays where the public could stop to admire the view, which, although interesting during daylight hours, was said to be outstanding at night.

Battersea Bridge was a favourite subject among the painters of the time and acquired a reputation among the public for magical cures. It was said that 'the confluence of airs to be met with halfway across, possess some strange curative and healthful magic existing nowhere else'. However, it was detested by those who relied upon the river for their living, including the ferrymen, the bargees and the tug captains, who found it difficult to negotiate. In spite of the massiveness of its beams, it was quite fragile and was put out of action several times during bad weather. It was strengthened in 1873, but in 1883 was declared unsafe for wheeled traffic and pulled down. It was replaced with a temporary wooden footbridge until the new iron bridge was completed in 1887.

After the building of the first Battersea Bridge, the southern shoreline altered completely and the whole area began to develop both industrially and residentially. Brunel chose to site his sawmill here because land was much cheaper than on the northern side of the river. Although prices had inflated since the opening of the bridge, there was still much to be gained by choosing Battersea in preference to a site closer to the City such as Southwark. There was the added advantage of being only 10 minutes' walk from his house in Chelsea.

In 1808, Brunel was awarded a patent for his design of circular saws (see Appendix 12F). Beamish stated that he took out two further patents for improvements in sawmills in 1812 and 1813, but only the one from 1813 has been traced to date (see Appendix 12H). Marc's efforts at Battersea did not go unnoticed. A writer in the contemporary periodical the *Gentleman's Magazine* stated that 'most of our readers must have seen or heard of the ingenious machinery at Battersea for sawing veneers with circular saws . . . the invention, and, in part, the property of Mr Brunel'.

In March 1810, Marc submitted his account for the Portsmouth blockmaking plant to the Navy Board, which immediately disputed it and ordered an investigation with the object of delaying final settlement. At the same time, Borthwick decided to go ahead with his planned reconstruction of the Leith sawmill. The Scot's long-delayed decision came at a doubly inconvenient moment for Brunel, for Maudslay was in the process of moving to new and larger premises in Westminster Bridge Road and for some time his output was limited. As a result of this, the shafting for Borthwick was made in Leeds by Fenton, Murray & Wood, but only after Farthing had been sent there to supervise and expedite the contract. Bryan Donkin (famed for his invention of paper-making machinery) was also contracted to make a waterwheel, but the bulk of the machinery was eventually made at Maudslay's new works. In June 1810, Marc travelled to Leith personally to inspect the site, but upon his return he discovered that very little progress had been made with manufacturing the plant. Meanwhile Borthwick, who had taken five years to make up his mind, was becoming impatient with the delays and lack of firm promises from his engineer. In August he wrote, 'I cannot put up much longer with your procrastination,' to which Marc replied, 'If you think you could obtain the result in a speedier way by applying to others, I am willing to give you the drawings and models for buildings without making any charge for them . . . in everything I undertake I consider my credit of much more consequence than the pecuniary advantage likely to be derived.' Borthwick relented and the whole of the apparatus was complete by late November of that year.

By the early months of 1811, Marc's affairs were beginning to fall into place. The Navy Board had paid him £17,000 (nearly three thousand pounds less than he had claimed), plus a further £1,000 for models. Borthwick's installation of the planned reconstruction of the Leith sawmill was almost complete, and the Board of Ordnance, after years of indecision, now finally approved Brunel's plans for Woolwich Arsenal. Things were looking up.

6. Woolwich

The town of Woolwich originally occupied both sides of the Thames, the Essex side being called North Woolwich. By the 15th century, however, only that part on the south bank continues to be mentioned in documents. The site was occupied well before the Roman invasion and there is evidence of a Roman cemetery; the old spelling of *Uuluuich* (before the *W* entered the language) appears in legal records as far back as AD918. It is mentioned in a grant of land by King Edgar in AD964 to the Abbey of St Peter of Ghent where it is spelt *Wulewich*.

From a small fishing village, it rose to prominence as a naval station and dockyard early in the 16th century, when King Henry VIII bought from Nicholas Partrich, grocer and alderman of London, and Marion, his wife, 'a messuage, salthouse and wharf at Woolwich', plus a small parcel of land. This was in order to increase the precincts of the rudimentary dockyard for the construction of his new flagship the *Henri Grace à Dieu* or 'Great Harry'. The fame of this ship so enhanced the reputation of the dockyard that it was enlarged to accommodate more and more shipbuilding including the *Elizabeth Jonas* (1559), *The Prince* (1619), *The Vanguard* (1631), *The Sovereign of the Seas* (1637) and *The Royal James* (1663).

Woolwich Dockyard was Admiralty property operated by the Navy Board. It had been a major employer of labour since the reign of Henry VIII. Woolwich Arsenal on the other hand was an Army establishment and quite independent of the dockyard. Founded in 1716, it expanded during the Napoleonic Wars into a vast industrial enterprise, employing over 10,000 men in wartime and nearly 6,000 during peacetime. It consisted of three great factories — the Royal Gun Factory, the Laboratory and the Royal Carriage Department. Together they covered an area totalling 600 acres. Although there was a certain amount of co-operation between the Dockyard and the Arsenal, due to their close proximity and the fact that the Arsenal was responsible for supplying ordnance to the Navy, they were nevertheless deadly rivals as far as prestige was concerned and fierce competitors in the fight for the allocation of government money.

Until the middle of the 19th century, Woolwich was renowned as a thoroughly unhealthy place in which to live and work. The Plumstead Marshes, which almost surrounded the town, were intersected with miles of open ditches containing extensive swamps of stagnant water, many of which had not been cleaned for over 50 years. Malaria and other waterborne diseases were widespread in Woolwich. The death rate among both workmen and prisoners was well above the national average, and long-term employees seldom lived beyond the age of 40. The situation remained unchanged, until eventually a new sewage system was opened in April 1865 by the Prince of Wales, who later became Edward VII.

In 1734 Woolwich Dockyard was extended, and again in 1784, but it still had no storage for timber and no accommodation for large numbers of workmen, including a great deal of convict labour. These prisoners lived mainly on the hulks (prison ships) moored in the river. Each hulk housed up to 600 men who were divided into classes according to their behaviour. Some of the convicts held were French prisoners-of-war, captured during the numerous engagements, but the majority were convicted felons awaiting transportation to one of the colonies. The cells were numbered consecutively, beginning with those on the lower of the three decks. The highest numbers were allocated to the best behaved prisoners, who were promoted from the decks below. No convict was permitted to be without an iron upon one or both legs, even if his work did not take him ashore. Conditions on the hulks were extremely harsh. The food was of a very poor quality, sanitation was almost non-existent, and drunken orgies were regularly reported in the press.

In 1829 the privilege of 'Watering Time' was abolished at Woolwich Arsenal. It had been the custom from earlier times to suspend work for half an hour in the morning and afternoon, to enable barrels of beer to be brought in from the local public houses to sustain the spirits of the workmen. Effectively it was the forerunner of the modern tea break. The authorities however felt that drunkenness among the workforce during working hours had become excessive. The privilege was being abused so it was withdrawn, much to the chagrin of the employees.

The dockyard at Woolwich extended to over 56 acres in total and the approaches had constant problems with silting. In 1803, John Rennie, with his assistants Joseph Whidby and William Jessop, conducted a survey on behalf of the authorities. They reported that Woolwich was uneconomic and that it should be closed. (Deptford and Chatham were also condemned.) In Rennie's opinion, Woolwich was too far from the sea: the water at low tide was so shallow that all ordnance and stores had to be unloaded at Northfleet before warships could proceed to the dockyards, and be reloaded there after repairs or refurbishment. There was also the problem caused by fresh Thames water entering the Woolwich installations which was detrimental to the wooden hulls of the ships: they deteriorated far sooner in fresh water than in salt. Rennie favoured the building of an entirely new deep-water dock at Northfleet. However, it was another 66 years before Woolwich finally closed, although the Admiralty did reconsider the idea from time to time.

The workers employed at nearly all naval dockyards were in constant dispute with the authorities and had been for years, but Woolwich had by far the worst reputation for unrest and revolt. This situation arose from a variety of reasons — Portsmouth's location on an island meant that the workers were almost entirely dependent upon the Dockyard for their living, whereas at Woolwich, dissatisfied journeymen were able to seek

Above: One of the back streets in Woolwich *c*1805. *Author's collection*

Left: Hulks and convicts at Woolwich *c*1800. *J. Williams*

Right:
Leg-irons worn
by convicts at
all times.
Author's collection

A view of Woolwich Dockyard c1800. Note the prison hulk centre picture.
Philip MacDougall collection

employment in other yards and businesses situated along the banks of the Thames.

Late payment of wages was the main cause of dispute, with the Navy Board delaying payment for anything up to 18 months on some occasions. This was not always the fault of the Board. The Treasury was, as always, reluctant to make the funds available until it could hold out no longer. There was also constant friction over the matter of 'chipping' and this applied to almost every naval dockyard and several of the privately-owned yards.

For centuries past, the dockyard carpenters had been permitted to collect the chippings produced during their work to take home for kindling, but by the 18th century this had grown into a major problem. Men now brought their wives and families into the yard to gather the wood, and vast quantities of good timber were cut up and taken out as 'chippings'; one carpenter even built himself a small shed within the dockyard in which to store his chippings until such time as it was convenient for his family to carry them away.

Numerous methods were tried to overcome the problem, but all were refused by the workers, who held the whip hand whilst the government was so badly in need of shipping. At one time, pilfering at Woolwich was so bad that the Navy Board introduced regulations to prevent the wearing of overcoats and wide trousers within the Dockyard, and labourers working in the Stores were not permitted to wear trousers at all!

All dockyards were corrupt from top to bottom, but Woolwich was notorious for the efficiency of its thieves. It was said that the Woolwich criminals could completely equip a ship with stores stolen entirely from HM's warships, and the Navy Board estimated that the loss through theft exceeded £1.5 million in 1814. Many of the managers of government dockyards were also proprietors of their own small yards along the river, mainly employed in the construction and repair of merchant vessels. The opportunities for diverting stores and labour from the Naval Dockyard were seldom neglected and systematic embezzlement on a huge scale was endemic among the managers and superior officials.

During the reign of George II, the ropemakers of Woolwich were in dispute with the Admiralty over their wages, and in consequence refused to take on apprentices. This so infuriated the king that he ordered eight of the ringleaders to be impressed into the Navy. As a result, the rest immediately volunteered for naval service, leaving the ropeworks completely idle. As ropes and cables were an essential part of naval equipment, the king 'graciously' relented and they all returned to work.

* * *

In 1808 Brunel was asked to submit plans and estimates for the erection of a sawmill at the Royal Arsenal. His appointment was suggested by Colonel Uppage, Inspector of the Royal Carriage Department, who proposed that an engine should be introduced into the factory for the cutting of timber. In a letter to the Board of Ordnance dated 25 April 1808, he stated that Mr Brunel, the Director of Block Machinery at Portsmouth, who worked a vertical sawmill there, was understood to be the most experienced person to consult and that he proposed to visit him. The Board approved of this suggestion and after a meeting, Colonel Uppage reported on the result of his interview with Marc Brunel. In this report he suggested that the Royal Carriage Department should be equipped with a 12-horsepower engine and machinery for vertical and other saws. On 20 May, Marc was asked to submit plans for a steam engine. He replied that he would forward his plans in due course, but that a 12-horsepower engine would be unable to cope with the load contemplated. Instead he recommended an 18-horsepower engine.

On 3 August 1808, Marc forwarded his terms for mechanising the Royal Carriage Department to R. H. Crew, the Secretary to the Board of Ordnance, at the same time suggesting that an extra four horsepower would be required for working lathes. He calculated that an 18-horsepower engine would be the minimum required, but recommended one of 20 horsepower which would cost an extra £100, but would be more economical in fuel since it would seldom be required to run at full capacity. The complete specification reads as follows:

'A 20hp engine with cast iron beam and water cistern; the whole to be fixed upon an entire cast iron foundation plate. Also two wrought iron boilers with fire apparatus, steam pipes, etc. The whole to be on the most improved construction.'

There follows specific details of the machinery — shafting wheels, saws, frames, clamps and a table. The 20hp engine together with all the machinery mentioned was to be erected ready for use for £6,000. If the 18hp engine was preferred, the figure would be reduced by £100.

It transpired that the Board was unable to proceed with the plan immediately owing to a problem with the proposed site. The unexpected delay annoyed Brunel, since he had been requested to give the project priority, and in so doing had refused several other enquiries. He wrote to the Board requesting remuneration for all the trouble and inconvenience he had been caused. The Board granted him the sum of £100 towards his expenses. By the end of February 1811, the siting problem was resolved, and Major-General Uppage, as he had now become, requested the Board to reopen the subject of the sawmill. Work commenced in May on a building to house the machinery and Marc was requested to submit an amended estimate for the erection of the plant. The new estimate was for £4,600, a reduction of £1,400, but he advised that an additional boiler would be required which, with its accessories, would cost £450. The total cost of the undertaking (including the erection of the building) would now be just over £9,000. Approval was finally given on 27 May 1812, and the work was put in hand immediately.

A month earlier, the Storekeeper and the Clerk of the Survey, in their corporate capacity as Respective Officers, complained that the site selected for the steam engine was in close proximity to their dwellings, making life intolerable for themselves and their families. They requested that the engine should be made to consume its own smoke which they understood to be possible at trifling expense. The Board agreed and issued the necessary instructions. In the summer of 1812, Brunel reported that his charges to date were £5,080, of which he had received only £1,000 on account. He also suggested certain improvements which would cost an additional £850. The Board agreed to the alterations and advanced him another £2,000, and in October 1813 a further payment of £1,500 was made. The following summer, Brunel applied for further remuneration, stating that his outlay had been £6,378 and that his profit had been only £16. The Board was unhappy over the affair and some protracted haggling took place over the final claim, but they were not long in reaching a decision. They refused to pay more than the amended estimate, but in view of the work Brunel had performed for the country, they awarded him a pension of £300 per annum. Following the fire at his Battersea sawmill which occurred in August 1814, Marc asked that his pension be commuted into a lump sum of £4,500, and this request was granted on 14 September.

Information concerning Marc's reorganisation of the Dockyard at Woolwich is difficult to come by, although he was responsible for several improvements and alterations there. This was in addition to the extensive rebuilding and re-equipping of the sawmill and timber yards of the Royal Arsenal. He made provision for civilian workers' housing close to the Dockyard, although the convicts were still accommodated on hulks in the river or in nearby prisons. He seldom refers to Woolwich Dockyard in his correspondence, so it is assumed that he was paid for the work on time and in a satisfactory manner.

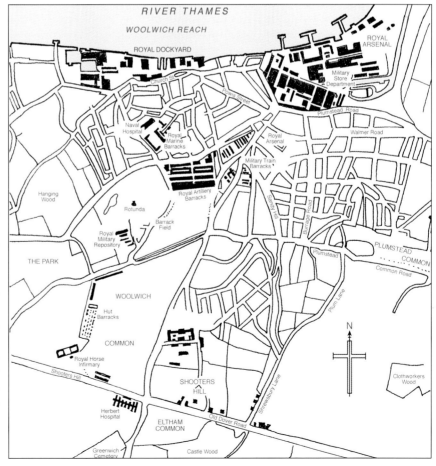

Map of Woolwich c1810 (not to scale). *J. Williams*

7. The Boot Factory at Battersea

In addition to running the Battersea sawmill, designing and erecting the sawmill at Woolwich Arsenal, and improving the Woolwich Dockyard, Marc had also begun experiments with steam navigation. Prior to that, in 1809 he had witnessed the remnants of Sir John Moore's gallant army disembarking at Portsmouth after the retreat from Corunna in the Peninsular War, where they had been overwhelmed by the French. Sir John was killed during the rearguard action which enabled the majority of the British troops to evacuate. Brunel saw the dispirited soldiers dragging themselves along the quay on festering feet, often bandaged in dirty rags or covered with open sores, with toes protruding from the remains of boots which had disintegrated in use.

It was certainly true that corruption among Army suppliers was as rife as those involved with the Navy. In 1807 Lord Dundonald began his campaign against corruption and reported such scandals in his *Autobiography of a Seaman*. In it he described how the troop ships were often in a wretched condition, with deficiencies common in anchors and stores, and the decks often being so leaky that, when it rained, the men below were soaked. Overcrowding was so common that soldiers with no bedding other than blankets were obliged to sleep on the deck. Brunel managed to obtain a few pairs of new boots from the Army suppliers and took them away for examination. He found that a layer of clay was sandwiched between the thin inner and outer soles to give the appearance of a heavy, substantial boot, which began to disintegrate as soon as it came into contact with damp conditions.

It had been seven years since the installation of the first of his blockmaking machines at Portsmouth and the Navy Board still withheld payment. The Board of Ordnance had shelved plans for the Woolwich Arsenal development (only temporarily as it turned out) and Marc now had a wife and three young children to support. He therefore decided to enter the bootmaking business privately, without involving government departments, much as he had already done with the Battersea sawmill. The boot and shoe workshop was erected close to the sawmill site. Marc initially employed 24 unskilled disabled ex-soldiers to operate a series of machines to produce good strong boots and shoes in nine sizes. Brunel's methods involved 16 different machines, all designed by himself and manufactured by Maudslay. Prices ranged from nine shillings and sixpence for a pair of shoes to one pound for a pair of Wellington boots.

Pairs of boots and shoes can be made either the same for both feet (straights) or for left and right feet individually. If they are produced as straights, separate lasts for each foot are not required and the number of lasts needed is halved. During the 17th and 18th centuries all women's shoes were straights, as were the majority of men's. Rights and lefts first appeared in any quantity in about 1790, when the invention of the pantograph lathe made the manufacture of mirror-image lasts possible. At first only the best quality boots and shoes were 'footed', purchased mainly by the aristocracy and wealthier classes; it was much later before the general public were able to afford such luxuries. During this period, both the Army and Navy were supplied with straights, and when Marc Brunel began manufacturing at Battersea he saw no reason to depart from the norm. Regimental standing orders for the King's Shropshire Light Infantry in 1800 included one that stated that soldiers should wear their boots on alternate feet on alternate days — obviously in order to extend their life. Examination of the drawings which accompanied Brunel's Patent Application No 3369 clearly shows that straights were produced by the Battersea plant (see Appendix 12G).

By 1812 it was already illegal to incorporate clay into the soles of boots and shoes, a fact recognised by the *Northampton Mercury* in May 1813, which stated that 'Shoemakers that have been in the habit of using clay in the bottom of shoes will be prosecuted'. The following week's edition of the newspaper reported that a George Neal, a journeyman shoemaker, had been convicted of 'wilfully spoiling the material of ten pairs of shoes by improperly making them up and putting a quantity of clay between the soles'. This practice must have been well known to the Army and Navy officials responsible for the purchase of such shoes, but they usually turned a blind eye.

Brunel's factory quickly became very successful and the Army was known to have secretly made a few tentative purchases soon after the plant went into operation. Visitors to the Battersea sawmill were invited to view the new boot factory and soon its fame spread. Lord Castlereagh, the Foreign Secretary, was particularly impressed; it was through his influence that Brunel was persuaded by the government to increase production to satisfy all the Army's requirements, and within a few months output had increased to 400 pairs a day. It must have been obvious to any astute businessman that production at this rate was not sustainable indefinitely, as one year's output would easily cover the needs of the entire British Army. But Marc Brunel was no businessman.

On 2 August 1810, Marc had been granted a patent for 'Certain Machinery for the purpose of Making or Manufacturing Shoes and Boots' (see Appendix 12G). On 12 March 1814 a further patent was granted for the invention of 'A New Method of giving additional Durability to certain Descriptions of Leather' (see Appendix 12I). This later patent was concerned with improving the strength and durability of the leather used in the soles. Since the products of the factory proved to be all that was claimed, the government issued a large order which was completed in the stipulated time and resulted in the troops being properly equipped

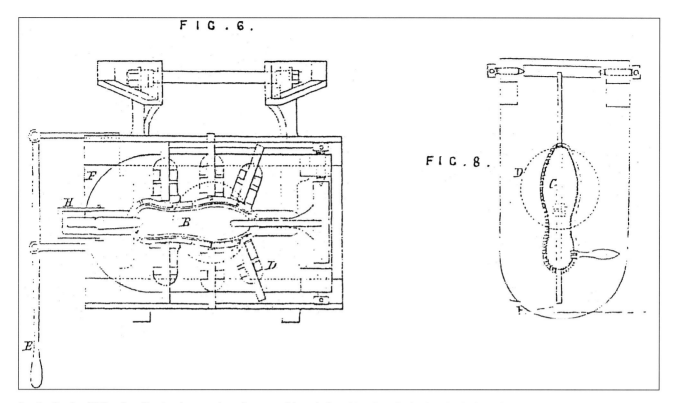

FIG. 6.

FIG. 8.

Marc Brunel's drawing clearly shows that straights were to be produced.

for the Battle of Waterloo. Production continued apace with verbal encouragement from government sources, but unfortunately for Brunel, following Waterloo, a peace treaty with France was concluded and it was decided that there was no longer a need for army boots on such a large scale. Marc was left with large stocks and few customers. Farthing had not been involved in this enterprise: had he been he would certainly have insisted upon a written contract.

Confident of the moral integrity of the government, Brunel continued in full production, incurring heavy liabilities connected with the plant in an effort to keep his disabled employees in work. Inevitably, financial difficulties arose from which he was unable to extricate himself. On 22 February 1819 he wrote to Nicholas Vansittart, the Chancellor of the Exchequer, drawing his attention to the condition of the soldiers as seen earlier at Portsmouth:

'I was prevailed upon and induced, therefore, to turn my attention towards supplying the deficiency; the great trouble and expense of which I took entirely upon myself. I invented and prosecuted a plan for making, by machinery, military shoes of a greater durability than those previously used in the army. As soon as the machinery I had invented was in a state to work, I applied to the Invalid Department for disabled men whom I proposed to be exclusively employed in it. When my plans were known to the principal officers in that department, directions were given to afford me every assistance I might require to accomplish my object. Having once the opportunity, when the Prince of Orange visited the Royal Arsenal at Woolwich in March 1813, to mention to several of HM Ministers then present, that I was engaged in forming an

establishment for making military shoes, I was commended for so laudable an undertaking. It was admitted that the shoes previously sent abroad were proverbially bad, and several officers then present added to these observations facts that excited a great deal of interest in the success of the scheme . . . My Lord Castelreagh, addressing himself to me, said, "When you are ready to make any number of them, let me know." I have since complied with his Lordship's directions but have received no answer.

'The Commissary-in-Chief, sensible of the importance of the undertaking, gave it every encouragement in his power, in taking all its produce in its infant state. Shoes were supplied to the 13th Regiment, and the report of them was favourable; but objections were made on the supposed impossibility of mending them. In actual service in the field, it is found that mending cannot be practised even with the common shoes; consequently the *intrinsic value of these shoes consists in their first durability*, and in the quality they possess beyond the ordinary shoes. In the course of 1814, HRH the Commander-in-Chief, being informed of the progress I had made, did me the honour to visit my establishment. HR Highness and the officers of his suite, passed the most flattering encomiums on the utility of the work. It was stated on that occasion, that shoes sent previously for the service of the army had been found *not to last one day's march*.'

He then went on to state how much of his own money had been invested in what he considered to be a very worthy enterprise and continued:

'The loss alone has fallen upon me, a loss amounting to upwards of £3,000 sterling; besides the encumbrance of premises for which I must pay £400 per annum, for they are still upon my hands. For this loss, sir, incurred in this way and on the grounds which I have detailed to you, I solicit a compensation which I feel I have a title to. The loss was incurred on an object for Government; the loss was incurred, as it were, with the very knowledge of Government; for I never should have gone on absorbing my own pecuniary resources, the fruits of my past services, unless buoyed up with the assurance and hope that Government would support me . . . I do therefore trust that I shall not be held to ask too much, in asking that the Lords of HM Treasury will be graciously pleased to take this statement into their favourable consideration, and order me a compensation of my losses, chiefly of the sum of £4,500, being that part of my former earnings from Government, which, as I before stated, I have absolutely sunk, together with much more of my own property in the establishment, of which I have now troubled you with the detail.'

Feeling that it was possible that no assistance would be forthcoming from those in authority at home, Brunel contacted the Prussian government in early 1819 and began negotiations for the supply of shoemaking machinery. He submitted a detailed specification of the machinery, including the design of the buildings required for the effective working of a great national establishment. There is no record of a reply, nor is there evidence that his proposals were ever put into effect by the Prussian government. This was most probably because it was normal practice for each Prussian regiment to train and employ its own soldiers to make the footwear for the corps.

June Swann, in her book *Shoes* (Batsford, 1982), implied that Brunel's products were of poor quality, stiff and uncomfortable to wear, and that after the end of the Napoleonic wars his products were dropped, being 'quite out of keeping with post war feeling'. New issues of boots, however, have always been subjected to hours of treatment with dubbin and polish before any comfort could be derived from them, so Brunel's boots were no different in that respect. It is true that Marc's methods went out of use for several years, but when a man named Crick began production in 1853, his methods were little changed from those of Brunel. Indeed, the machinery he used was in many instances so similar in design to Brunel's that many experts believe that it was copied directly from Marc's drawings.

Marc's misfortunes had been compounded in August 1814 by the almost total destruction by fire of his Battersea sawmill. Another major fire was raging in the City at the same time, and only three fire engines could be spared for Battersea; it was low tide and very little water could be pumped from the Thames, which at that time was wider and shallower than it is now. In consequence the mill burned for more than three hours. At the time, it was noted that the greatest exertions were made 'to preserve the stock of wood and veneers, the greater part of which were saved, at the risk of the lives of those who strenuously exerted themselves upon the occasion'. Despite their efforts, everything else was consumed by the flames, with the exception of the steam engine and one wing of the main building. A writer for the *Gentleman's Magazine* described how, in two hours, 'these most valuable machines, which, in point of execution and perfection, exceeded everything we know . . . presented the awful sight of a heap of fragments; and the fruits of six years of exertion and ingenuity, attended with an expence of above 20,000*l*., were destroyed'.

Brunel was at Chatham when the news was brought to him that night. To most people it would have been a catastrophe, but Brunel was rather more sanguine. Once he had established that nobody had been hurt, he turned to Josiah Field who was with him at the time and said, 'I can make better machinery now.' Furthermore, in a reply to a Mr Edgworth's letter of condolence shortly afterwards he answered, 'The misfortune is not without its consolation, as I shall now have the opportunity of carrying out many improvements which I have often contemplated.' This he proceeded to do with rather naïve optimism, apparently unaware of the financial strain the business was now called upon to bear.

Chatham Dockyard

8.

In 1807, Bentham's office of Inspector-General of Naval Works had been abolished. Bentham was subsequently appointed a Commissioner of the Navy and in this capacity continued to have considerable influence over naval matters. He had long been concerned about the inefficiency of operations at Chatham Dockyard, where thousands of man-hours were consumed annually simply in the moving of timber from one place to another. Although Rennie had earlier dismissed Chatham (and several other dockyards) as being unsuitable for development, Bentham had overcome the problem of silting there, which would soon have made Chatham unusable for large vessels, by the invention of a new type of steam dredger which could remove the silt faster than it built up. As a consequence, Chatham continued to be available for building the Navy's largest ships.

Bentham designed a lighted cover over a slipway to enable work to continue during bad weather, which also protected the timber whilst it was being seasoned. The newly sawn scantlings had to be stacked outside for at least two years before being taken inside to the shipwrights for processing. Whilst this procedure resulted in better and more durable ships in the long term, it required the storage of vast quantities of timber in enormous seasoning yards. In 1810 over 12,000 tons per annum of Russian oak were being handled at Chatham Dockyard by a combination of men and horses.

When Brunel was approached by Bentham, he calculated that the cost of dragging this timber around the yard was in excess of £4,000 a year. It was then all sawn by hand in sawpits with a pair of sawyers to each pit. After careful consideration, Marc formulated plans which he put before members of the government and the Admiralty. He stated that: 'The imperfection of the various mechanical contrivances . . . led me to direct my views to the invention of such machinery as should be the means of obviating these difficulties, and I foresaw that the field would be open to me of rendering service to the naval establishments of the Kingdom of a magnitude much exceeding that which had been derived from my improved system of making blocks.'

Through Bentham's influence, Brunel received an official request from the Admiralty in January 1812 for plans embracing a

Part of Chatham Dockyard showing the position of the sawmill and canal entrance. *Redrawn by Dr Philip MacDougall from charts held by the Public Records Office*

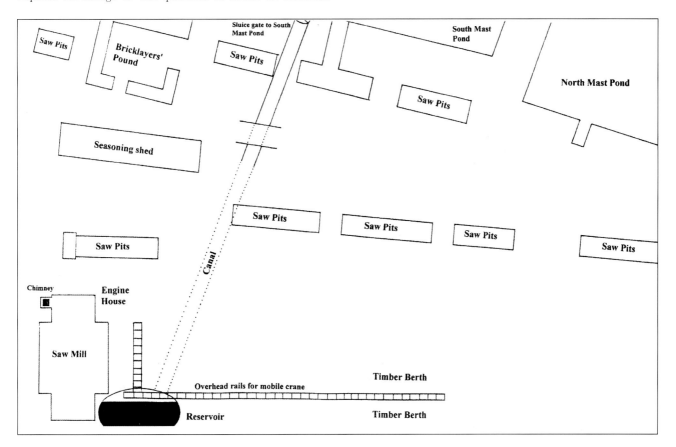

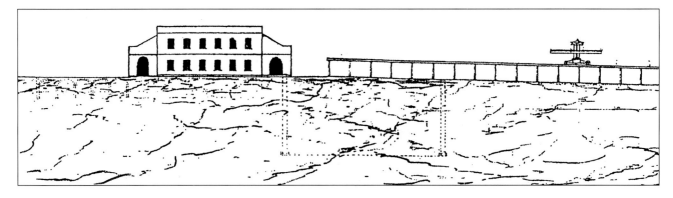

Timber mill and mobile crane at Chatham Dockyard. Drawing by Marc Brunel dated June 1817. *Courtesy Public Records Office. PRO ADM 140/99*

log handling and sawmilling plant for the dockyard, and by this time he had decided how best to transport the timber. The fact that Marc's revolutionary proposals for Chatham were accepted within eight months was most unusual. Bentham's influence coupled with Brunel's reputation, which was now growing apace in influential circles, was decisive. He was already deeply involved in the shoemaking enterprise, but this does not seem to have deterred him; his ability to deal with several matters at the same time was never in doubt.

Chatham Dockyard employed about 150 sawyers who were normally paid at the rate of four shillings and two pence per 100ft sawn, working in a sawpit. This was an oblong trench over which the timber would be secured, allowing it to be cut using a two-handed saw, one man down in the pit, the other above. The top man or 'top dog' as he was known, was the more skilled of the two and was responsible for setting and guiding the saw. He was the more highly paid. The man below in the trench was known as the 'under dog' and it was he who had the more unpleasant job which must have entailed getting covered in large amounts of sawdust each day. (It is interesting to note that the phrase 'underdog' originates from this working practice.) In an average day a pair of sawyers could expect to saw about 220ft and would earn jointly about 55 shillings per week. This meant that Chatham, with its 75 pairs of sawyers, was involved in an annual expenditure of approximately £11,000. Brunel was confident that this figure could be reduced to £2,000 to cover wages and maintenance of the machinery.

According to Dr Philip McDougall in his Research Paper written for the Chatham Dockyard Historical Society, part of which is reproduced below, the sawmill that Brunel designed for Chatham consisted of eight saw frames that each carried an average of 36 saws. They thus produced 1,260ft of scantling per minute. Potential savings on sawing alone were enormous, but in addition Brunel planned further savings by revolutionary methods of timber handling across the yard. Prior to the erection of Brunel's woodmill, all log timber coming to Chatham was landed at the dockyard wharf before being dragged by horses to a convenient place for stacking. According to Brunel, in any one year, 'there is

required at least 6,000 goings and comings of teams of horses, merely to lay the timber for survey — 6,000 times to and from the stacks — at least as many more times one hundred yards in aiding the lifting on the stacks'.

From the timber stacks, the logs, once surveyed, would have to be removed to the saws and then to a new stacking area once they had been sawn. All this movement, when the cost of wages and the use of horses was included, amounted to a further £4,000. Brunel proposed extending the use of the steam engine to be installed in the mill to assist in the movement of the timber across the yard. The process would begin with the construction of an underground canal that linked with the River Medway and connected with a stacking and surveying area sited close to the mill workshops.

The building which Marc designed consisted of a sawing hall with a pitched roof and two open sides, supported by columns at regular intervals, which was flanked by two three-storey wings. One of these housed the boilers, whilst the other supported a large cast-iron water-tank on its roof with sloping sides and an open top. This building occupied a site between the old and new dockyard walls on rising ground overlooking the Medway. In front of the wing supporting the water-tank, an oval, brick-lined shaft was driven downwards about 60ft into the chalk and from its base a tunnel was excavated towards the South Mast Pond — a lagoon linked to the Medway.

When completed, the canal and tunnel provided a waterway through which logs could be floated from the Medway wharf to the base of the shaft beside the sawmill. Within the shaft Marc installed a counterpoise which could be filled with water from the tank above. This raised a tree trunk to the surface where it was grabbed by a crane which travelled on a gantry, and it was then laid on a carriage running on iron rails. The carriage was then winched up a gentle incline and delivered to the sawing floor in front of one of the saw frames. When the log had been converted into planks, these were again loaded on the flat carriage and returned to the travelling crane which lifted them onto a railway truck. This ran on a 860ft-long railway down an incline and into a timber seasoning shed. On both sides of this track were plots for stacking sawn timber. After the timber had been removed and stacked, the truck could be drawn back up the slope by a rope and windlass operated also by the steam engine.

The positioning of the sawmill on a high point of land was

crucial to the successful working of Brunel's entire scheme. Much of the movement of timber, both in its rough and converted stages, relied upon the combination of gravity with the gently sloping sides of the hill. This area of rising ground just outside the dockyard wall had always been regarded as an encumbrance, but Brunel immediately saw the advantages it afforded. 'That hill must be bought,' was his statement to one of the officers of the yard. When the officer replied that this would involve removing a mass of earth, Brunel replied, 'Remove! Take away that noble hill — the most valuable bit of ground in all the yard! No, no! but buy it — buy it as quickly as possible.' It was upon and below this hill that the sawmill and termination point of the canal was subsequently constructed. In addition, the slopes to the east were to carry the rails used by the mobile crane.

Unfortunately when work began in August 1812, Brunel was not given sole charge of the building works. Moreover, he was often overruled in areas where he was the undoubted expert. The canal tunnel, for example, which he had designed with an elliptical inclined arch, was changed during his absence to a vertical segment design that proved to be much weaker. As a consequence, this, combined with the poor workmanship of the bricklayers, resulted in the tunnel collapsing over a distance of 45ft, killing one man and injuring 10 others.

A further issue that brought Brunel into contention with the authorities was that of the 120ft-high chimney attached to the mill. Owing to its foundations being set in a combination of clay and chalk, it began to settle irregularly. On 27 October 1813, Brunel expressed concern as to its safety. He informed Edward Holl, the architect of the Navy Board, that he had called a halt to its progress. This caused considerable annoyance. As a result, George Parkin (Master Shipwright) and P. Hellyar (Assistant Master Shipwright) wrote to the Navy Board one week later complaining, 'We beg leave to state that the chimney is, in our opinion, in a state of perfect safety; that it may be proceeded with without fear. In referring to Mr Brunel's remarks, we have to observe that the chimney stands firm on a firm foundation.'

To this letter the Navy Surveyor and member of the Navy Board Sir Robert Barlow replied: '. . . from what is therein set forth, added to my own observations on the state of the building, I should recommend proceeding to the completion of it (the chimney), agreeably to the original design'.

The work was accordingly restarted. On 18 February 1814 Brunel made the following entry in his journal, 'Observed that the chimney cracks very much, and continues to bulge out. Mr Vinall is now sensible of it, and proposes to prop it by means of buttresses; and I propose, in addition, some wrought iron ties, which may be buried into the brickwork.' As a result of these alterations, the chimney was secured. The use of iron ties, then unique in concept, was another example of Brunel's innovation and vision. The practice of using iron ties became commonplace in later years.

Despite the various problems, the mill finally became operational in June 1814. This was a remarkable achievement, considering the revolutionary nature of the project and the vast

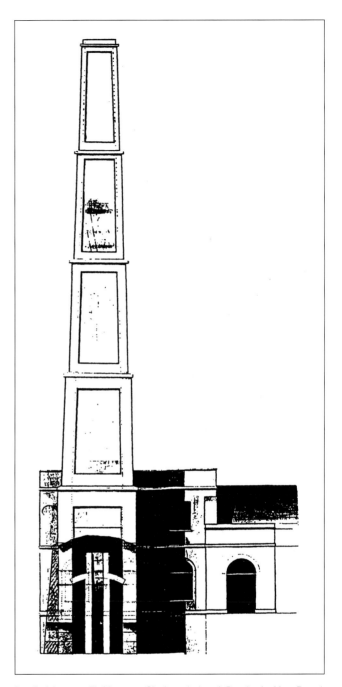

Detail of the sawmill chimney at Chatham dockyard. Drawing by Marc Brunel dated June 1817. *Reproduced by courtesy of the Public Records Office. PRO ADM 140/99*

amount of reorganisation of the day-to-day schedules that this involved. It brought massive changes to the Chatham Dockyard, with the numbers of sawyers and horse teams drastically reduced over the following months. A book published in 1817 on the history of Rochester, referred to the sawmills as follows: 'At the north-east extremity of the yard some new works have lately been constructed, commonly called the sawmills, projected and executed by that modest and persevering mechanic, Mr. Brunel, who has effected as much for the mechanic arts as any man of his time.' It went on to describe the workings at the yard and commented on how the operation of raising the timber was 'worthy of observation; and the steady, though quick motion with which it ascends is truly astonishing'. The author then described how he had witnessed a baulk measuring 60ft in length and 16in square being raised up by 60ft in the space of 60 seconds. The tone of his writing suggests that he was suitably impressed by Brunel's design.

Even the vertical saws which had been unchanged for centuries past were redesigned by Marc Brunel. Previously a saw cut only on the down stroke, the following upstroke was non-productive but was subject to considerable friction from the sides of the cut. Marc, by a simple stroke of genius, altered the action of the mechanism so that on the return stroke the saw drew back from the cut a fraction of an inch, thus enabling the upstroke to be friction free. This tiny alteration alone saved almost one third of the power needed.

Furthermore, Brunel's other great invention was also brought to Chatham. This was a set of blockmaking machines, identical to those at Portsmouth, and sited in the upper floor of the woodmill. It ensured that pulley blocks could not only be produced economically at Chatham, but that in the event of a fire at Portsmouth, the Navy's requirements could be met from the Chatham mill instead.

* * *

Many of the problems encountered during the constructions at Chatham arose through the employment of convict labour. Men were drafted into jobs such as bricklaying without ever having the slightest idea of the skills required and not wishing to acquire them.

The prison officers in charge of them were equally ignorant. It is doubtful whether anyone in charge of the convicts could read a plan — if they could read at all — and many of the prisoners (a proportion of whom were French prisoners-of-war) resented being put out to such hostile employment and could not be expected to attack it with enthusiasm.

Immediately Brunel's proposals were accepted by the Board, he appointed two assistants, Mr Bacon and an ex-curate named Ellacombe to supervise construction and control the labour force. They did this admirably for some time, until the Navy Board dismissed Ellacombe for reasons of economy during Marc's absence in France. Brunel was furious; he had selected two very able and honest people to work under him and, once again, the Navy Board had interfered with his carefully calculated arrangements. The Board relented following Marc's strong protestations, but by then Ellacombe had had enough. He refused to resume his duties at Chatham, preferring the relative peace offered by a return to the clergy. Marc wrote to Ellacombe, 'May you, my good friend, be as great an ornament to the Church as you have been in that most arduous career in which you leave your very sincere friend, with one of his lights out.' Fortunately Bacon's services were retained and he remained in charge of the plant for many years after its completion. Brunel visited Chatham some years later in May 1823 and was very pleased to find that everything was in the highest order. He observed in his diary that the flying crane was 'doing its duty with as much precision as the first day; the same with the Lifting Machine'.

The Chatham sawmills continued in production, albeit on a reduced scale, until well into the 20th century. Eventually the Brunel sawmill was replaced by a new sawmill sited on the St Mary's Island dockyard extension. The canal was filled in and the crane rails taken up. The old sawmill became the dockyard laundry and was finally handed over to the Chatham Historic Dockyard Trust which is its current owner. Great Western Railway enthusiasts should note that it was Marc Brunel, not Isambard, who originated the broad gauge. The rails on which the crane ran at Chatham were exactly 7ft apart.

9. Disasters Leading to the Debtors' Prison

By 1812 Farthing had withdrawn from the Battersea sawmill, although relations between the two ex-partners remained cordial. Nonetheless the shrewd and trustworthy associate had gone and the commercial management of the enterprise was in new and untried hands. At the end of Farthing's time at Battersea, the mill was yielding an annual profit of £8,000 and seemed set to enjoy many years of continued prosperity.

During 1813, Brunel was almost exclusively occupied with the Chatham installations, but in the following year he was approached by the Duc de la Rochefoucauld to submit a specification for a steam engine for the *Conservatoire des Arts et des Métiers* in Paris. This transaction was abandoned after long-winded wrangling about payment, but Brunel learned a lot from the researches he was forced to undertake.

Marc had a double-acting steam engine manufactured by Maudslay which was installed in a specially adapted packet boat, the *Regent*. This vessel was built to ply between London and Margate, and there was great excitement among the younger Brunels, who assembled on a pier below London Bridge to wave to their father as the newly equipped paddle-steamer bore him away on her maiden voyage. It was a great success, although the landlord of the York Hotel at Margate was less than enthusiastic. He, like many of the local inhabitants who had connections with the sailing packets, saw Brunel's invention as a potential threat to their livelihood. Consequently he refused Brunel a room for the night. Beamish noted that in a letter he received from Brunel many years later, Marc described how once again he had arrived at the York Hotel in Margate and commented that it was at '*this same hotel* that in 1814 I was refused a bed because I came by a steamer, and every one of the comers met with a very unfriendly reception'. The *Regent* continued in service between London and Margate for many years without incident, and was the forerunner of the famous Eagle Steamers which covered the same route (extended to terminate at Clacton) until the outbreak of World War 2.

Brunel was confident that steam-powered tugs could tow large sailing ships out of and into harbour in spite of unfavourable winds and adverse tides, but when the idea was presented to the Admiralty (see Appendix 2), the Navy Board issued a notice in reply

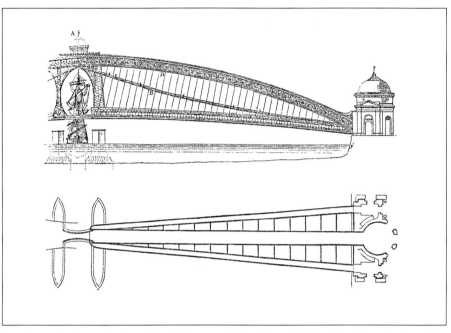

This sketch, thought to be by Brunel, for a bridge over the River Neva at St Petersburg, shows the plan and elevation of half the arch, and the suspended roadway which could be broken in the centre for the passage of large ships. It also indicates the ties ('B' and 'B'), which were to give the necessary stability and to reduce movement between the abutment and the centre ('A'). This design was copied to a large extent in the building of Tower Bridge much later. *Reproduced from Richard Beamish's* Life of Brunel *(1862)*

stating that it was their duty to discourage the employment of steam vessels, as they considered the introduction of steam would 'strike a fatal blow at the naval supremacy of the Empire'. The logic behind this statement is difficult to understand; perhaps their Lordships felt their personal incomes to be endangered if this newfangled idea was encouraged. However, within 20 years the employment of steam tugs to manoeuvre sailing ships through the dangerous Medway channel into Chatham Dockyard became normal practice; prior to that it was not unusual for ships to wait up to three days for a favourable combination of wind and tide before attempting to berth at Chatham.

Brunel's nautical schemes had to be abandoned temporarily after the Battersea works burned down. Although insured against fire, it is unlikely that he would have recouped his losses entirely and there was certainly a considerable consequential loss in addition. To his dismay, his bank statement now showed a balance of £865. The previous October it had stood at £10,000. The loss of Farthing had proved to be a major disaster.

The Battersea mill was rebuilt and re-equipped and was in full production again by 1816. The reconstruction, however, had been financed to a large extent by the sale of the remaining stocks of boots and shoes at a considerable discount. Visitors to the new mill were convinced that they were seeing the beginnings of a great and wonderful epoch. 'I beheld the planks of mahogany and rosewood sawed into veneers the sixteenth of an inch thick, with a precision and grandeur of action which was really sublime,' wrote Sir Richard Phillips. 'The same power at once turned these tremendous saws, and drew their work upon them.'

Brunel was badly in need of a shrewd and honest partner to manage the enterprise for him, for he was already busy with new schemes. The first of these was a knitting machine which was patented in 1816. The machine was found to be very efficient in producing cylindrical knitting suitable for the production of woollen stockings, but unfortunately for Brunel, the low cost of Lancashire-produced cotton goods made Marc's woollen hose unsaleable and the idea had to be abandoned.

A fourth child was born to Sophie in the month following the fire, a daughter named Harriet. Some accounts insist that Harriet was actually the fifth child because the hair of a child in a locket engraved 'S.B.' is preserved. This could refer to a child who died in infancy or the locket could have been a gift to Sophie from Marc containing a piece of hair from one of her surviving offspring. Little Harriet barely survived a year; she died on 23 October 1815, adding greatly to Marc's distress.

In 1817 Marc went to France and took his family with him. He was employed by an English company tendering for an installation to supply Paris with pure water. The scheme had the support of Louis XVIII who regarded Marc as 'a good Frenchman lost', but it was violently opposed by the Paris water porters who felt their livelihood threatened. The project was eventually abandoned in spite of royal support, and the Brunels returned to England.

When the French ship sailed out of Calais for Dover, Marc could not have guessed at the ordeal that they were about to be subjected to. As they cleared the shelter of the harbour, the violence of the southwest gale became apparent. All thoughts of steam engines, water works and knitting machines must have been driven from his mind; the survival of his family was all that occupied him for the next few hours. The captain set a direct course for Dover, due to its proximity, despite the fact that all his men were advising against it. Desperate pleas also followed from the terrified passengers, including Brunel, to get the captain to run with the wind, the obvious course of action in such tumultuous seas. But the captain refused to listen to reason until, seeing his folly, the crew finally refused to obey his orders. The ship was in severe danger of foundering when one of the passengers, in a last attempt to rescue the situation, offered the captain a bribe to hand over command to Brunel. After a short but acrimonious discussion, agreement was reached and Marc was able to sail the vessel safely into the port of Deal, 10 miles east of Dover.

* * *

Shortly after this potentially disastrous incident, some more positive news was forthcoming from the Russian Ambassador in London, Prince Lieven, who represented Alexander I, Czar of Russia. Brunel had already come across the imperial family in 1814 during a visit of various crowned heads to England. Beamish relates how 'his Imperial Majesty' placed with his own hand a diamond ring on Brunel's finger as a symbol of the value and esteem that Russia held for him, and their willingness to secure his services if they had need of him in the future. The Ambassador now requested Brunel to design a bridge to span the River Neva at St Petersburg. This project was particularly appealing to Marc since it involved solving several difficulties which he had not previously encountered. The width of the river exceeded 800ft at the proposed spot, and to this challenge was added the difficulties with the climate, for the Neva was covered with ice for most of the winter. But by far the greatest problem was that of communication. Marc spoke fluent English and French but no Russian; the Russian Ambassador spoke English but no French, so during negotiations they could converse with little difficulty. However, Brunel realised that the Russian workmen he would have to control knew neither English nor French. It was very likely that they were illiterate and would certainly have been unable to read an engineer's plans. A suspension bridge of such proportions had never before been attempted, and, after much deliberation and calculation, Brunel came to the conclusion that in this situation it would be impracticable. This was predominantly because the Neva was a major waterway carrying very large ships and the roadway would have to be carried higher than the masts of the tallest ship at high tide.

Marc also considered constructing a tunnel. It needed to be larger and longer than that at Chatham which carried the subterranean canal, but tests revealed that he would have to bore through soft alluvial strata compressed only by the weight of the water above. This would be tricky because until then, his only experience of tunnelling had been at Chatham, which involved only 400ft of tunnelling through stable chalk. After much deliberation the tunnel idea was abandoned.

The final plan, submitted through Prince Lieven, was for a bridge with a span of 880ft — half as long as that over Sydney harbour — resting on massive masonry abutments on either bank. The centre section of the roadway would be hinged so that it could be raised to allow the passage of tall ships whose masts could pass easily below the high arch which linked both units. It is interesting to note that Brunel's principle was employed in 1884 in the construction of Tower Bridge in London, another fact which is seldom acknowledged. The Neva bridge was to be of timber construction and the great central arch would be prefabricated and floated onto its abutments by means of four pontoons. (This method was used over 40 years later by his son, Isambard, to erect the arched tubular girders for the Tamar Bridge which still carries the railway into Cornwall.) The scheme was finally rejected by the Czar because the Treasury could not face such a costly expenditure. The project was halted pending more prosperous times.

This was a shattering blow for Brunel, who had expended a considerable amount of time and energy on the project. He had made several models and had also been on the point of travelling to St Petersburg to supervise the work. He now found himself in desperate circumstances financially. Profits from the boot factory

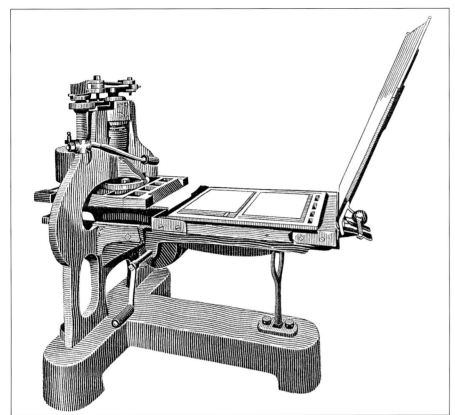

creditors who were pressing him. The Council's Rate Books for the property list several different names which seem to be unrelated. There are also entries under the name of 'Burnet', which could be a misspelling, and others under 'Mudge' which, being a far from common name in the area, could mean that one of Sophie's kin had been brought in to control the business.

Marc Brunel's financial problems were not entirely of his own making but his failure to engage the services of an efficient partner or manager to replace Farthing had certainly not helped. One of Brunel's main weaknesses however was his unwillingness to concentrate his energies on existing enterprises. Instead he constantly chased after new ideas and inventions which drew heavily on his capital and time. Although he regularly worked a 15-hour day, it was usually divided between several different subjects and at no time was he concerned with commercial administration or the exploitation of his inventions beyond perhaps an initial approach to a prospective developer. This was clearly a fundamental flaw in his character.

During this period Marc became interested in the printing industry. This had progressed little since the days of Caxton and was still largely a series of labour-intensive manual processes. New ideas and innovations were constantly being proposed and tried but the basic printing machine was still the 'Stanhope Press', which was a cast-iron version of the wooden press used in Caxton's day. Designed in the early 1800s by Lord Stanhope, the 'Stanhope' was capable of an output of 300 impressions per hour in the hands of a team of skilled operators. *The Times* was equipped with a large number of these presses.

Brunel had previously had dealings with the engineering firm of Taylor & Martineau of Whitecross Street, London. Taylor's brother was a London printer who was co-operating with Koenig (a German/American manufacturer of printing machinery) in the design of a new type of rotary cylinder press. The first demonstration at *The Times* in 1811 was unsuccessful, but Marc realised its potential and began to explore ideas concerning the production of curved stereo plates which he knew would be required once the press had been perfected. The newspaper's proprietor, John Walter, took an active interest in the work but realised that there was no possible way of using the invention before

were negligible, the Battersea sawmill was just beginning to recover from the effects of the fire but was making only a nominal profit. Marc asked a city banker friend named Sansom to look into his position. Sansom's report was most alarming as the following excerpt reveals: 'The more I investigate, the more I feel it necessary for your safety that you should sift to the bottom of every cash transaction. It was a most extraordinary jumble which you have certainly not understood, and I should have wondered if you had. I should hardly have been more surprised than I am if one of your saws had walked to town.'

After taking several legal opinions, Sansom wrote again in January 1817: 'If you have ever been ill in your life, and have depended upon medical advice, fall down on your knees and bless God that you had fewer doctors than you have had lawyers about you. If that had not been the case, you might have been making sawmills on the other side of the Styx, or inventing a steamboat for the Old Charon.' He continued sympathetically saying, 'Your conduct has from the commencement of our treaty been in the highest degree honourable, liberal, and friendly; and, although I am very anxious for my friend, I can sometimes hardly help regretting that I could not myself embark with you. Your business at Battersea, I can very clearly see, is to be made a very lucrative one, and if I were your partner, I would answer for showing you a very different balance sheet for the year 1818.'

It is possible that, following Sansom's advice, the Battersea site was registered in the names of Brunel's managers or relatives around that time to secure it against possible impeachment by the cylinder press was perfected, so withdrew his support. Brunel however was convinced that the future would eventually lie with the cylinder press, and in 1820 he filed a patent entitled: 'Certain Improvements in Making Stereotype Plates' (see Appendix 12L). When in 1814 the improved version of the Koenig machine was installed at *The Times*, the foresight of Marc Brunel was justified. The new machine was capable of producing 2,000 copies per hour, and the modern wide-circulation newspaper was born. However, Brunel's stereotyping system was not used. At first special wedge-shaped type was fixed to the curved surface of the cylinder, but this was unsatisfactory; it was difficult for the compositors to handle (no mechanical composing machines had yet been invented) and if the type was loosened by the vibration whilst printing, the whole arrangement could collapse and cause considerable damage to the machinery. Eventually it seems that John Walter sidetracked Brunel's design and adapted Marc's ideas without actually infringing the patent or paying a licence fee to the inventor.

* * *

In 1820 Marc's daughter Sophia was married to Benjamin Hawes, the elder son of Marc's great friend, also called Ben, and they took up residence in Barge House, Lambeth, the home of the Hawes family. Sophia had grown into a remarkable young woman. Her grasp of engineering was so great that she was known as 'Brunel in petticoats'. Since this profession was barred to her it was lucky that

she possessed equally well-developed artistic skills. Her husband Benjamin eventually became the MP for the new Borough of Lambeth. He launched the Fine Arts Commission, and later still, when Under Secretary of State for War, was responsible for sending a hospital designed by Marc's son Isambard Brunel to the Crimea under pressure from Florence Nightingale. Meanwhile, early in 1820 Isambard was sent to France to improve his education at the college of Caen. He transferred to the Henri Quatre Lycée in Paris in November of that year where, in the opinion of his father, it was thought he would obtain the best mathematical education.

Above: The first unsuccessful Koenig cylinder press demonstrated at *The Times* in 1811. *Reproduced by kind permission of Koenig & Bauer Aktiengesellschaft, Würzburg*

Left: The first double-cylinder printing machine by Koenig & Bauer installed at *The Times* in 1814. *Reproduced by kind permission of Koenig & Bauer Aktiengesellschaft, Würzburg*

In December 1818, Brunel had filed a patent specifying 'A New Species of Tin Foil, capable of being Crystallized in Large, Varied, and Beautiful Crystallization' (see Appendix 12K). He was relying on decorative tinfoil for his financial salvation but was almost a century in advance of his time. The manufacture was put in hand at Battersea under the control of his latest partner, a Mr Shaw. The attractive wrapping material quickly became popular. The cost of developing the process was comparatively modest and it seemed reasonable to expect a rapid improvement in the state of his finances.

But by the end of 1820 Brunel was getting deeper and deeper into debt. The sawmill and the boot factory had not come up to his expectations (mainly through Shaw's dishonesty and mismanagement) and the manufacture of decorated tinfoil had been widely pirated despite the protection supposed to be offered by the patent. In a fit of near-desperation Marc invented a device which he hoped would be an instant success. After his experiences with Shaw's management of the 'Battersea concern', he went back to his former partner Farthing and gave him instructions to market a portable duplicator which he called 'A Pocket Copying Press' (see Appendix 12M). It was manufactured by Taylor & Martineau, and produced copies in reverse, similar to the mirror images left on blotting paper, but could be easily read from the reverse side of the thin paper used. It was an immediate success and seemed set to overcome Brunel's pressing financial problems.

In the New Year of 1821 Marc received an enquiry for a bridge to link the island of La Croix with Rouen. He set to work with enthusiasm for he welcomed the possibility of revisiting the scenes of his childhood. To see once again the quays and docks which had fascinated him when a lad and to be able to converse with his workmen in the language common to both, was an attractive proposal. His elation was short-lived; within days of despatching his plans to France, his bankers Sykes & Company were declared bankrupt and went out of business. Upon learning this, his creditors became more demanding. His business enterprises on the whole were very shaky, and although the Pocket Copying Press was selling well it had not covered the initial outlay and profitability was still some way into the future. The final blow fell when the French authorities rejected his plans because he was not a member of the Government Corps of Engineers.

On 18 May 1821 Marc was arrested for debt. He was tried and committed to the King's Bench prison in Southwark. (Some accounts erroneously state that he was imprisoned in the Marshalsea.) He was accompanied by his devoted Sophie and spent the following 88 days incarcerated there. It was to be the lowest point of his life.

* * *

The King's Bench prison originated in the days when the king, as fount of all justice, personally tried and judged selected delinquents seated upon the great marble bench in the hall of the Palace of Westminster. It was transferred to Southwark following the closure of the infamous Star Chamber. When it was burned down in 1780 by the Gordon rioters, together with the Fleet and Newgate prisons, it was rebuilt with a brick façade closely resembling the nearby Marshalsea prison, except that it was considerably larger. It was primarily a debtors' prison but included a section set aside for naval miscreants awaiting trial or sentence, as indeed was part of the Marshalsea.

During the reign of George II and for decades afterwards, London was the main centre of crime in England, with over 300 receivers of stolen goods operating mainly in the Dockland area. Most of the contraband was stolen from ships moored in the Thames during the process of loading and unloading by barges and lighters, before the system of docks had been widely developed. It was calculated at the time that the total value of goods stolen annually exceeded £700,000. A large proportion of this was being purloined from His Majesty's stores and ships of war, and to counteract these practices severe penalties were imposed. A large and constantly increasing criminal class led to overcrowding in the prisons, which became unsanitary dens of iniquity, populated by men, women and children without segregation.

Although there had been some improvements since 1780, the three major London prisons were still operated for profit and were far from pleasant. Newgate still confined men and women together, and in every prison corruption was rife from the Prison Governor down to the lowliest turnkeys — men who could let a prisoner out on daily leave in return for payment. According to a report on the conditions of London prisons, written in 1814 and presented to Parliament, the receipt of money on the part of the turnkeys 'opens a wide field for every species of abuse' (see Appendix 11). It appeared that all prison employees were experts at making the most profit from their unfortunate charges and applied the rules for their own benefit. The report detailed conditions inside the King's Bench prison, describing how the building was poorly lit and extremely dirty. It apparently smelt not only from the open sewers but also from the piles of excrement heaped up behind the prison and open barrels of urine awaiting collection. It was noted that scavengers were paid by the Marshal to clear the excrement away, but they made a profit from this practice, including selling the urine to the woollen trade, and only collected the material when the quantity repaid the labour of removal. The report recommended that the scavengers should be made to do their duty and remove all the dirt from the prison on a weekly basis. Furthermore it was advised that at least one lamp should be placed in each stairway to improve the lighting. Conditions had improved very little by the time of Brunel's incarceration.

Several sources state that both Marc and Sophie were imprisoned for debt but this is not so. Only Marc was convicted, Sophie accompanied him voluntarily as wives often did during this period in order to sustain and support her husband in his misfortune. Their eldest daughter Sophia also attended him and was a great help in preventing him from falling into a deep depression at a time when no solution to their problems seemed in the offing. It appears that they had a room to themselves in the prison. They were debtors but not destitute. Marc never took advantage of the rules to leave the prison in spite of the fact that he was considerably depressed by the place. He could quite easily have spent his days at home and returned at night to his cell for a small daily payment to the turnkey, but had he done so, his pleas to friends and supporters would have been necessarily diluted and his incarceration extended indefinitely. The Brunels still had a small income or savings enough to support them in the day-to-day necessities and to retain the services of a servant, but it was insufficient to pay off the debt for which they were suffering.

Their son Isambard was now 15 and safely away in France improving his education. Some accounts say that he was supported at this time by friends of Marc, but it is more likely that his father had foreseen the coming events and taken steps to protect his son from the troubles at home. Father and son were extremely close and enjoyed very similar thought processes. Richard Beamish once described how his chief reacted prophetically to a drawing of Navier's first suspension bridge across the Seine, which collapsed soon afterwards. 'You would not venture, I think, on that bridge unless you would wish to have a dive.' And on another occasion as they walked past a new warehouse in Deptford, Marc tugged at his sleeve, exclaiming, 'Come along, come along; don't you see? . . . it will fall!' And that night it did fall. Isambard inherited this same ability to assess the stability of structures. 'That will fall,' he told a group of his schoolmates, noting the gathering storm and the faulty construction of a partly built house. Next morning only a heap of rubble remained. Even at this stage in his young life, Isambard's commercial instincts were proving more acute than his father's: he had laid a wager on the collapse!

Very little is known about the time Marc and Sophie spent in prison, a period they always referred to as 'the Misfortune'. This is because many of Marc's papers and parts of his diary relating to that time were destroyed after his death. Glimpses of how Marc felt can be detected from later journals when he was referring back to this period, which he variously described as 'a great reverse', and 'a serious disaster of fortune', and the entry for 18 May 1839 stated 'this day in 1821 I had a serious most serious (sic) difficulty to contend against in my affairs'. From the writings that have survived, it is apparent that at times Marc was desperately depressed and disgusted at the treatment he had received from government departments, and he was persistent in his correspondence with anyone he thought able to assist him to obtain his discharge. Many friends came to visit him in the prison, and probably did their best to get his case reviewed, but unfortunately with little effect. Among those who visited on several occasions was Marc's old friend Dr Wollaston. On one occasion he was accompanied by Admiral Sir Edward Codrington who described the conditions the Brunels had to endure: 'the small room, in one corner of which sat Brunel at a table littered with papers covered with mathematical calculations, while, seated on a trestle bed in the opposite corner, sat his wife mending his stockings'.

All the fair-weather friends discreetly disappeared when the family found itself on the brink of ruin; Sophie, however, never wavered or criticised. She never once left Marc's side. As the months dragged on, the feeling of helpless isolation began to take hold. 'My affectionate wife and myself are sinking under it,' he wrote to Lord Spencer, 'we have neither rest by day nor night. Were my enemies at work to effect the ruin of my mind and body, they could not do so more effectually.'

He wrote again a little later, 'It is now ten weeks, that I am in this cruel position. I have called to my aid all the forces of my soul; but I feel that I cannot longer support that which may compromise my name in the eyes of the world.' In another letter to Sir Edward Codrington dated 9 June, Marc thanked him for his visit that morning: 'I take this opportunity of returning my sincere thanks with those of my unfortunate and afflicted partner, for your kind visit this morning. As a Father and a Husband you must, indeed, have felt at finding Mrs B. in such a miserable abode. Indeed, her health, I fear, is sinking for the want of those comforts which we have lost.'

In a fit of indignant desperation, Brunel resumed his correspondence with Alexander I, the Czar of Russia, who had regretfully suspended the earlier negotiations for the Neva bridge. It now seemed possible that the Russian Treasury was prepared to sanction the project. He declared himself ready to transport his family to St Petersburg, to take Russian nationality and to work under the protection of the Czar, whose enlightenment and liberality seemed to be in complete contrast with the callousness of the British government. How the details of this correspondence leaked out is unknown, but the consequences were almost immediate. No sooner was it rumoured that the unfortunate Mr Brunel was about to shake the English dust from his feet, the authorities bestirred themselves. Now the 'National Interest' was involved. Lord Spencer and Marc's son-in-law, Ben Hawes, reported a feeling of anxiety at the Admiralty. When the Duke of Wellington (another of Brunel's stalwart friends) asked the Prime Minister what his penny-pinching Chancellor of the Exchequer was doing to ensure that Mr Brunel's services were retained for Britain, the embarrassment was unmistakable.

Lord Spencer initiated an active campaign. He felt that, however lacking Brunel might be in commercial instincts, his peculiar genius for invention was too good to lose to another country. The Duke of Wellington threw his weight behind the campaign and promised to press the government for an award to cover the debts. Mr Bandinel of the Foreign Office, Admiral Sir Edward Codrington and Mr Arbuthnot took charge of the negotiations. Notwithstanding the efforts of his friends, these negotiations seemed unnecessarily protracted to Brunel, as shown by a letter he wrote to his friend Dr Wollaston:

'If I had been guilty of any crime against the State I could not be treated with more severity, not to say cruelty, than I am. After the most unequivocal assurance of relief with profession of liberality, nine successive Board days have gone by without producing the least effect . . . Good God, when will this end! Day after day without any favourable issue. My worthy friend Bandinel is ill. I am indeed wretched, much beyond I was in the first fortnight of this confinement, when my whole property was, as it were, gone from me. I then, feeling confidence in my own exertions, had made up my mind for the future. I am at present so completely overcome as to be at times alarmed as to the state of my nerves.'

Though this arbitration seemed to the prisoner to take for ever, Lord Spencer was pressing Vansittart to make a grant and had finally obtained his agreement on 18 June. The Crown's solicitors however advised this to be kept secret, lest the creditors, upon learning that the government was prepared to assist, should increase their claims. In consequence, the Chancellor wrote to Brunel to say that he had been informed of a report that Marc, upon being released, would go to Russia and if such were the case, the government would not release him. Brunel replied, 'I must starve, or get employment here, or go to Russia, but if I see honourable and permanent employment here, you may be assured that I shall not be wanting in zeal, but shall devote my future services and talents for the benefit of my country.'

A memorial presented to the Treasury set out that there were 'strong publick grounds for the Government to avail themselves of Mr Brunel's extraordinary talents'. So a grant of £5,000 finally resulted in Brunel's release. This was on the understanding that he would abandon any thoughts of going to Russia. It was accompanied with an observation from the government that 'the step was taken more in liberality than in absolute justice, and they had a right to hope for the benefit of his future services'. On 10 August 1821, all formalities now completed, Brunel was released from the King's Bench prison after a confinement lasting nearly three months.

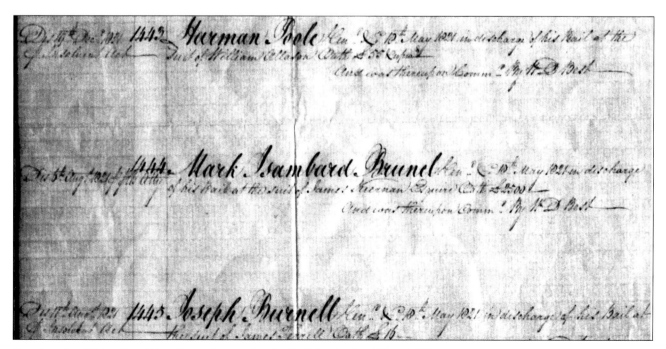

The entry in the official register of Marc Brunel's discharge from the King's Bench prison, 6 August 1821.

10.

Back in Harness

In the year following his release from prison, the government asked Marc to supply plans for a sawmill on the island of Trinidad in the West Indies. Brunel also filed a Patent for Improvements in Marine Steam Engines (see Appendix 12N). This was a subject that had occupied his mind to a large extent during his time in King's Bench and one which he felt had helped to maintain his sanity when all seemed lost. This coincided with a request from the French government to design and supply two suspension bridges for the Île de Bourbon (now known as the island of Réunion in the Indian Ocean). Each bridge was to have two 8ft 9in teak roadways which would be given lateral support by eight bracing chains, making the bridge capable of withstanding the 100mph hurricanes which are a regular occurrence on the island. The smaller St Suzanne river bridge would have a span of 131ft 9in. Each of its roadways would be carried by three suspension chains which would pass over iron towers rising from stone piers on either bank. The bridge over the River du Mât was to have two 131ft 9in spans, one either side of a central mid-stream iron tower.

Sturge's Bowling Ironworks Company at Bradford was the initial choice for the work, but the contract was eventually awarded to the Milton Ironworks near Sheffield. This, however, proved to be so unsatisfactory that Brunel ended up personally supervising the operation, involving numerous journeys to Yorkshire over the two years the contract took to complete. In addition to the inefficiency of the works at Milton, there was a further hindrance due to a particularly severe winter. In his journal of 15 January 1823, Marc recorded that the cold was so intense that cast iron measuring 15in by 10in was broken. By the end of April the bridges had been erected in Sheffield by the contractors. They were then inspected by Monsieur Sganzin on behalf of the French government and accepted.

Now they had to be dismantled and taken to London for shipment to their destination. But before this could be done the contractors demanded £500 payment above the agreed figure, for what they referred to as extra work. Naturally, Brunel resisted, and a protracted correspondence between solicitors ensued. The bridges were finally dismantled in July under the supervision of Brunel's trustworthy superintendent Thomas Mathews. He had supplied regular reports of the progress of the work, together with evidence of the numerous efforts made to elude his watchful eye. These included the weighing of the ironwork when Mathews' back was turned or when work had stopped for the night. The weighing of ironwork was one way of ascertaining whether holes caused by poor casting had been filled with clay and painted over — an activity which certainly contributed to the collapse of the Tay Bridge in 1879. 'I told 'em,' reported Mathews, 'of all the jobs I had ever been at, I never saw such goings on as these.'

Eventually, in mid-August, the dismantled bridges arrived in London at the East India Wharf packed in wooden cases; but by now

Brunel had lost all confidence in Milton and he insisted upon the cases being opened and reinspected. It was as well that he did, for it was discovered that the under-chains were 800ft shorter than the specified length. Not only that, but 200 of the flat links were missing. The bridges could not therefore be loaded aboard until the end of October, and in this operation yet more defects were discovered which caused even more delays. It was not until 29 November 1823 that the bridges finally left Gravesend for their destination.

* * *

In August of the previous year Isambard had returned from France after a period of three years. The final year had been spent under the tutorship of Abraham Louis Breguet, the great Swiss chronologist, who had inherited the mantle of England's Thomas Mudge. Breguet was very impressed with Isambard. He wrote, 'I think it is important to cultivate with him (Isambard) the happy inventive tendencies which he owes to nature or to education and which it would be a pity to see wasted.' Sophie, with bitter memories of imprisonment still fresh in her mind, was not keen for her son to follow in his father's footsteps but was eventually compelled to accept the inevitable. So Isambard, at the age of 17, joined Marc in his offices in The Poultry, Cheapside, where he took part in designing the various projects in hand. These included a suspension bridge over the Serpentine, a swing-bridge for Liverpool Docks, plans for a cannon-boring mill for the Dutch government, and designs for paddle-tugs on the Rhine.

Before his imprisonment Marc had begun to put his affairs into better shape with the assistance of Sansom. The inefficient Shaw was replaced by two brothers named Hollingsworth and by a man named Mudge, who was probably related to Sophie through the marriage of her sister Elizabeth to Thomas Mudge Junior. Brunel now divested himself of the boot and tinfoil enterprises and retained only a half share in the Battersea sawmill, the Hollingsworths and Mr Mudge taking the other share. This meant that his new partners, like the admirable Farthing, could be expected to behave responsibly since they would be reluctant to squander their own investments. On the face of it this appeared to be a wise decision.

Isambard was often left in charge during the frequent absences of his father, who was in great demand as an expert witness in patent actions. Marc also travelled about the country investigating the numerous approaches that were being made to him. It is known from his diary entries that he often combined these trips with sightseeing visits to the local attractions in the area. His visit to York in March 1823 included some time spent at York Minster. Marc's engineering eye could not help but notice how the flying buttresses on the south side were needed only for the roof and not for the

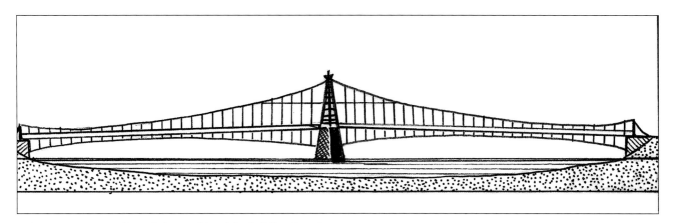

Reproduction of a drawing by Marc Brunel of a bridge to span the River du Mât on the Île de Bourbon (now Réunion). *J. Williams*

arches as a support. Although he considered the building fine, the nave apparently could not compare with those to be found at Amiens and St Ouen. Rather amusingly, Brunel also noted that he had never seen choristers so filthy in their smock frocks as they were at York, and interestingly he noted that there was only one person in attendance at the afternoon service.

The following month, on his way to Birmingham on business, Brunel stopped off at Blenheim for a visit. His diary entry recorded the following comments: 'It would be useless to attempt a description of the magnificence of that place, the vistas and the extent . . . It is truly grand . . . The house is a heavy monument, but in a Stile (sic) corresponding with the age it was built in: the Paintings very valuable the Library very rich on choir Books and the architectural ornaments costly but heavy.' On that same journey he noted that Warwick looked a neat town and that the castle was very fine, although there had been no time to view it internally.

Whenever possible Brunel called upon his friend Lord Spencer at Althorp, where he was always warmly welcomed. On one such visit, on a bitterly cold day at the end of January 1823, Marc noted in his diary the inadequacies of the heating system at the house: 'Made some observations on the mode employed for heating the Library and other Rooms at the house. It is evident that the arrangement is very bad.' He then recommended the Spencers to contact a Mr Silvester. Marc himself would communicate the plans of the flues and stoves to him as soon as possible. Upon his return to London he contacted Mr Silvester, who possessed a thorough knowledge of the nature of heat, and it was this gentleman's belief that, 'in view of the good basements at Althorp, he could easily produce such temperature as would be found equal in all parts of the apartments by heated air conveyed in horizontal flues', a modification of the system widely used by the Romans. An interesting aside is that during this same visit, Marc promised to obtain ducks from Rouen for Lady Spencer's lake. At that time the 'Rouen' breed of duck was almost unknown in England and very sought after. A large bird, it was originally developed for its table properties and much used in cross-breeding with the English Aylesbury, but it was also renowned for its beautiful plumage and striking colour patterns. Lord Spencer, aided by Marc Brunel, was one of the first to introduce the breed into England. Unfortunately Rouen ducks no longer inhabit the lake at Althorp.

Another request made to Marc was for a swinging footbridge at Liverpool, which had to be designed to allow for the passage of quite large ships into the dock, and at the same time to incorporate safety features unknown until then (see Appendix 3). Marc's design fulfilled all the requirements, and in consequence he built a model which Isambard took to Liverpool and demonstrated to the Dock Committee who gave it their approval. However, a delay occurred when the Finance Committee was approached to provide the funds. At about the same time, the management of the docks was completely reorganised and brought under the control of a new department. This resulted in Marc's efforts being sidetracked, as the newly appointed surveyor put the work in hand using Brunel's model as a pattern. Once again, Marc had been cheated; the hours of investigation, the construction of models and the preparation of detailed drawings and specifications went unrewarded. No doubt he could have sued the authorities at Liverpool, but since control had been moved several times since he was first approached, it would have meant a difficult and protracted law suit. In any case, he had little faith in the British legal system after his recent experiences.

Marc seldom declined work, especially if it included difficult commissions that would stretch his imagination and skills. In 1824 he undertook a study for a canal across the isthmus of Panama, 57 years before de Lesseps began work on the Panama Canal. Unfortunately finance was not forthcoming for the venture. He designed a suspension bridge over the Thames at Kingston, and two smaller bridges for the Huddersfield Canal Company. For the Grand Surrey Canal Company, he designed new coal docks, and at Bermondsey new cargo docks for a group of financiers. There was also a sawmill for British Guyana, and an underground aqueduct to supply Hampstead with Thames water from Hammersmith. Of all these, only the sawmill was built, under contract by the Board of Ordnance.

In the spring of 1825 Marc collaborated with Professor Gay-Lussac in an improved process for the refining of tallow to make candles. Joseph-Louis Gay-Lussac (1778-1850) was a Member of the Institute of France and had been appointed Professor under the patronage of Napoleon. He discovered boron in 1813 and the properties of iodine in the same year. He was an acknowledged

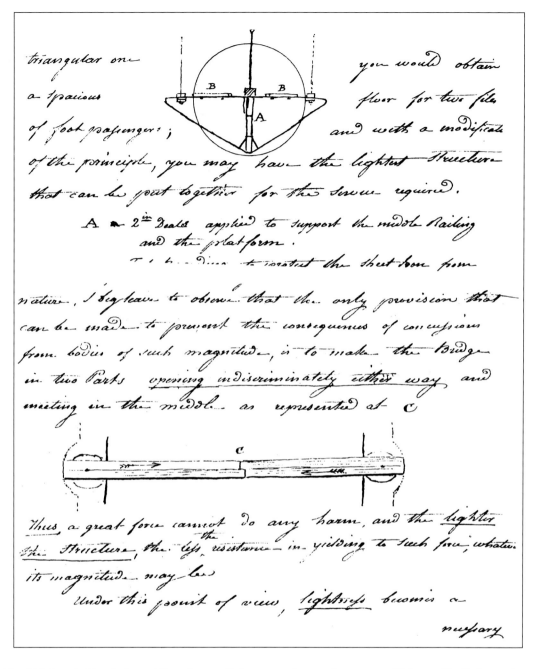

expert on the nature of acids which is probably why he and Brunel joined forces. Theirs was an attempt to improve the refinement of tallow which at that time was extensively used in the manufacture of soap as well as candles. Nothing more is known about this project but it is perhaps more than a coincidence that Price's Candle factory, one of the largest in the world, was established on a site close to the Battersea sawmill.

Brunel was experimenting at the same time with rope-hauled railways in France and London, and a diving-bell crane which was intended for use in the Bay of Vigo where Spanish treasure was thought to lie beneath the sea. He was also working on a plan for a canal to link the harbour at Fowey on the southern Cornish coast

to Padstow on the northern coast. This would eliminate the hazardous passage around Land's End where many ships foundered in bad weather. Unfortunately, the project was not considered to be viable because it was thought the ships would use it only in bad weather. Eventually it was abandoned as being too expensive but Isambard, who worked with his father on most of these commissions, gained valuable experience from it.

In the autumn of 1825 Marc submitted plans to the Chester Bridge Committee for a rubble version of the Grosvenor Bridge across the Dee which Thomas Harrison planned to construct in stone. Rubble was defined as brick or stone set in mortar and reinforced with iron. Brunel estimated the total cost would be in the region of £10,000. Despite this very reasonable proposal, Harrison's design was adopted instead, and the bridge was completed at a cost of five times Brunel's estimate. It was also during this period that Marc received a letter from the Duchess of Somerset requesting a design for a bridge over the River Dart at Totnes. 'You are always so obliging,' she wrote, 'that in the midst of your important business I venture to trouble you upon a concern in which I would ask your advice, well knowing that upon the subject in question *no* opinion is more valuable; indeed were you not so deeply involved *underground*, I should propose your putting yourself into one of the coaches and coming to us to Berry.' Brunel in reply sent a design for a graceful bridge of two arches constructed in stone, but there is no evidence that his design was put into effect. The present bridge at Totnes is credited to another architect and the design is quite different.

In 1826 Marc was engaged to design a passenger embarkation pier for the port of Liverpool. There is scant information regarding this enterprise in the Liverpool archives, probably due to the several changes in the administration which affected the footbridge. Nonetheless, there is evidence that a small floating stage, approached by a movable bridge, was installed near the Princess Dock. This was almost certainly Brunel's floating pier and remained in use until about 1847 when the larger St George's Pier was built to the design of Joseph Simpson.

Various books contend that during this period Brunel was engaged upon the design of a new cemetery at Kensal Green in association with Augustus Welby Northmore Pugin (1812-1852), the celebrated Victorian architect. This is disputed by the records of the cemetery company and other sources. The confusion has arisen over the similarity of the names of Augustus Pugin and his father Auguste Charles Pugin. The latter, as a young French architectural draughtsman, escaped from Paris during the Revolution and came to England in 1792. Here he enrolled as a student at the Royal Academy Schools in London at the age of 24. Soon afterwards, in Wales, he was appointed as a draughtsman to the architect John Nash. Nash in 1796 returned to London with Pugin, who by now was sufficiently qualified to exhibit at the Royal Academy. In February 1802 he married Catherine Welby who was related to a wealthy landowning family in Lincolnshire. Their son, Augustus, was born in March 1812. Auguste was a friend of Marc Brunel but so far as is known never co-operated with him on any venture. As a member of the committee of the cemetery company, Auguste Charles took an active interest in its organisation until his death in 1831, two years before it was opened. The company invited architects to suggest designs for the new cemetery and offered a reward for the best entry. A total of 48 entries was received, but not one from Marc Brunel or either of the Pugins. Auguste was actually one of the judges. Although the prize was awarded to Henry Edward Kendall it was never proceeded with. The final design was mainly the work of John Griffith, one of the committee members.

John Nash suggested the younger Pugin to Barry when part of the latter's designs for the new Houses of Parliament was rejected. As a result, Augustus Pugin became responsible for much of the decoration, furnishing and planning for the new building. At the time of the formation of the cemetery committee, however, Augustus would have been only 18 or 19 years old, and it is inconceivable that the staid committee members would have considered accepting one so young and relatively unknown into their ranks. The only connection that exists between the Kensal Green cemetery and Marc Brunel is that the Brunel family grave is located there with a simple, elegant monument in white marble designed by Sir Marc himself.

Few of Marc's inventions improved his financial situation but they frequently made fortunes for others who exploited them, often illegally. Although the Battersea works yielded £2,500 in 1822, Brunel was shamelessly cheated by the Hollingsworth brothers. They often undertook cash transactions and pocketed the receipts. Mudge, too, proved to be utterly useless. The majority of Marc's income at this time came from the royalties from his steam engine patent and the several sawmill inventions. Although always prepared to consider and often implement new suggestions and offers put to him, Brunel would dismiss projects which he considered impracticable or dangerous. The proposal to construct a ship canal between London and Portsmouth was one such example. He saw immediately that such a waterway would involve the construction of several tunnels of lengths never before attempted. In addition, a large number of locks and some aqueducts to cover the variation in the elevations of the countryside through which the canal must pass would be needed. He calculated that the cost of the construction would be astronomical and the operational charges so high as to deter its commercial use.

Another project with which he refused to be associated was the application of locomotive steam engines to the propulsion of carriages on common roads. Railroads were in their infancy, and the turnpikes used by stagecoaches were in a very poor state of repair. Many accidents were caused by potholes and other hazards, and coaches regularly overturning. Brunel felt that until the roads were improved, the bridges strengthened and the routes generally straightened, the increase in speed which the steam locomotive offered would be offset by a rise in fatalities. A few steam-propelled road vehicles were designed and produced by other inventors including Richard Trevithick, but they were never popular, although Foden steam wagons continued to be manufactured into the 1920s. At one time Marc was approached by the directors of the Thames Tunnel Company to design an elevator to operate in both the Rotherhithe and Wapping shafts by which vehicles could be lowered and raised as required, but when they insisted upon the use of a steam engine for the power unit, Marc refused the commission because he feared it would scare the horses and result in accidents. Brunel's concerns about the conditions of the roads in England and Wales continued to bother him in later years, as can be seen in a letter he wrote in 1834 concerning the postal services (see Appendix 6). The condition of the roads was one of many subjects that held an interest for Marc, but it was not all-consuming; the project that was to take up all his energies — mentally, physically and emotionally — was about to begin.

11. The Tunnel

The Rotherhithe Shaft and the Great Shield

In 1176, the first stone bridge across the Thames in London was begun. The man responsible was Peter of Colechurch, the Curate of St Mary Colechurch in Cheap. It lasted for 650 years and replaced a series of wooden bridges, one of the earliest of which was mentioned by Dion Cassius at the time of the Emperor Claudius in AD44.

Peter Colechurch's bridge took 33 years to complete and spanned the Thames in 19 arches, supported by 20 massive piers which occupied 700 of the 900ft of waterway at low tide. These great piers were needed to support what amounted to a small town on the bridge itself. It was about 36ft wide and had more than 100 buildings upon it. In the centre was a two-storey chapel dedicated to the martyr Thomas Becket, which could be entered from the bridge level or by steps up from the river. The money for the construction of the bridge was raised by a tax on wool, which gave rise to the old saying that London Bridge was built on woolsacks.

Peter Colechurch did not live to see his bridge completed. He died in 1205, by which time the work was almost finished; he was laid to rest in the crypt of the chapel on the bridge. Many modifications, repairs and accidents were to befall the bridge over the coming centuries. Shortly after completion, on the night of 10 July 1212, a fire which broke out at the Southwark end attracted a crowd of helpers and spectators who massed on the bridge. Unfortunately a fresh wind sprang up and carried sparks to the opposite end, setting that alight also, and the crowd was trapped. It is recorded that 'above three thousand bodies were found severely or partly burned, besides those reduced to ashes and unidentifiable'.

During the following 68 years, little or nothing was done to repair the damage, so much so that in 1280 the bridge was in a 'ruinous condition and in danger of falling down'. In 1281 five of the arches were carried away by a combination of ice floes and flood water, and emergency repairs were put in hand. At noon on 14 January 1437 the stone gate with the tower on it at the Southwark end fell down, together with two of the arches of the bridge.

On 13 February 1633, between 11pm and midnight, a fire broke out at the premises of a Mr Briggs, a needlemaker at the northern end of the bridge, which raged unabated until mid-morning and destroyed 43 houses on the bridge. Many of these houses had not been rebuilt at the time of the Great Fire of 1666. The damage caused to the bridge by the Great Fire cost more than £15,000 to make good, but would have been much greater had the fire not been confined to the northern end. Once the repairs had been completed, building leases were quickly taken up, and within about five years the line of houses and shops was once more complete on both sides of the roadway. Another fire on the night of Wednesday 8 September 1725 destroyed about 60 houses at the Southwark end of the bridge and damaged the first and second arches. It so damaged the old Traitor's Gate that it had to be demolished and rebuilt from the foundations up.

By the late 18th century the East and West India Docks, London Docks and St Katharine Docks occupied the northern shore by Wapping, and the opposite shore housed mills, warehouses and a vast assortment of factories and businesses. During the 1820s more than 4,000 wagons, carts and drays crossed the bridge every working day, and more than half of these turned down Tooley Street through the alleys and back streets of Southwark towards the industrial area of Rotherhithe. Every one had to pay a toll to cross London Bridge and it was quipped that it cost more to carry skins from Wapping to Rotherhithe than across the Atlantic from Hudson's Bay. Progress across the bridge was so slow and unpleasant, due to the congestion, that many foot passengers preferred to cross the river by wherry, thus providing a regular livelihood for the 350 watermen, who naturally opposed any improvements to the bridge or plans for new bridges.

* * *

The first attempt at a tunnel under the Thames was made in the late 18th century by Ralph Dodd. He proposed a 900yd-long tunnel between Gravesend and Tilbury fort. It is probable that Tilbury was chosen because Daniel Defoe's tile and brick works had been sited there earlier, when the abundant clay in that area had been found to be ideal for his purpose. (Daniel Defoe was not only the author of several books, he was also a businessman, a spy for one government and an agitator against another.) Dodd thought that the river here flowed over a clay base, below which was a substantial layer of chalk through which his tunnel could be bored. After drilling to a depth of 146ft, however, they had still not encountered the chalk, and after a fire in the engine house the venture collapsed.

In 1802, Robert Vasie, a Cornish mining engineer nicknamed 'the Mole', proposed a shorter tunnel between Rotherhithe and Limehouse. In 1805 the Thames Archway Company was formed and empowered by Act of Parliament to build a tunnel based on Vasie's plan, 'capable of taking horses and cattle, with or without carriages, and foot passengers'. Problems were immediately encountered when the initial driftway (which was only 5ft high, and 3ft wide at the base, tapering to 2ft 6in at the roof) was overwhelmed with water which poured in at such a rate that the pumps were unable to cope.

With costs mounting, the directors halted the work whilst they consulted John Rennie and William Chapman who disagreed as to the best means of continuing. So on the advice of Vasie and Davies Giddy (a mathematician who later changed his name to Davies

Above: Some idea of the congestion over London Bridge can be gathered from this illustration from the Illustrated London News. *Illustrated London News Picture Library*

Right: Richard Trevithick (1771-1833). A portrait by J. Linnell, 1816. *Science Museum/SSPL, London*

Gilbert and became MP for Bodmin), the board called in Richard Trevithick, who at that time was dredging gravel from the Thames with his newly invented steam dredger — the first of its kind in the world. Trevithick's reward was to be £1,000 when the work was completed, and he was confident that his earlier mining experiences would overcome any difficulties that might arise.

Richard Trevithick was born at Illogan in Cornwall in 1771. The son of a mine manager, he was an inattentive and rebellious pupil at school, and his standards of literacy were the despair of his teachers. However, he had a natural grasp of basic mathematics and learnt a great deal about the practical problems associated with mining by observation and discussion with the miners who worked with his father. He acquired a vast knowledge of the steam engines and pumps which were essential to the efficient working of every mine, and was thus able to devise improvements which were to benefit the entire industry.

At that time nearly all Cornish mines were equipped with

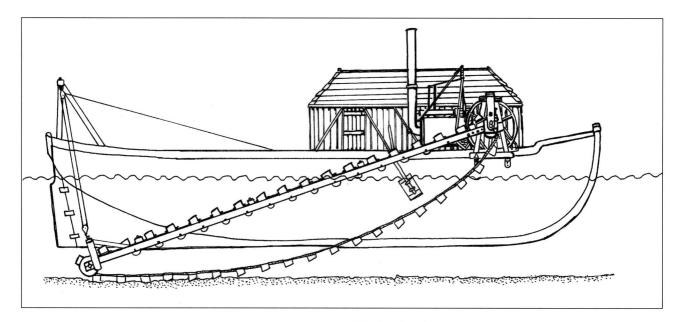

Trevithick's dredger. Illustration based upon *Rees's Cyclopedia* (1819). The chain of buckets delivered ballast into barges alongside. *J. Williams*

James Watt's engines, for the use of which Watt charged a royalty based upon the savings made in the coal used against a Newcomen engine of the same power. The mine owners accepted this at first as a reasonable reward for using the improved engines, but when the patent was extended until 1800, giving it a life of 31 years in total, they began to consider alternatives. The first of the Watt engines had been installed in 1780, and by 1795, when Trevithick was 24, there were quite a few of the old Newcomen engines still in operation so he would have been familiar with both types.

Being 6ft 2in tall at a time when Cornishmen in general were seldom taller than 5ft 8in, Richard was a very big, strong man who was soon renowned for his enormous strength and sense of fun. There are dozens of stories told about his antics. One such example was when he picked up another well-built miner at a dinner, upended him and placed his boots against the ceiling.

By the age of 36 Trevithick had established for himself a formidable reputation and was more than a match for the hesitant directors of the Thames Archway Company. A letter to his friend Giddy reads: 'Last Monday I closed with the tunnel gents. I have agreed with them to give them advice and to conduct the driving of a level through to the opposite side . . . to receive five hundred pounds when the drift is halfway through, and five hundred pounds more when it is holed out on the opposite side. I have written to Cornwall for more men for them. It is intended to put three men on each core of six hours' course. I think this will be making one thousand pounds very easily . . .'

In August 1807 Trevithick began work on the driftway — a small, initial tunnel intended to be used as a drain for the main tunnel to be constructed later. Only one miner could work at the face at a time, passing the spoil to others behind. When a shift ended and a new operator had to take over, it was only with great difficulty that they could pass in the narrow passageway. The lighting was by means of links (flares) or primitive oil lamps which would have considerably reduced the amount of oxygen available. The further they proceeded under the bed of the river, the worse conditions became.

Trevithick engaged miners from Cornwall who were used to working in cramped conditions, but they were unaware of what they were to be asked to endure. Even so, they had the continual support of their chief who regularly inspected the progress at the workface. This was in spite of the fact that his great height and massive build caused him tremendous discomfort in the claustrophobic drift which necessitated him crouching during his entire period underground.

On 23 December 1807 there was a severe risk of collapse when the drift broke into quicksand at 950ft from the shore. A month later, at a distance of 1,040ft from the Rotherhithe bank, the river again broke through and the men had to scramble for their lives back through the drift. Trevithick was the last to leave and was almost drowned, the water being up to his neck before he reached safety. Various remedies were tried, but when the drift had advanced only another 12ft, the directors panicked and the project was abandoned.

The board now offered £500 to anyone who could produce another (workable) plan and as a result, 49 schemes were submitted to Dr Charles Hutton and William Jessop for their consideration. These arbiters, a mathematician and an engineer respectively, found none to their liking and reported as follows: 'Though we cannot presume to set limits to the ingenuity of other men, we must confess that, under the circumstances which have so clearly been presented to us, we consider that an underground tunnel, which would be useful to the public and beneficial to the adventurers, is impracticable.'

This marked the end of the Thames Archway Company. The numerous setbacks and near-disasters which attended these attempts at a Thames tunnel made the raising of capital for any new venture very difficult. There was one more attempt, in 1816, when a Mr R. F. Hawkins proposed a new method. His confidence was such that he patented it. Two shafts of brick and cement were to be sunk in the river 200ft from each shore to a sufficient depth. From these shafts the excavations were to proceed in both directions. The bottom of the shafts were to form wells from which the water would be pumped by engines fixed to the top of each shaft; the roofs were to be made of cast iron. No mention is made of coffer dams being sunk into the riverbed so the method he intended using to construct his shafts is not clear. There is no record of any actual attempt being made to construct a Thames tunnel by this method, so Mr Hawkins was probably unable to obtain backing for his patent; certainly no more was heard of it.

* * *

Marc Brunel had long been interested in tunnels, as shown in his original plans for a tunnel beneath the River Neva in Russia. In 1818 he had patented a method of 'Forming Drifts and Tunnels under Ground' (see Appendix 12J), and now set to work designing a new type of tunnelling equipment. One day while working at Chatham Docks, he spotted a piece of keel timber which had been removed from one of the ships that was being renovated. It was punctured by a series of holes and in one of these he discovered a living ship-worm, *Teredo navalis*, which was claimed to have sunk more ships than all the cannon in existence.

The worm operates by boring into the wood with a proboscis at its head which is protected by hard shell, digesting the wood particles as they are produced, then excreting the residue in the form of a thin paste with which the tunnel behind it is lined. Upon exposure to the air, the lining hardens and thus provides complete protection to the worm. According to the author of an article named 'Tunnel' in the Edinburgh Encyclopaedia, which was written during the first half of the 19th century, it was Mr Brunel himself who had informed the writer that he had taken inspiration from this worm. Whether this is true or not, there is no doubt that the 'shield' Marc designed used the same principle and has become the accepted method of tunnelling up to this day. The Channel Tunnel between Folkestone and Calais was constructed using a modified 'Brunel Shield' and the development of London's underground railway system would have been impossible without it.

Brunel's idea for a tunnel under the Thames was, like Vasie, to sink a shaft on the Rotherhithe (south) bank, from the bottom of which a horizontal tunnel would be driven under the river. The shaft would be 50ft in diameter to allow for the vast quantities of spoil to be drawn up to the surface, and would in due course be matched by a similar shaft on the Wapping side of the river. These enormous shafts would be converted, after the completion of the tunnel, into access points with gently rising stairways for the use of pedestrians wishing to cross the river on foot. Approach roads for wheeled traffic were planned to follow, but these were never developed, although the tunnel was constructed with dual carriageways and footpaths for pedestrians. The importance of a new mode of communication between Rotherhithe and Wapping was stressed by contemporary writers, who pointed out that, although these two locations were only 1,200ft apart geographically, they were actually four miles apart travelling by road, via London Bridge. It was also noted that many foreign goods were being brought into the various docks on the north side, the majority of which needed to be taken south of the river, and at present the goods were almost entirely conveyed by land. The benefits of building a tunnel were therefore seen as highly favourable.

Marc, the inveterate letter writer, wrote to every person of influence he could contact. They included MPs, cabinet ministers, Dukes, Duchesses, Lords and Ladies, engineers and architects — in fact anyone he thought able to advance his cause and desirous of investing money in it. After months of diligent effort he was rewarded. Several people began to show interest and on 18 February 1824 a general meeting was held at the City of London Tavern. A total of 2,128 shares of £50 each were at once subscribed for. George Hyde Wollaston's name headed the list for 500 shares, followed by that of his distinguished brother Dr Wollaston. A committee was formed which included the names of men known for their mechanical and professional achievements: Timothy Bramah (son of Joseph Bramah the lockmaker), Bryan Donkin (the inventor of the first automatic paper-making machine), Gray, Martin, Ritchie, B. Shaw, W. Smith MP, W. Taylor and G. H. Wollaston. The city surveyor, Mr Montague, was appointed to survey and estimate the value of property to be acquired on either bank, and the firm of Joliffe & Banks was to make the necessary soundings and borings to enable Brunel to plan the details of his enterprise.

On 25 June 1824 the Bill incorporating the company received the royal assent. On 20 July a meeting of shareholders of the Thames Tunnel Company took place at the City of London Tavern, with W. Smith MP in the chair. In the report submitted to the committee, it was announced that 'the result of *thirty-nine borings* made upon two parallel lines across the river has fully confirmed the expectations previously formed, there having been found upon each a *stratum of strong blue clay of sufficient depth to ensure the safety of the intended tunnel*. The ground on the Surrey side of the river, near to Rotherhithe Church, was also bored, and a deep well being sunk on the north side, for a parochial purpose, gave a result of the most encouraging nature'. The report continued, 'in compliance with instructions from the subscribers at the original meeting, your directors have made arrangements with Mr Brunel for the use of his patent, for which they have agreed to pay him 5000*l.* when the body of the tunnel shall be securely effected and carried sixty feet beyond each embankment of the river, and a further and final sum of 5000*l.* when the first public toll under the act of parliament shall have been received for the use of the proprietors'.

Brunel was appointed engineer to the company with a salary of £1,000 per annum for a period of three years, 'the whole of which sum the directors have agreed to give him in case the work should be accomplished to their satisfaction at an earlier period'.

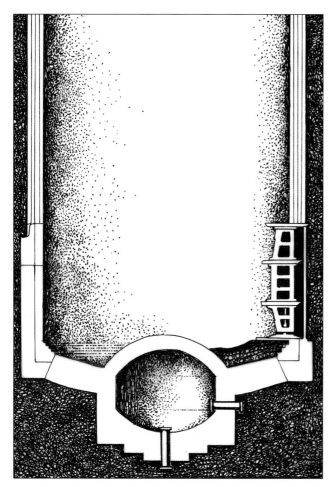

Section through the Rotherhithe shaft showing the drainage cistern and the shield. *R. Beamish*

his offices and his family from No 29 The Poultry to Bridge Street in Blackfriars — a much poorer area than the one he had been living in. His attempts to sell the house at Chelsea failed, so he was forced to lease it to tenants on an annual basis. Nearly all his time was now occupied in perfecting his plans and in this he was ably supported by Isambard, who was now 18 years old and fast becoming a reliable assistant.

On 16 February 1825 work began on clearing the ground for the construction of the first shaft. The site chosen was close to Cow Court, Rotherhithe, 141ft back from the river bank. The problems the earlier tunnellers had encountered, strengthened by the advice of the geologists, resulted in Brunel boring no lower than 14ft below the riverbed to the roof of his tunnel. In the light of later accidents, it was agreed that 40ft would have presented fewer problems but as it was, complete confidence was put in the surveyors' reports and the work put in hand without the slightest trepidation.

Brunel based his plans upon the results of the trial bores supplied to him which are set out below:

'*Stratum 1,* to a depth of 9ft. Brown clay.
Stratum 2, to a depth of 35ft 8in. Loose gravel with a large quantity of water, 26ft 8in thick.
Stratum 3, to a depth of 38ft 8in. Blue alluvial earth, inclining to clay, 3ft thick.
Stratum 4, to a depth of 43ft 9in. Loam 5ft 1in thick.
Stratum 5, to a depth of 47ft 6in. Blue alluvial earth, inclining to clay, mixed with shells, 3ft 9in thick.
Stratum 6, to a depth of 55ft. Calcareous rock in which are embedded gravel stones, and so hard as to resist the pickaxe, and to be broken only by wedges, 7ft 6in thick.
Stratum 7, to a depth of 59ft 6in. Light-coloured muddy shale in which are embedded pyrites and calcareous stones, 4ft 6in thick.
Stratum 8, to a depth of 60ft. Green sand with gravel and a little water, 6in thick.
Stratum 9, to a depth of 68ft 4in, green sand, 8ft 4in thick.'

After considering the results of all these borings and surveys, Brunel made several adjustments to his plan of operation. The main change was the abandonment of the idea of a horizontal tunnel in favour of one descending slowly to the centre of the river, where the channel used by shipping was deepest and the riverbed therefore liable to most disturbance from passing traffic or anchoring.

Brunel employed a second draughtsman named Pinchbeck to help complete the drawings of the Great Shield and intended seeking tenders from Bramah & Company, Donkin & Company and Henry Maudslay. However, this was rejected by the directors because the principals of the two former companies were directors of the tunnel company. Tenders were therefore obtained from Sturgess & Company of Bradford and Henry Maudslay. It was a great relief to Brunel when it was found that Maudslay's price was the lower. Maudslay and Brunel had worked together on the

At the same meeting it was agreed that the company would erect turnpikes at either end of the tunnel as soon as it was completed, and would charge the following tolls:

'Foot passengers 2d (two pence).
Six-horse carriages 2/6d (2 shillings and sixpence).
Three- or four-horse carriages 2/0d.
Two-horse carriages 1/0d.
One-horse carriages 6d.
Horse-drawn wagons and carts 4d.
Wheelbarrows 2½d.
Horses, mules or asses without carts 2d.
Not more than 1/0d for every score of cattle.
Not more than 6d for every score of calves, sheep or lambs.
Not more than 6d for every score of geese, ducks or turkeys.'

Redundant watermen, who were taking on average 3,700 passengers daily, were to be compensated to an extent to be settled by a jury, and the authority vested in the company would lapse if the tunnel had not been completed within seven years.

In order to be closer to the tunnel workings, Brunel transferred blockmaking machinery, the circular saws and the bootmaking plant, among others, and each respected and admired the other. Maudslay was duly awarded the contract.

The day that witnessed the laying of the Thames Tunnel's first brick by Marc Brunel on 2 March 1825 is described in his son Isambard's journal. Isambard related how he had got up very early that morning to go to Rotherhithe with his father. All were hard at work up until the last minute as was usual in such cases. Isambard described how a stage was erected, and that the crowd began to gather very early. He then wrote, 'My father laid the *first brick* I the *second* Armstrong and Smith followed.' In Marc's account of the same event he described how a large concourse of people assembled to witness the ceremony and that church bells were rung. A 'sumptuous collation' for over 200 guests followed, punctuated by speeches and toasts. There was a large model of the tunnel in sugar which adorned the centre of the table, and a bottle of wine was reserved at the end of the feast to be used at the banquet planned to celebrate the opening of the completed tunnel.

The construction of the shaft was in itself a major undertaking. It had never before been attempted on such a large scale by entirely untried methods. The plan was to construct a circular metal ring, 50ft in diameter and 40in high, cast in 48 segments bolted together. The lower edge was chisel-shaped to cut into the ground. At the top was a 10in-wide inward facing flange to which a ring of timber 1ft thick and 3ft wide was bolted. Upon this ring, a brick wall was built 2ft 6in wide. As the wall progressed upwards the metal ring was forced into the ground by the increasing weight. The soil within the ring was excavated by men and machine at a steady rate, varying between 3in and 20in per day, controlled mainly by the speed of the bricklaying at the head of the shaft.

The tunnel itself required over seven and a half million bricks and thousands of tons of 'Roman' cement, a quick-setting cement that Brunel had previously used at Chatham with excellent results. (This cement has no connection with Rome; it is a hydraulic cement made from calcareous nodules from the London clay.) A rapid-setting cement was essential for work in the tunnel where pressure would be exerted against the newly erected walls by the great jacks in order to advance the shield. The use of this type of cement was disputed by Bryan Donkin and Timothy Bramah, who conducted a series of experiments with other types to disprove Brunel's theories. They all failed, and so contracts for Roman cement were issued to three firms — Turner & Montague, Francis & White and Wyatt Parker & Company.

Negotiations with Pritchard & Hoof, the contractors who had built the Medway canal, having fallen through, it was decided to use directly employed labour. On 25 August 1825 William Armstrong was engaged as the resident engineer at a salary of £300 a year plus accommodation and (eventually) an allowance for coals and candles. In November, a clerk of works named Litchfield was recruited, and in December a smith named Redman agreed to make over a thousand small screw-jacks for the shield, and duly set up his workshop near Cow Court. At about the same time negotiations began with a craftsman named Stewart, who applied

for the post of chief brickmaker at the tunnel company's own brickworks, which it planned to establish on nearby land. Unfortunately the contractors delivered only 53,000 of the initial order of half a million bricks, and Brunel was forced to obtain supplies elsewhere at an increased cost since the company's brickfield was not yet in production. The additional expense was another stick in the hands of Bryan Donkin, who continued to agitate about the use of Roman cement at every opportunity, and often in public, to the annoyance of Brunel.

The extraordinary 'sinking tower' was a public spectacle which fascinated everyone. The Duke of Wellington, Lord Somerset and General Ponsonby visited it on 22 April 1826, and climbed down inside the structure to watch the operations. After the Duke came several members of the aristocracy, including the Duke and Duchess of Cambridge, the Duke of Gloucester, Prince Leopold, Lord Spencer, the Austrian Ambassador, Prince Lieven, the Duke of Northumberland, and Mr Robert Peel. 'The crunching sound produced by the entrance of the iron curb into the gravel . . . being reverberated from the walls of the tower, had,' as Richard Beamish wrote, 'a striking, not to say startling, effect; while it tended to exalt the impression which the magnitude of the operation was so well calculated to inspire.'

By early May the structure was down into the 27ft-thick stratum of gravel, and a battery of hand-pumps was rigged at the top of the tower to remove the water which flooded into the excavation. 'Seventy-two men employed in pumping,' wrote Brunel on the 9th, 'thirty-six by day and as many by night, and thirty-two more for heaving up the gravel, all of which would have been saved had the steam engine been at work, and that in the course of a week or ten days at the utmost.'

Ten days later the rim of the curb reached the stratum of 'strong blue clay', but water continued to pour in. After another frustrating week, Brunel decided to call upon Maudslay to remonstrate about the delay in the supply of the steam engine which was on order. He was received by Joshua Field who by now was the partner in charge of administration. When Brunel suggested that the directors of the company might require compensation for the delay in producing the steam engine, Field replied that, 'knowing Mr Maudslay as I do, and you should by now, I am convinced that if he became aware of such intentions he would not hesitate to lay by and set the Company at defiance at once'. Brunel realised that this would mean the immediate cessation of all work on the steam engine and perhaps as importantly, the happy partnership which had developed between them over many years would be threatened if not broken for ever. He returned to Rotherhithe and finished sinking the shaft in less than seven weeks without the use of a steam engine.

Maudslay was renowned as a testy individual who did not suffer fools gladly, not that Brunel was anybody's fool. Henry Maudslay had clawed his way up from being a mere labourer at an ordnance factory to becoming the most famous engineer of the time and the inventor of machine-tools which had revolutionised British manufacture. He had every right to be self-opinionated and

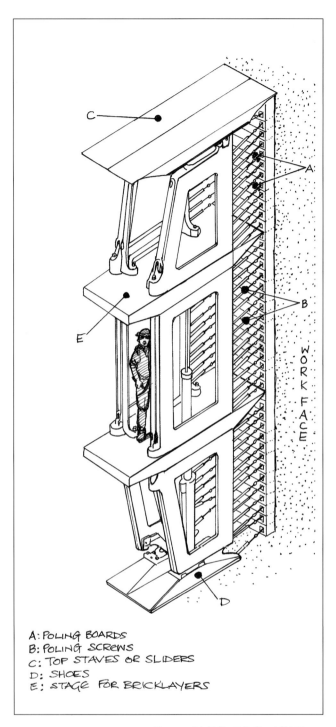

A: POLING BOARDS
B: POLING SCREWS
C: TOP STAVES OR SLIDERS
D: SHOES
E: STAGE FOR BRICKLAYERS

Above and right: Brunel's Great Shield. Note [*right*] the frame can now be advanced whilst the poling boards remain supported. *Reproduced from Richard Beamish's* Life of Brunel *(1862)*

arrogant. He was a perfectionist. He knew it and so did Brunel, and each appreciated that the other was supreme in his class. The delay in the supply of the steam engine was due to problems encountered in its manufacture, and Maudslay would not release it until it was perfect or as near perfect as he was able to make it.

The shaft was finally completed on 21 November 1825, and the Great Shield's 12 frames were assembled at its foot, ready to advance towards the northern shore. It was a huge relief to Brunel after all the problems which had arisen during the preceding months. The new steam engine had overcome its teething problems, the ground beneath the clay was said to be standing exceedingly well, and the strata above appeared to be sound and free from water for 32ft.

On 2 November, during the last stages of completion of the shaft, Marc was taken ill and confined to bed for a week by his doctor. Now aged 56, he had been working a seven-day week ever since the work was started, except for two days spent in Brighton in October with Sophie and his daughter Emma. Pressure of work affects people in different ways; Brunel dealt with all the problems, solved them, had a breakdown and was back at work a week later. Armstrong, the resident engineer, was approaching breaking point, although still active and dependable, but young Smith, Brunel's junior, was badly affected and had become erratic. He would soon be dismissed. Isambard, however, seemed never to need rest and was completely

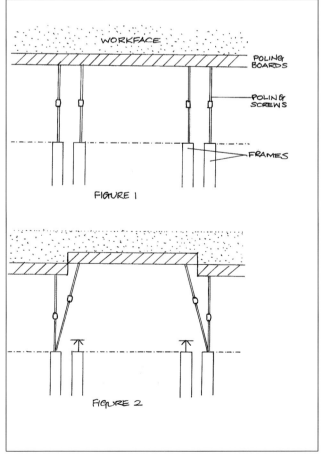

68

Section
showing the
scale of the
tunnel.
J. Williams

untroubled by the stress the others endured. His unbridled energy was a constant source of bewilderment. He was now 19, and spent his occasional leisure time at the opera, the art galleries, or racing on the Thames with William Hawes in their jointly owned skiff. On returning home in the evening, he would raid the library for books such as Hume's *History of England*, or the publications of the Royal Institution of British Architects or the Institution of Mechanical Engineers. The following year he was elected a member of the Royal Society — a rare honour for one so young.

* * *

To understand the problems which arose later and the methods used to overcome them, it is necessary to provide at least an outline of the construction of the Great Shield together with its operation. The shield consisted of 12 frames of which one is illustrated on page 70. Each frame was 21ft 4in high and 3ft wide, divided into three compartments or cells approximately 7ft high. There was therefore a total workface area of 756sq ft being worked at the same time, with one miner to each cell operating independently. There were 36 cells in total. Each frame could be moved independently of its neighbour for a maximum of 9in.

The unworked ground in front of the shield was supported by oak shuttering called Poling Boards, each 3ft long, 6in wide and 1in thick (A). These were held in contact with the soil by Poling Screws (B), which were small screw-jacks produced on site by the smith Redman, and which fitted into recesses in the cast-iron frame. On the top were the Top Staves or Sliders (C) and on the base, the Shoes (D). The large propelling jacks have been omitted from the illustration for the sake of clarity, but can be seen in illustration GS3.

The floors of each division formed stages for the bricklayers (E),

who, by working with their backs to the miners, allowed mining and bricklaying to be carried forward simultaneously. The frames were numbered from 1 to 12 and were identical, except for numbers 1 and 12 — these were fitted with side panels of similar construction to the top staves, which served to retain the walls of the excavation until the brickwork was completed. Each of the staves covered an area of the workface and could be slid forward as the mining progressed. The only section of ground left unprotected at any time was that between a stave when moved forward and the brickwork behind, and that was only for the time necessary for the bricklayers to add the next course of bricks. Tails were subsequently added to ensure that no ground was unprotected until the brickwork was filled in.

Each miner operated the following procedure. The top poling board in his cell was removed and the exposed ground excavated to a depth of 9in (the length of one brick). The top stave and the poling board were then pushed forward into the cavity and secured there by extending the poling screws. The board immediately below was then removed and the procedure continued until all boards in all three cells were 9in forward. The bases of the poling screws were now transferred one by one to equivalent positions in the adjacent frame where a second indentation in the casting had been made to accommodate them. When all the poling screws had been relocated, the frame could be propelled forward into its new position by the big jacks without releasing the pressure of the poling boards on the workface. The poling screws could now be contracted, brought back into their correct positions and the work proceed as before.

At first sight this may seem to be a most elaborate procedure, but it ensured that the workface was never left unprotected even when the shield was in motion. To compensate for any lateral pressure, friction rollers were set between the frames to permit easy movement. All the frames could not move forward together: frames

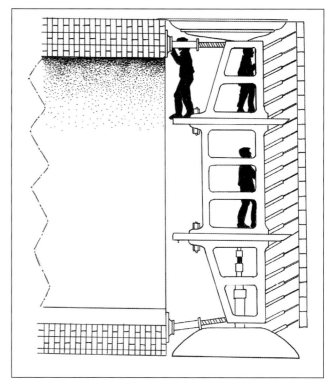

Left: Section through the Thames Tunnel showing the method of working. *Reproduced from Richard Beamish's* Life of Brunel *(1862)*

Below: Detail of the Great Shield in operation. *Reproduced from Richard Beamish's* Life of Brunel *(1862)*

1, 3, 5, 7, 9 and 11 would move first whilst the remainder were used to support the relocated poling screws. Once these were in their new position, the remaining six frames could also be taken forward. A movable wooden staging was erected behind the frames to facilitate the extraction of spoil from the workface and the supply of bricks and mortar to the masons.

The genius of Brunel's design lay in the fact that every part of the shield was capable of being removed or repaired without the slightest danger to the rest of the structure. All the component parts of the frames were bolted together and could be dismantled and replaced whenever necessary. J. Saunders, who wrote a chapter on the Thames Tunnel in Knight's volume on London, referred to Brunel's shaft in the following terms: 'This seems to our eyes, uninitiated in the wonders of engineering, not one of the least marvels of this altogether marvellous work.'

The character of the soil encountered during the first 500ft of the excavation for the shaft was described by Richard Beamish. Commencing at the top, 'Two feet of very strong blue clay, decreasing as the dip of the tunnel exceeded that of the stratum, and which at a distance of about 250 feet no longer appeared. Six feet of pure blue silt for 300 feet. Six feet of blue silt with a great abundance of small bivalve shells very minutely broken, a few only perfect. These shells were sometimes found in a thick layer, leaving the silt pure. Sometimes a number of strata — shells and silt alternating — presented themselves. A layer of about an inch thick of indurated sand, has been constantly found in the bed of silt. Two to three feet of stone, a sort of bastard gypsum, not at any time continuous, but coming out in large lumps. And, lastly, three to nine feet of gravel intermixed with green silt, which formed the foundation.'

Here then was a totally different type of ground from that which had been suggested in the report of the surveyor and announced by the directors on 20 July 1824. In place of 'a stratum of strong blue clay of sufficient depth to ensure the safety of the intended tunnel,' a variety of strata was encountered, varying in density. According to Knight's book, *London — Pictorially Illustrated*, the favourable report by the surveyor 'induced Mr Brunel to go to work in a somewhat bolder way than he had otherwise intended'. It is obvious that a

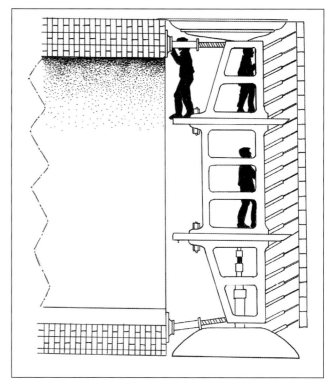

machine which was designed to operate in homogeneous blue clay was scarcely calculated to contend with a friable sand or sand so impregnated with water as to have become absolutely fluid.

* * *

On 28 November 1825 the shield began its progress. Two teams of miners, each consisting of 36 men with complementary teams of bricklayers, worked eight-hour shifts. They soon ran into trouble, however, when the shield encountered water in such quantity as to halt operations until repairs could be effected and the flow staunched. Meanwhile Marc had fallen ill and was confined to his bed. Many leeches were applied to his head, and with their aid or in spite of it, he managed to recover.

12.   **Tunnelling**

The First Phase

By 23 December 1825 the first section of the double archway had been completed and the shield entered into undisturbed ground free from water. However, by late January 1826, when 14ft of the brickwork had been completed, water burst in with considerable force and the pumps were overwhelmed. The steam engine had to be stopped and work suspended as the water rose 12ft in the shaft. When work was resumed on 3 February, such was the influx of water that the miners refused to re-enter the workings until they could be assured of their safety.

In mid February, water and gravel again broke into the workings, but by now the miners had more confidence in the protection offered by the shield; they stayed in their positions and — by the introduction of lead piping behind the frames — drew off the water to relieve the workface. By the 25th of the month, the workface was free from water and on 11 March the shield had, without injury, passed through ground which no previously known system of tunnelling could have penetrated.

The importance of this achievement was not lost on the public who continued to heap praise on the man who had invented and designed the Great Shield. One of these was the printer and publisher Charles Knight of Ludgate Hill, who was a regular visitor to the tunnel workings. In 1843, he published his great work, *London — Pictorially Illustrated,* which included a detailed description of Brunel's enterprise. However, the chairman of the company, William Smith, was not so impressed. He thought that the shield was an unnecessary luxury and the construction of the brickwork false in principle. On several occasions he tried to get Brunel dismissed from his position as Chief Engineer and one of his own cronies substituted. He was continually undermining Brunel's attempts to get government support for the venture, and was of the opinion that the work could be completed more cheaply and rapidly by a combination of simpler methods using contract labour. Smith's continuing hostility distressed Brunel and almost brought about a second breakdown in his health. Fortunately the success of the shield enabled him to bear with equanimity the barbs to which he was constantly exposed, and he soon returned to his normal cheerfulness and optimism.

Unfortunately the resident engineer was less able to shrug off the jibes and petty complaints. Early in April Mr Armstrong broke under the strain at the same time as Brunel himself was confined to bed. The whole direction of the undertaking was now thrown upon the shoulders of Marc's son, Isambard, a young man scarcely 20 years of age who had now to spend nights as well as days underground supervising and directing the work.

By the middle of May, over 100ft of the tunnel had been completed. The machinery for raising the spoil to the top of the shaft was working well, and the men below were becoming more efficient and confident by the day. The directors (led by Smith and Donkin) had refused Brunel's request for a drain on the grounds of economy, and had substituted a pipe which now proved to be totally inadequate for the purpose. This decision eventually added considerably to the cost of the work. On 22 May, the top plate of No 1 frame suddenly collapsed with a startling report. There were no signs of unusual pressure, and the mishap was thought to have arisen through a sudden change in temperature to which the upper part of the excavation was often subjected.

By the middle of the summer, Mr Armstrong, in view of his illness, decided to resign from his position as resident engineer, and young Isambard was officially appointed in his stead, with Richard Beamish as Isambard's assistant. This gives an indication of how much trust now lay between father and son. Indeed a diary entry from June 1826 reveals Marc's thoughts on his son's suitability, commenting that 'Isambard is the most efficient inspector we have. He is constantly in the work.' Another entry details how Isambard gave instructions as to the better disposition of the staves of the Great Shield, to which his father made the following comment: 'Isambard's vigilance and constant attendance were of great benefit. He is in every respect a most useful coadjutor in this undertaking.'

The directors, at the instigation of Smith and Donkin, decided that the bricklayers should be paid by a system of piecework, but the men soon found that it was impossible to earn a living wage under the new arrangements. They therefore began to skimp the work and put pressure on the miners to speed up their efforts in order to make more brickwork available sooner. Unfortunately, against his better judgement, Marc was forced to allow the experiment to continue with disastrous results. The speeding-up process required the length of the poling screws to be doubled, and these now needed the application of physical power, possessed by few of the miners, every time the shield was moved forward. The poling board at the top of a cell was 2ft or more above the miner's shoulder, and when the soil was removed in front of the board, the workface was about a foot further forward. With his arms extended to the utmost, the miner was in the least favourable position to resist any pressure from the exposed part of the workface. In addition, under considerable pressure from the company directors, Brunel reluctantly agreed to increase each forward movement of the frames to 18in (the length of two bricks), a decision which went against all his mechanical expertise and earlier experience. Both Beamish and Isambard thought that, had the progress been limited to the planned 9in advances and the payment of the bricklayers by piecework abandoned, many of the future problems would have been avoided or at least reduced in scope.

On the morning of Friday 8 September 1826, water was seen to be falling from the tails of Nos 7 and 8 frames. This was checked by stuffing the gaps with oakum. In two hours, diluted silt was observed and during the night it burst in with considerable force. By three o'clock the next morning, when Beamish relieved Isambard at the workface, the force was so great as to resist the united efforts of three men to keep the stuffing in place. It was not until six in the evening that the silt was all washed away and clay began to seal the gaps. It was now Saturday evening, and the utmost vigilance was required during the whole night to keep the men at their posts so that work could be suspended as usual on Sunday. By 11am on Sunday, they were finally satisfied that the workface was sufficiently secure to leave for the rest of the day.

The following day witnessed water and silt occasionally bursting through the back of No 6 frame when any attempt was made to move forward. Marc Brunel was constantly in attendance with advice and suggestions. Timbers were introduced in front where the ground was more solid and capped with clay; they were then forced up by powerful screw-jacks. But upon an effort being made to move the frames forward, water appeared in front in such quantity as to threaten the destruction of the workface. To relieve the pressure, borings were made through the brickwork of the central pier and pipes inserted at the back of Nos 6 and 7 frames. By late evening, after considerable labour, the objective was achieved and the water flowed with great velocity, promising to relieve the pressure and thus to prevent the further dilution of the silt and clay. The place was now a hive of activity; the din of the workmen and the crashing of the water as it fell 22ft on to the iron plates on the floor below was deafening. Suddenly the water ceased to flow. The men stopped work. Not a sound broke the intensity of the silence. 'We gazed on one another with a feeling not to be described,' says Beamish. 'On every countenance astonishment, awe perhaps, was depicted, but not fear. I saw that each man, with his eyes on Isambard Brunel, stood firmly prepared to execute the orders he should receive with resolution and intrepidity. In a few moments — moments like hours — a rumbling, gurgling sound was heard above; the water resumed its course; the awful stillness was broken; life and activity once more prevailed; and the works proceeded without farther material interruption.'

By five o'clock on the morning of Tuesday the 12th, everything appearing quiet and peaceful, the superintendents retired to rest. It was now clear how important it was for a regular staff of efficient supervisors to be constantly in attendance for, however admirable the conduct of the workmen, it was obvious that without the guiding and controlling power of an officer, many would have abandoned their posts. The first difficulty, commencing on the 8th, had required 53 hours of consecutive superintendence; the second had required 20 hours, and these were not isolated cases. Beamish's journal records 24 hours on the following Thursday and Friday, and 21 hours on the Saturday and Sunday.

The demands on Isambard were still more onerous for, with the exception of a few hours, he never left the workings until the 13th, taking sleep only by snatches on the staging. His father was duly worried: 'I am very much concerned at his being so unmindful of his health,' Marc wrote. 'He may pay dear for it.' Prophetic words.

The promptitude and courage exhibited by Isambard were not confined to underground operations. A few days before the first irruption, the feed-pipe of one of the boilers of the steam engine burst. To stop the pumps would have caused great inconvenience and possibly danger. Isambard and Beamish were at the top of the shaft and ran to the engine house as they noticed a change in the tone of the engine. Isambard immediately understood the nature of the incident. Seizing some packing material and a length of timber, he jumped upon the boiler and applied the packing to the fissure with the aid of the timber. He intended jamming the other end of the post against the roof of the engine house, but when he attempted to do so, the roof lifted, so he hung onto the timber, applying all his weight to the fissure until relieved by others who rushed to his assistance. By Isambard's valiant effort, time was gained to allow the auxiliary boiler to be brought into service and the steam engine continued its work without interruption.

The increasing demands now being made upon his son's physical powers determined Marc to ask the directors for permission to engage extra assistants. In justification he laid before them a written detailed report of the events of the last 20 days, which he had already communicated verbally to the board. To his surprise he was charged by the chairman with misleading the directors, on the grounds that this was the first they had heard of the problems. Fortunately this view was not shared by the rest of the board, who acceded to Brunel's request. They appointed Mr Gravatt, the son of Colonel Gravatt of the Royal Engineers, who was a civil engineer with practical experience in Bryan Donkin's factory. A little later, Gravatt was joined by Mr Riley, a young man also experienced in the profession, but with a constitution quite unsuitable for the demands shortly to be made upon it.

On 22 October, Isambard was taken ill, and on the 29th Beamish was also struck down. For the next 10 days neither was able to resume active superintendence. Marc's entry in the *Thames Tunnel Minutes of Occurance* for 30 October states, 'Want of Assistants! inconvenience from it. All of them ill!' The responsibility now devolved entirely upon Marc and Mr Gravatt, one or the other having to remain all night in the frames, and Mr Gravatt's shift on more than one occasion exceeded 38 hours.

By the end of the year the legs, heads and top staves of nearly all the frames had been renewed. Many of the operations connected with the removal of the top staves and head were not only dangerous, but involved considerable delay. In consequence the progress in 16 weeks reached an average of only 7ft. The main problem was the lack of a drain (which the directors had deemed an unnecessary expense), for which handpumps were a very poor substitute. The valves were continually being choked and the leathers damaged. The bottom of the excavation was therefore alternately full of water or empty, although never dry. There was no choice but to continue working in water whilst the costs increased by £150 per week.

To all this was added another unexpected and unwelcome problem. The directors decided to recruit labourers from Ireland in an attempt to reduce costs and speed up operations. These men, although honest and hard-working under the normal open-air

conditions to which they were accustomed, found conditions in the tunnel so new and incomprehensible that their energies seemed to completely desert them — except in flight. Any unusual activity among the miners at the workface, any sudden gush of sand, or rattling of gravel upon the frames would immediately result in them abandoning their posts *en masse* in a headlong rush to the shaft.

An incident occurred while the shield was being restored after the first irruption. The entry in Isambard's journal states that 'At two o'clock in the morning . . . Kemble, the overground watchman, came stupefied with fright to tell me that the water was in again. I could not believe him — he asserted that it was up the shaft when he came. This being something like positive, I ran without my coat as fast as I could, giving a double knock at Gravatt's door on my way. I saw the men on the top, and heard them calling earnestly (down the shaft) to those who they fancied had not had time to escape . . . Miles had already in his zeal thrown a long rope, swinging it about, calling to the unfortunate sufferers to lay hold of it . . . I instantly flew down the stairs. The shaft was completely dark. I expected at every step to splash into the water. Before I was really aware of the distance I had run, I found myself in the frames in the east arch. Nothing whatever was the matter, but a small run in No 1 top, where I found Huggins (foreman) and the *corps d'elite*, who were not even aware that any one had left the frames.'

* * *

On 27 January 1827, Mr Riley, the second assistant, was taken ill with fever. By 5 February he had become delirious and he died on the 8th at the premature age of 24. Marc Brunel's journal records, 'Isambard, Gravatt, Beamish ill — Munday and Lane, foremen, very ill'. On the evening of the 10th, whilst superintending the operations of the bottom boxes and seeing that all was made safe for Sunday's rest, Beamish reported that 'a haze rose before my eyes and, in the course of half an hour, I had lost the sight of my left eye. The active treatment to which I was subjected, while it prevented me from resuming my duties until the 7th of March, proved only partially successful in restoring the vision.' The constant stench and disgusting effluent from the filthy Thames water in which they had to operate, was doubtless the principal cause of illness among those engaged in work underground.

By now the miners at the workface had become accustomed to small movements of the ground and minor slips were often ignored. Progress had increased to more than 13ft per week, the record for one day being 3ft. The number of men employed had increased from 180 in October to 467 in March. This included the extra Irish labourers required to remove the excavated spoil, and those manning the pumps. Of the regular miners and bricklayers, about seven per cent were on the sick list. Saunders, writing in Charles Knight's great work on the history of London, related an amusing incident which occurred one night while Isambard was on duty supervising the works. It was around midnight that he and a few others who were with him heard the cry, 'The water the water! Wedges and straw here!' Silence followed. Naturally Isambard

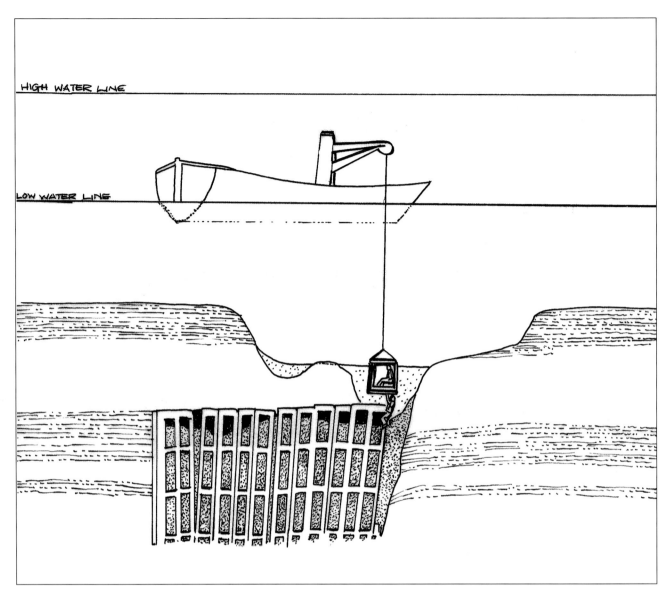

Mapping the hole in the riverbed after the first irruption. *Reproduced from Knight's London — Pictorially Illustrated of 1842*

rushed to the spot where the sound had come from. He found the men safe but all were fast asleep. It became obvious that one of them had been dreaming of a new irruption in the tunnel and had cried out in his sleep.

On 26 February 1827, the directors permitted members of the public to visit the works. Upon payment of one shilling they could proceed down the western archway for about 300ft. This was a rather sly attempt to derive some income from their investment, but it caused Brunel considerable anxiety. The men employed underground were now all experienced and well drilled in procedures for rapid evacuation should the necessity arise. To place the public in such a position could result in uncontrolled panic and consequent loss of life. Furthermore, fleeing visitors could easily block the exits and make it difficult or impossible for the miners to get to safety. Despite his concerns, visitors were admitted and very soon the tunnel became one of London's main attractions. It was quite common for between 600 and 800 people

to pay to view the works every day. Indeed Marc himself decided to take his family to the tunnel to see how much progress had been made. His diary entry for that day reads as follows: 'I went to the Tunnel. The arch being well lighted up, and the whole walk completed, a few visitors were admitted. The coup d'oeil was splendid. Mrs Brunel, Emma, Sophia and her three little children were the first. It gave me great pleasure to see the whole of my family in the new scene.'

Progress was still being made, but the ground in front of the shield began to show signs of impending trouble. So diluted had the silt become that, early in April, the miners in the upper boxes had, on removing a poling, to scrape away the spoil with their hands, while a man stood behind ready to supply clay which they rammed in and pressed against the saturated ground with the aid of the poling screws. In spite of every effort, whole faces would come down, the diluted silt oozing out, leaving a cavity in front of the polings which, having no pressure against them, fell.

It appeared that the workface was in such a condition as not to be trusted and orders were given to have straw, clay and oakum at hand in the top boxes. Four 6in pumps were at work in the back of the west frame and a double pump at the east, together with a horizontal pump which was worked by the engine. It was now that the need of an efficient drain was felt in earnest. In the west corner, so little of the original base of the river was left that, on 21 April 1827, stones, brickbats, pieces of coal, bones and fragments of glass and china came down into the frames.

After many applications from Brunel to the directors, the hire of a diving bell was conceded. The one used by the West India Docks Company was borrowed and immediately began a survey of the bed of the river. It was found that the ground over No 1 frame was depressed to a considerable extent and the soil so loose that an iron rod could be pushed down to the frame. Subsequently an iron pipe was thrust through the ground and direct communication established between the frame and the diving bell. (Some gold pins were passed up to the divers in the bell and presented to friends of Sir Benjamin Hawes as a memento of this extraordinary event.)

Meanwhile a steening (a type of tank) was prepared, and with the aid of Isambard and Mr Gravatt, sunk from the bell and deposited over the western frame. After the water had been pumped out it was filled with concrete and proved to offer a big improvement, although there was still some leakage. Some of the old watermen believed that very little gravel still remained in that part of the river, 'it were all dredged away for ballast, and some ships' anchors must have penetrated the clay from time to time'. During the time that the work was going on, one of the workmen lost his shovel in the river — an occurrence which will be referred to again later. It is interesting to note from Marc's journal that on 25 April, Sophia and Mrs Baldwin 'went down with the diving bell'. It is not clear whether he was referring to his wife or his daughter in this case. It is nevertheless surprising that he allowed it to happen at all as it would have been a precarious adventure for anyone inexperienced in its use.

By May Day 1827 540ft of the tunnel had been completed, but the difficulties encountered appeared to be on the increase. The situation was aggravated when the men went on strike for an increase in wages in spite (as Beamish bemoaned) of earning between three shillings and threepence and three shillings and ninepence a day. Even more damaging was the fact that they picketed the tunnel entrance to prevent others from entering. With the aid of the foremen and a few faithful employees, the senior staff had no option but to make all secure and close down the workings.

The strike lasted eight days, and the miners returned to work only when they were convinced that their claims were not going to be met. All were re-engaged, with the exception of the known ringleaders who were dismissed. The strike however proved more serious than was at first thought; the ground had settled down and had become more diluted. The sliding plates and screws had rusted and the distorted condition of some of the frames required much greater power than had been needed previously to move them forward. Consequently, considerably more supervision was required, often extending to 27 hours without relief for the principals.

On 11 May, the sheaf of a ship's pulley block was excavated from the ground at No 5 frame, and a piece of brass and an old shoe buckle from No 9. On the following day, the shovel which had been lost in the bed of the river on 2 April was discovered among the spoil taken from No 6 frame. On the same day, during an attempt to work down the top face of No 6 frame, the ground came in with great force, not diluted, but in masses. The top box was filled and cleared and again filled, and it became necessary to timber the back and sides of the box and leave it. The yellow mottled clay which overlaid the silt was broken, and, subjected to the pressure of a column of water of about 30ft at high water, forced itself through apertures into which barely two fingers could be inserted, where it would swell out to the size of a man's head. Marc Brunel, perceiving that it was almost impossible for the miners to prevent the workface collapsing as soon as the top polings were removed, directed that the forward movement of the frames should be restricted to 7in at a time; but by then the damage had been done.

The following description of what happened next is taken from Richard Beamish's report — an eyewitness account of an event which came close to causing the entire project to be abandoned there and then.

'At two o'clock on the morning of the 18th of May I relieved Isambard Brunel in the superintendence of the working. At five o'clock, as the tide rose, the ground seemed as though it were alive. Between Nos. 6 and 7 frames there were occasional bursts of diluted silt, which subsided, however, as the tide ebbed. The men who came on at six a.m. exhibited extreme reluctance to go to their work, hanging about the engine house, where arrangements had been made for them to dry their clothes. The day passed on with not more than the usual amount of alarm; but, as the flood-tide returned, the same general disturbance of the ground was observed as had occurred in the morning.

'The visit of a dear friend (Lady Raffles) with a large party, about five o'clock p.m., did not tend to allay a strong feeling of apprehension which took possession of my mind. No sooner had she taken leave than I prepared myself for what, I was satisfied, would prove a trying night. My holiday coat was exchanged for a strong waterproof, the polished Wellingtons for greased mud boots, and the shining beaver for a large-brimmed south-wester.

'The tide was now rising fast. On entering the frames, Nos. 9 and 11 were about to be worked down. Already had the top polings of No. 11 been removed, when the miner Goodwin, a powerful and experienced man, called for help. For him to have required help was sufficient to indicate danger. I immediately directed an equally powerful man, Rogers, in No. 9 to go to Goodwin's assistance; but before he had time to obey the order, there poured in such an

75

overwhelming flood of slush and water, that they were both driven out; and a bricklayer (Corps) who had also answered to the call for help, was literally rolled over on to the stage behind the frames, as though he had come through a mill sluice, and would have been hurled to the ground, if I had not fortunately arrested his progress. I then made an effort to re-enter the frames, calling upon the miners to follow; but I was only answered by a roar of water, which long continued to resound in my ears. Finding that no gravel appeared, I saw that the case was hopeless. To get all the men out of the shield was now my anxiety. This accomplished, I stood for a moment on the stage, unwilling to fly, yet incapable to resist the torrent which momentarily increased in magnitude and velocity, till Rogers, who alone remained, kindly drew me by the arm, and, pointing to the rising water beneath, showed only too plainly the folly of delay. Then ordering Rogers to the ladder, I slowly followed.

'As a singular coincidence, I may here remark that this man, Rogers, who showed such kindly feeling and devotion, had served with me in the Coldstream Guards.

'As I descended from the stage, the water had so risen in the tunnel, that all the loose timber near the frames, the cement casks, and the large boxes used for mixing the cement, were not only afloat, but in considerable agitation. The light was but barely sufficient to allow me to grope a way through these obstructions, which, striking against my legs, threatened seriously to arrest my progress. I felt that a false step could not be retrieved, clad as I was, and with heavy boots quite full of water. After a short struggle, I succeeded in gaining the west arch, which, having been appropriated to visitors, was comparatively free. The water was perceptibly rising; it had already reached my waist; still I could not venture to run, feeling that a stumble might still prove fatal. If I could only gain the barrier which limited the ingress of visitors, I should be clear of the floating timber which must be there arrested!

'As I approached this barrier, the sight of some of our most valued hands cheered me. Not understanding the cause of procrastination, they could not withhold their expressions of impatience, Mayo and Bertram swearing lustily at my apparent tardiness. Arrived at the barrier, four powerful hands seized me, and in a moment placed me on the other side. On we now sped. At the bottom of the shaft we met Isambard Brunel and Mr. Gravatt. We turned. The spectacle which presented itself will not readily be forgotten. The water came on in a great wave, everything on its surface becoming the more distinctly visible as the light from the gas-lamps was more strongly reflected. Presently a large crash was heard. A small office, which had been erected under the arch, about a hundred feet from the frames, had burst. The pent air rushed out; the lights were suddenly extinguished, and the noble work, which only a few short hours before had commanded the homage of an admiring public, was consigned to darkness and solitude. It only remained to ascend the shaft, but this was not so easy. The men filled the staircase; being themselves out of danger, they entirely forgot the situation of their comrades below. For the first time I now felt something like fear, as I dreaded

the recoil of the wave from the circular wall of the shaft, which, if it had caught us, would inevitably have swept us back under the arch. With the utmost difficulty the lowest flight of steps was cleared when, as I had apprehended, the recoil came, and the water surged just under our feet. The men now hurried up the stairs, and though nearly exhausted, I was enabled to reach the top, where a new cause of anxiety awaited us. A hundred voices shouted, a rope! a rope! save him! save him! How any one could have been left behind puzzled and pained me sorely. That some one was in the water was certain. With that promptitude which ever distinguished Isambard Brunel, he did not hesitate a moment. Seizing a rope, and followed by Mr. Gravatt, he slid down one of the iron ties of the shaft. The rope was quickly passed round the waist of the struggler, who proved to be old Tillett, the engine man. He had gone to the bottom of the shaft to look after the pumps, and being caught by the water was forced to the surface, from which he would speedily have disappeared, but for the presence of mind and chivalrous spirit of his officers.

'The roll was now called, when, to our unspeakable joy, every man answered to his name; and we were thus relieved from the painful retrospect that must have followed any sacrifice of life.

'To convey the intelligence of the disaster to Mr. Brunel was our next object. Taking a pony belonging to Isambard, I hastened to town; but Mr. Brunel having gone to dine with a friend, it was not until ten o'clock at night that I was enabled to execute my painful mission.

'After the first shock, and upon receiving the assurance that no life had been lost, it was marvellous to witness the elasticity with which he took part in the preparation of a letter to *The Times*, which should convey to the public correct information as to the nature of the catastrophe, and in determining the mode by which he proposed to overcome the effects of the misfortune.

'At one o'clock I threw myself on my bed; and, breathing forth a grateful thanksgiving for the protection which a beneficent Providence had extended to me and to those placed under my charge, I sunk, utterly exhausted, into a profound sleep, having been twenty-three hours under unusual anxiety and continuous activity.'

Beamish was very fortunate to have survived. The Brunel Archive Collection held at the University of Bristol still has in its possession the *Thames Tunnel Minutes of Occurance* book. The first entry is on 13 August 1825 and the last entry finishes abruptly on 18 May 1827 — the day of the flood as referred to above. The book is heavily water-stained but nevertheless remarkably legible.

It is interesting to note that following this particular accident, Marc records in his diary that in a sermon given by the Rotherhithe curate that day, the curate referred to the accident as a fatal one and was therefore a 'just judgement upon the presumptious aspirations of mortal men'. It appears that Marc did not hold much sympathy for these views. Immediately after this statement he wrote the words, 'The poor man!' It may be assumed that he was referring to the curate.

More Problems

The next day, 19 May, it was decided to examine the bed of the river again from the diving bell. A hole was found extending from about the centre of the excavations eastwards for a considerable distance. In some parts the sides were vertical and no gravel was found even on the undisturbed bed of the river. This confirmed the opinion of the watermen that all the gravel in that part had been dredged out. A number of coal barges had reached this spot the night before, but moved on because the anchors would not hold. Further examination showed that the Great Shield was still in place. By almost dropping out of the bell, the occupants were able to place one foot upon the back of the top staves and the other on the brickwork of the arch. To fill the hole, Brunel directed that bags filled with clay which had hazel rods run through them should be lowered over the hole in such quantity as to form an arch. The clay was put into bags to prevent it being washed away, and the rods were to allow for interlocking. So effective was this operation that, five days later, the pumps were able to begin the task of draining the tunnel. As a further security, under more pressure from some of the directors and against his opinion as to its effectiveness, Brunel had a raft of timber, measuring 35ft square and 2ft deep, loaded with 150 tons of clay, lowered over the frames.

The pumps were now brought into full operation and the water level rapidly fell in the shaft. By six o'clock on 31 May, it was possible for a boat to enter the western arch. However, before much of an examination could be made, the waters began to rise once more. By nine o'clock, the pumps were overpowered and the shaft again filled. An examination of the river-bed revealed that the western end of the raft had been tilted up by the strong ebb tide and the eddy which resulted had washed away most of the loose ground at that end. This confirmed Brunel's opinion as to the inefficiency of a rigid covering like a raft, and he directed that it be raised and abandoned. This was done but only after a great deal of toil and effort against a tide running at four miles an hour and with the aid of seven chains fixed to eye-bolts on the raft.

Marc now reverted to the use of clay in bags, but before these were positioned, a number of iron rods were laid over the cavity with the aid of the bell. They rested between the brickwork and the top of the shield to form a grating to support the clay bags. The frequent use of the diving bell resulted in many narrow escapes being experienced. Links in the chain by which the bell was suspended sometimes fractured; occasionally vessels collided with the bell barge and once, when Isambard was below in the bell and Beamish on the barge attending to the signalling, the foot board of the bell was observed floating on the surface of the water close by. A man named Pinckney, who accompanied Isambard inside the bell, was confident that he could step out of the bell and stand upon the ground immediately below. Unfortunately the ground gave way

and he managed to scramble back to safety only by grabbing hold of Isambard's foot. During the struggle the board broke away and floated upwards. There was great relief on the barge when signals from the pair below provided evidence that all was well.

Clay and gravel continued to be applied to the breach without intermission. When about 19,000cu ft of these materials had been deposited in the hole, pumping was resumed. Brunel was forced to consent to yet another experiment in the form of a tarpaulin spread over the repair: by the use of anchors and chains this was eventually fixed in position, but proved to be as useless as the raft.

By 25 June, the shaft was completely free from water, as was about 150ft of the tunnel. However, since the tunnel descended in a gentle gradient to the centre of the riverbed, there was a considerable quantity of water at the frames, where water was still within 3ft of the top of the arch. When Beamish, accompanied by two other men, Woodward and Pamphilon, took a boat into the tunnel and came to within about 120ft of the waterlogged frames, they encountered a great mound of soil and debris brought in by the irruption. Beamish left the boat and, carrying a lantern, scrambled over the heap into the frames on hands and knees.

To his amazement he discovered that the frames had resisted the whole force of the river and were still in place with their top staves level. The cells were about a quarter filled with silt and clay, and the bags which now formed an artificial riverbed above, protruded in front and behind the frames, looking as though ready to burst. All the cells opposite the western arch were completely filled with silt and clay, and a steady stream of water flowed in the east corner.

The admiration and near-reverence which Beamish and the other superintendents held for Marc Brunel and his son Isambard shines through the pages of Beamish's book *Memoir of the Life of Sir Marc Isambard Brunel* which was published in 1862, 13 years after the death of Sir Marc. Unhappily this hero-worship was not apparent among the senior directors of the tunnel company, who continued

The Bell barge in the river. After a drawing by Clarkson Stanfield RA.
J. Williams

interfering with the engineer's plans at frequent intervals, much to his chagrin and their cost.

The newspapers of the day carried full reports of the progress and problems of the tunnel, and rejoiced when it seemed possible that the work could be restarted after the disaster. Some of the directors now expressed a wish to investigate the extent of the damage for themselves. Brunel strongly opposed any such expedition, but was powerless to stop it. Two of the directors — Mr Martin and Mr Harris — proceeded down the shaft accompanied by Mr Gravatt and two miners named T. Dowling and Samuel Richardson. They boarded a small boat with the intention of travelling down the western arch. The following account is taken from Beamish's book:

'Richardson, though not called upon by Mr. Gravatt, insisted on getting in at the stern of the boat when he had shoved it off. The weight proved too great. The water came in; and though one of the gentlemen was requested to go forward that the boat might be balanced, neither was willing to stir, until Mr. Martin, feeling the water inconvenient, and forgetting that the boat had made considerable progress down the archway, diminishing rapidly the headway, suddenly stood up, struck his head against the top of the arch, and fell backwards upon the others. The boat filled, and the party were at once plunged into water about twelve feet deep. The only swimmers were Mr. Gravatt and Dowling. The directors clung to Mr. Gravatt, who could only release himself by diving. Swimming out, he quickly returned with a second boat. Meantime, Mr. Martin and Mr. Harris had succeeded in getting hold of the plinth in one of the side arches and there supported themselves until relieved by Mr. Gravatt; but there was no account of Richardson. A drag was procured, and his body at once found. It was conveyed to my bed as the nearest; and though every means was resorted to that science dictated, all proved in vain; the vital spark had fled — the silver cord was broken. The shock to Mr. Brunel was great, far greater than the announcement of the irruption of the river. For that a remedy was at hand; but who could call back the departed spirit? It must be ever remembered that one of the most striking characteristics of Brunel's inventions was the means provided for the protection of life; and notwithstanding all the difficulties by which the operations of the tunnel were beset, no life had yet been sacrificed where the necessary care had been practised.'

On 11 July Marc Brunel was laid low by a serious illness and was indisposed for several weeks. During this period, Isambard was in control, assisted by the faithful acolytes Gravatt and Beamish. Conditions in the tunnel worsened considerably and the foul atmosphere continued to cause many casualties. The ventilator in the central pier had been choked by the irruption and the timber replacement was still incomplete. All the men reported a black deposit in their nostrils and mouths. Headaches, sickness and giddiness became endemic. A newspaper report described how the impurity of the air would so overcome the labourers that 'the most powerful had frequently to be carried out insensible'. Richard

Beamish suffered an attack of pleurisy which kept him off work for nearly six weeks. Many of the key men were taken ill at this time and one (known as 'Old Greenshield') died. In consequence the work was painfully slow.

Progress on the tunnel was being followed by people all over the country. Those living in Scotland were given regular updates by *The Scotsman* newspaper, which eagerly informed its readers of all the latest goings-on at the workface. On 8 September 1827 it reported that 'no stronger proof could be adduced of the great anxiety and interest entertained by the public generally in the success of this great undertaking, than the fact that £62 was received, during the last week, as admission money from visitors, great numbers of whom were scientific foreigners, whose visits are made almost weekly'.

In his journal of 30 September, Marc reports, 'How slow our progress must appear to others; but if it is considered how much we have had to do for righting the frames, and for repairing them — what with timbering, shoring, shifting and refitting, all executed in a very confined situation — water occasionally bursting upon us — the ground running in like slush — it is truly terrific to be in the midst of this scene. If to this we couple the actual danger — magnified by the re-echoing of the pumps, and sometimes by the report made by large pieces of cast iron breaking — it is no exaggeration to say that such has been the state of things. Nevertheless, my confidence in the Shield is not only undiminished; it is, on the contrary, tried with its full effect.'

On 1 November 1827 there was an unusually high tide. The water rose 3ft above the Trinity high-water mark and 9in higher than had been recorded for over 20 years. The greater part of Rotherhithe was flooded, and it was only by prompt action that sufficient clay was obtained to form a bank around the top of the shaft to prevent the tunnel being completely submerged. This was no ordinary tide; the water rose and fell three times that morning between 10 and 11.30am. First it rose and fell 9in; then it rose 15in and fell 12in; and lastly it rose 4in before falling in the normal manner. But the frames held and the pumps continued to cope with the increased flow. Writing about the Thames Tunnel in Knight's work on London, Saunders commented on all the difficulties that Marc had had to face, declaring that, although Marc had great confidence in the shield that he had designed and in his own resources, nevertheless it was impossible that he could have anticipated 'the all but overwhelming amount of obstacles that he has actually experienced, principally from the character of the soil, and the extraordinary influence which the tides exercised even at the Tunnel's depth'.

By Saturday 10 November, the situation in the tunnel was so much improved that Marc was persuaded to invite several dignitaries and officials to a celebration dinner under the river (see Appendix 14). The band of the Coldstream Guards played, and the side arches were adorned with crimson drapery. The tables were lit with candelabra served by portable gas. 50 people were present but Marc was not well enough to attend in person, so Isambard represented his father. Numerous healths were drunk and many speeches of congratulation made by politicians, Foreign Office

Relative position of the tunnel beneath the Thames. *Reproduced from Knight's* London — Pictorially Illustrated

officials, members of the aristocracy and even the Equerry to HRH the Duke of Gloucester, besides friends and associates of the Brunel family.

In the adjoining arch over a hundred workmen were also invited to celebrate. At the end of the evening, following the traditions of their craft, they presented Isambard with a spade and pickaxe. How the celebrants coped with the damp conditions and fetid air is not recorded, but the evening closed on a cheerful note with congratulations all round.

* * *

The numerous repairs which the shield continued to require meant that progress was slow during the months of October, November and December. As the year 1827 approached its close, the west side of the excavation became increasingly troublesome. No 2 frame had become inundated by masses of soil and debris and early in the New Year some of the rocks which had been previously excavated, taken up and thrown back into the river came down again into No 2 frame. Unfortunately the person in charge did not deem it necessary to refill the depression in the bed of the river which resulted, and so no bags of clay were applied as was normal. On Saturday 12 January, Beamish was above ground dealing with official paperwork when a shout issued from the shaft: 'The water is in — the tunnel is full!'

He rushed to the workmen's staircase which was blocked by men scrambling to escape, so he went to the visitor's staircase where he was met by the rush of water. He managed to grab Isambard who had been swept up by the great surge of the wave. Isambard was barely conscious having been badly battered by the debris amongst which the sudden burst of water had thrown him. Even so, he was most concerned about the men who had been with him at the workface only seconds earlier. 'Ball! Ball! — Collins! Collins!' he managed to murmur to Beamish before losing consciousness, but there was nothing to be done. They were gone.

Later, Isambard gave an account of the disaster as he saw it develop at the workface. Collins and Ball were working in No 1 frame when it was decided to open the ground at the top of the frame. When the side shoring was removed, the ground began to swell, and within a few moments had burst in through a hole between 8in and 10in in diameter. This was immediately followed by an overwhelming torrent of water and sludge. Collins was thrown back from the frame by the force of the water and all the efforts of Ball to replace the shoring were in vain. So rapid was the influx of water that, had the three not left the stage immediately, they would have been swept off like leaves on a flooded stream. A rush of air suddenly extinguished the gaslights and they had to struggle on in total darkness. When less than 20ft from the stage, they were thrown about by baulks of timber which were in great agitation in the boiling waters which now reached up to Isambard's waist.

With great difficulty Isambard managed to extract his right leg from something heavy which had fallen upon it and struggled into the east arch where he called to Collins and Ball. Receiving no reply and with the waters now up to his chest, he struck out for the workmen's staircase. Unfortunately the day shift was coming down whilst the night shift was still going up, and in the confusion the staircase was completely blocked. He tried to reach the visitor's staircase in the west arch but was caught up in the torrent and propelled into the shaft where Beamish was able to rescue him. Three men who, finding the staircase choked, tried to ascend a long ladder which they found in the shaft, were swept out of the shaft and under the arch by the recoil of the wave and were lost. The ladder and the lower flight of the staircase were broken to pieces. Isambard was found to have severe internal injuries as well as damage to a knee joint and was confined to his bed for months. He later recounted that his knee was so injured that he could scarcely swim or climb the stairs. During his long convalescence he referred to the accident in his journal: 'I shan't forget that day in a hurry, very near finished my journey then.' Frustration at being laid up at such a critical time is obvious from another journal entry. Following some musings on what might happen to his family, Isambard declared, 'damn all croaking, the Tunnel must go on, it shall go on'. Unfortunately he never again entered the workings as part of the engineering team. His term of office at the Tunnel had covered less than two years (see Appendix 4).

During the weeks that followed, Marc Brunel was subjected to a barrage of suggestions and advice as to the best way of repairing the damage and sealing the hole, but it invariably turned out, to quote the engineer's words, 'that the ground was always made to the plan, not the plan to the ground'. Out of over 500 suggestions, there was not one of any practical value. Isambard's protracted illness resulted in Marc taking upon himself the duties of resident engineer again, and it was during the long nightly vigils the two spent together that Marc narrated to Beamish the story of his earlier life and experiences which Beamish later included in his book.

The funds of the company were now almost exhausted. They had no option but to suspend operations. In July the frames were blocked up pending an appeal to the country for the finance necessary to complete the work. It is clear that Isambard was most anxious for his father's health at this stressful time. Two months earlier he had written in his journal, 'My poor father will hardly survive the Tunnel my mother will follow him I shall be left alone . . . what will follow I cannot guess.' All the problems and the continuing uncertainty connected with the tunnel during this period prompted *The Times* to refer rather wittily to the project as, 'The Great Bore'. A public meeting was held on 5 July 1828 at the Freemason's Tavern. It was attended by HRH the Duke of Cambridge, the Duke of Wellington, the Duke of Somerset, the Earls of Aberdeen and Powis, C. N. Palmer MP, W. Smith MP,

J. Masterman MP, Robert Dundas MP, Sir John Sinclair and Sir Edward Owen among others. The Duke of Wellington proposed a series of resolutions, expressing his confidence that the work would eventually be crowned with success. He praised the skill and ingenuity of the engineer in command, whom he had supported on previous occasions and whose abilities he admired. After a long and detailed discourse concerning the advantages to be gained from the completed tunnel, its effect upon the commerce of London, the interest of the rest of the world and the disappointment which would be felt world-wide if the project were abandoned, he put the following resolution to the assembled company:

'Resolved, That this meeting do earnestly invite the public at large to support the plan proposed for the completion of the work, and to subscribe their names for debentures which are issuable towards it, in sums of 20*l.* and upwards; and for donations, &c.; and that books be now opened.'

The resolution was carried unanimously. In proposing a vote of thanks to the Duke of Wellington, the Duke of Cambridge expressed his hearty concurrence with the views stated and added that in the course of the long residence which he had had abroad, it was his pride to notice the favourable opinion which was entertained towards the final success and completion of the Thames Tunnel. The pride of Englishmen was at stake in the undertaking. He therefore called upon all those present to support it by liberal subscriptions.

A sum of £18,500 was immediately subscribed. The Duke of Cambridge, the Duke of Wellington, Mr Maudslay and Mr Palmer were eager donors, having put their names down for £500 each. Despite their aid, the debenture scheme failed. All works halted on the site and in August 1828 the tunnel was bricked up.

Subsequently a mirror was placed at the end of the visitor's arch. This arch was stuccoed and being lit with gas, continued to be an object of attraction and interest among members of the public for several years. Writers and critics of the day were more scathing. Brunel's great project had failed and they wanted to make known their opinions on the structure. Thomas Hood was a humorist and satirist, known for his skilful use of puns. In his *Ode to M. Brunel*, Hood was merciless:

Other great speculations have been nursed,

Till want of proceeds laid them on a shelf;

But thy concern was at the worst,

When it began to liquidate itself! . . .

I'll tell thee with thy tunnel what to do;

Light up thy boxes, build a bin or two,

The wine does better than such water trades;

Stick up a sign — the sign of the Bore's Head;

I've drawn it ready for thee in black lead . . .

14. The Long Pause

Putting the tunnel and its problems behind him, Brunel resumed his profession as a civil engineer and undertook surveys for several organisations. These included the Grand Junction Canal Company, the Oxford Canal Company, Cork Jail bridge and the projection of a bridge over the Vistula at Warsaw. In addition, he was often employed in examining mechanical constructions for which patents had been secured, and in giving evidence in Courts of Law for or against their validity.

By now his fame had spread far and wide. In 1827 the City of Rouen had elected him an honorary member of the *Société libre du Commerce et de l'Industrie de Rouen*. In 1828 he was elected to no fewer than four institutions: *l'Académie Royale des Sciences, Belles Lettres, et Arts de Rouen;* the Royal Academy of Sciences of Stockholm; the *Société Royale d'Agriculture et de Commerce de Caen;* and the *Société libre d'Emulation de Rouen*.

In November 1829 he became a Member of the *Société française de Statistique universelle* under the Presidency of the Duc de Montmorency. Four years later he was awarded a Silver Medal by *l'Académie de l'Industrie de Paris* in recognition of the benefits his efforts had conferred upon practical science.

Meanwhile, the directors of the Thames Tunnel company were considering every plan and suggestion put before them to enable work to be recommenced. With the collapse of the debenture scheme it was obvious that government funding would be essential for the success of any plan they decided to adopt, and there was no guarantee that this would be forthcoming.

Of all the 500 proposals received from members of the public throughout Europe and America, only one finally gained the backing of the directors, who recommended its adoption during a meeting on 3 March 1829. They regarded it as the least impracticable of all those received and proposed it in the following words:

'That the plan was totally and essentially different from that of Brunel.

That it was a matured plan.

That it was accompanied by the opinion of several eminent scientific men.

And that the price would be considerably less than one half, and probably a still smaller proportion of the antecedent cost.'

In order to obtain the backing of the government, the chairman (W. Smith) approached the Duke of Wellington with the plan in the hope of securing his support. The Duke was not impressed. He felt that it was not the want of a plan which had caused the tunnel work to be suspended, but the lack of money. He was also very concerned that the work already completed should be protected in the event of any plan being adopted. The chairman

assured him that the completed part of the tunnel would be secured by the erection of an iron door at the start of the drift by which the work was to be carried forward from the present termination. This would be used to completely seal off any irruption of water, gravel and soil into the existing tunnel. To this Wellington replied that if the water was shut out the men would be shut in.

While these discussions were taking place, Lord Spencer sent a letter to the Duke of Wellington in which he recommended that the government should advance the sum of £300,000 for the completion of the tunnel according to the original design. His Grace replied that 'his colleagues and himself thought best to postpone laying the affairs of the company before Parliament till next session, when he hoped to have the benefit of his Lordship's assistance in so doing'.

Several months passed during which no 'scientific men' had been produced to back the new plan, and the directors had refused to submit that plan to Brunel, who was always willing to offer his assistance to enable them to judge the feasibility of it or any other proposals. Eventually Brunel sent a statement to the directors reviewing the arguments put forward for the adoption of new methods of completing the work. After some preliminary observations, he stated that:

'Taking, then, the proceedings of the meeting altogether, — the report, the discussion and the resolution, — the arguments used by some of the supporters of this resolution were: —

'1st. That my plan was unnecessarily expensive.

'2nd. That the cost of the Tunnel when compared with what I had calculated it at, showed that my estimates were not to be relied on.

'3rd. That it was therefore necessary to change the plan of operation, and adopt a cheaper one than that before the directors; which, backed by contractors, under adequate security, gave a just ground to expect the completion of the work at a considerably less sum than it would cost on my plan.

'In the observations which I deem it my duty to make, I shall follow the order in which I have now stated what I understand to have been the arguments used at the meeting alluded to.

'1st. Then, as to the expense of constructing a tunnel on my plan.

'It is well known that in the year 1808, a company was formed to effect a passage technically called a driftway, or heading, across the river at Limehouse; it was carried to the distance of 1,040ft. It was *an experiment* to ascertain the cost and practicability of making a larger tunnel for general traffic, and to which it was to be the drain; and was carried on in the old method by the best miners, and superintended by a well known practical and able engineer, Mr. Trevithick. Every particular of this undertaking together with the accounts of the expenditure were published.

'Now I shall proceed to make an exact comparison between the expenses incurred on that occasion and the present; so as to obtain from two experiments on a large scale the true comparative cost of constructing a tunnel on the old system (supposing it to be possible) and on mine; or, in other words, I shall ascertain from the published documents of the Thames Tunnel, and the archway company of 1808 –

'1st. What each paid for excavating and removing a cubic yard of earth.

'Expenses incurred in carrying on the works' . . . '

There followed detailed figures for Trevithick's Driftway (5ft high and only 2ft 9in wide) and the Thames Tunnel (22ft high and 38ft wide), showing the cost per cubic yard of the former to have been 12 pounds 17 shillings and sixpence against four pounds five shillings for the latter. Brunel continues:

'The difference is great, but the causes are easily explicable. In a heading, as in the case of the archway company, the miner has at every step to construct and fix in place a framework to support the ground; and if the soil is at all unfavourable, this operation is attended with considerable delay and risk.

'Whereas in the Thames Tunnel this support is completely effected by means of the shield, leaving the miner nothing to attend to but the more immediate working of the ground; and by the space and protection afforded to the workman, we are enabled to take full advantage of that subdivision of labour so necessary to economy, and by the facility and certainty with which the ground is supported without interfering with the operations of the miners, the bricklayer, the pumper, and the labourer can proceed simultaneously, and without impeding each other.

'The plan therefore which I have adopted, and which till now has been held out as the cause of our success in a work of acknowledged difficulty, is not, as stated in the report, an apparently ingenious but really expensive plan; but has proved, what it was intended to be, simply a more safe, easier, and more economical mode of securing the ground during the excavation than the old one.

'To pursue this comparison, — for it is not in my power to make another, — and it would be absurd to take ordinary canal tunnelling as a measure of the expense and difficulty of the tunnel under the Thames, — and passing through its loose alluvial bed,– it is worthy of remark that we have completed and secured the work as we proceeded, and that in spite of the difficulty we have encountered, we have never lost a single foot of what has been completed. We have a substantial structure, the strength of which has been proved beyond a doubt; whereas with regard to the driftway or heading of 1808, nothing remains but the recollection of it.

'Dr. Hutton and Mr. Jessop, — two very high authorities who were consulted on the resumption of that work, after the irruption of the river, — gave it as their opinion, that although they could not pretend to say what means might hereafter be suggested, they considered, 'that effecting a tunnel under the Thames by an underground excavation in the old mode was impracticable'. Yet it is to this old mode that you are now called upon to return, under the promise of *economy, security, and despatch.*

'With what probability I leave you to judge, as the work with which I have made the comparison of expense, was conducted by an engineer of honour, talent, and great experience; and failure and abandonment of the experiment, and the subsequent opinions of competent judges, which I have quoted, will give an idea of the practicability.

'Now with respect to expenditure merely, I beg to observe, that the company had a committee of works, and a committee of finance; and I am not aware of any proposition having proceeded from them by which any material saving was effected. Two modes were at different times suggested to me by the directors, with a view to economise our funds; one was to get the work contracted for at a fixed price; and the other was, to pay the men by piece work. With regard to contract work I have placed what observations I have to make upon it under another head. Piece work was tried; but we soon found that the work was hastily and imperfectly done, particularly when difficulties increased, and great waste of materials and time ensued in consequence. We were often obliged to do the same thing over twice, and, in consequence, it was given up. In fact, in a work like the tunnel, there must be no inducement held out to the workmen to conceal difficulties in the vain hope of avoiding them; or to hide defects in their work in a situation where inspection must be imperfect.

'As to the estimate: On this subject I must remark, that when weight is laid on the expense of the works under my management, the proprietors ought to have been informed at the same time of the following facts: viz. that the tunnel, subsequent to the formation of the original estimate, was increased one third in its dimensions; and the brickwork, consequently, from a rod to nearly a rod and a half per foot run.

'This alteration, made in concert with the directors for the greater convenience of traffic, was a very material enlargement of the original plan, and necessarily induced a corresponding addition to the estimate. This was the first cause of deviation from the original estimate.

'The next was the discrepancy between the *real* state of the ground, and that expected from the local information I had gathered together, after long continued enquiry, and equally from the result from the first series of borings instituted by the company, and conducted by parties entirely unconnected with myself and uninterested in the future proceedings of the undertaking. Besides all this, I am free to avow that the difficulties and uncertainties of the undertaking have exceeded all anticipations.

'The estimate for the future stands on peculiar grounds; it is an application of the actual expenses of the past to the future; it is no calculation founded on data which may prove false, but an abtract from the books of the company. With regard to the proposal to finish the works by contract under security: I object strongly to contract work for an underground situation like the tunnel. It is quite impossible to insure a sound and substantial structure.

Work by contract is always done with as cheap materials and in as slight a way as can be admitted by the specification. The opinion of the committee of works, and confirmed by the directors, is already recorded in an elaborate report on this subject in March, 1827, 'in which the danger of the work being slighted, instead of being performed carefully and so as to ensure permanent durability,' is assigned, amongst other reasons, for the determination to abandon all idea of contract work.

'Contract work is but piece work on a larger scale, where the objections to the one apply, at an increasing rate, to the other; it would be, in fact, most injudicious to apply either contract work or piece work to the tunnel. They may be usefully applied where the eye can follow them; but with us, where the consequences of a failure in any part might be fatal and irremediable, and would only be discovered when too late, there must be no inducement to slight the work for either profit or speed.

'And when it is considered that the tunnel is intended to stand the wear and tear of the traffic of the metropolis, it is of the utmost importance that not a doubt should exist in the public mind as to its solidity; and it has acquired that confidence in the plan on which it has been carried on.

'With regard to security for the performance of a contract, it is evident that it should be such as to compel a contractor to finish the work at all hazards, and finish it substantially, whatever difficulties may arise to swell his expences beyond his calculation. Such securities must, moreover, cover the full value of the work done; because it is only by the completion of the remainder, that this can be rendered profitable; and it may therefore be totally lost by any serious accident occurring to the new work which might be too expensive to be worth remedying.'

* * *

It was not until 30 March 1830, following 12 months of vexatious negotiations and disappointments, that the alternative to Brunel's plan was put before the experts. Professor Peter Barlow, Mr James Walker and Mr Tierney Clark were asked for their opinion as to its practicability. Their report was laid before a special general meeting of the company on 22 June 1830.

They had interviewed Marc and Isambard Brunel, Mr Gravatt and several of the miners; they had perused various plans, the section of borings, the journal and other documents; they had written to Messrs Pritchard & Hoof (the proposed contractors) for specific information as to the mode in which they intended to proceed. They did not receive a reply. They then gave their opinion that 'the plan submitted to them is by no means adapted for overcoming difficulties of the nature of those which have already been encountered, and which are likely still to present themselves. That there is therefore no probability of the tunnel being completed in the manner proposed; and that it would, to say the least, be little better than a waste of time and of money to attempt it.'

After stating that it was unnecessary to trouble the board with the reasons which had caused them to come to this conclusion, the report concluded, 'Few professional questions have ever come before us which have admitted of a more decided solution than this did, after we had been informed of the nature of the ground and of the plan proposed.'

Resulting from this clear and explicit report, a resolution was passed to the effect that 'no other plan be used than that of Mr Brunel, and that an application be made to obtain the necessary funds from Government'. This was a triumph for Marc. According to Beamish, on 2 July 1830, the tunnel company members were not a little surprised to learn that, upon a proposition being made by Lord Durham in the House of Lords to transfer to the tunnel company a loan then solicited by a Canadian company, the Duke of Wellington replied that a loan had already been offered by the government but had been refused by the tunnel company directors!

Despite the advice and opinions expressed by the most eminent engineers of the day, the chairman continued to press for the work to be completed by contract labour, without the shield and without Brunel in charge. It seems incredible that the chairman of the company, who had appointed the committee to enquire into methods of completing the work, should then completely ignore their conclusions and throw his weight behind a system which had already been partially tried and found to be severely wanting. But contract work often has its advantages — particularly to the person responsible for awarding the contract, in this case Mr Smith. It was suspected by some of the other directors that Smith was to receive a percentage of the value of the contract. One at least believed that the contract had already been verbally agreed before the meeting of the board took place on 22 June and that it only awaited the final signatures.

Smith's continual opposition to Brunel over this and several other matters came as the last straw. (He had been the chief opponent when Brunel had wanted to construct a drain in the tunnel, and was also the loudest voice clamouring for the introduction of piecework underground.) In the words of Beamish, 'Mr Brunel now resigned his appointment, feeling no longer able to contend against the spirit of hostility which had been exhibited towards him by the chairman.'

Shortly afterwards Marc Brunel complained of an extraordinary stiffness, coupled with a nervous irritability and it is thought that this was probably a minor stroke. He did make a full recovery and continued to work at full pressure, in spite of being advised by his doctor to slow down and take life at a more leisurely pace.

15. Alternative Interests

On 13 February 1830 a deputation of gentlemen consulted Marc Brunel regarding the erection of a bridge over the Avon Gorge at Clifton in Bristol. During the discussion he explained to them 'how the lateral agitation may be prevented, and how the effects of the wind might be counteracted'. As a result, towards the end of the year it was agreed to offer a reward for the best design. Among the submissions appeared the name of 23-year-old Isambard Kingdom Brunel, who had the enormous advantage of his father's assistance. Marc had already had experience with his Bourbon, Totnes and other bridges, and so was well qualified to advise and suggest ideas for the Clifton Bridge.

On 12 January 1831, Marc recorded in his journal, 'Devised this day for Isambard's bridge, a new mode for carrying the heads of the chains, and sent him a drawing of it.' On the 20th, he wrote, 'Engaged this day on the mode of passing the chains of the bridge over the heads, with all the combinations necessary for repairing, and likewise for the compensation against dilatation', and on the 26th he wrote, 'Engaged on the Heads for the Suspension Bridge also on the Land Fastenings'. Throughout February, March and April, Marc was almost continually engaged in perfecting the details of the bridge and making drawings to send or give to Isambard. To his delight, on 19 March Isambard was appointed Chief Engineer to the Clifton Suspension Bridge. Marc's joy, however, was tempered somewhat by the death of his old friend Dr Wollaston, who had been one of his staunchest supporters throughout the problems encountered with the directors of the tunnel company, and also earlier during Marc's incarceration in the King's Bench prison.

Admiration for the tunnel operations and the manner in which the problems had been tackled and overcome was not restricted to Britain. The feeling on the Continent was that the directors had made only feeble efforts to complete the undertaking. A company in France, the Paris Association, offered to purchase the entire stock from the shareholders for £250,000. This was promptly declined by the chairman and the senior directors, thereby strengthening the suspicion that bribery was intended to be part of the eventual contract.

For the remainder of the year, Marc was mainly engaged in designs for the Liverpool Docks, and in perfecting those which had already been accepted for the Clifton Suspension Bridge. He was however pessimistic about the latter ever being completed. He wrote in his journal on 20 April 1832, 'the trustees of the Clifton bridge had a meeting: though disposed to give their money gratis, I augur but indifferently of such liberality. They have resolved to draw a prospectus, and to go round with it to invite the public to subscribe. It may fairly be inferred that the project is sinking in public estimation. Coupling the state of things (at Bristol) with the prospect of the trade with the West Indies, we may pronounce at once and unhesitatingly that the scheme of Clifton bridge is gone by.'

* * *

The success of the experimental combination of iron ties with bricks and cement in the building of the shaft of the Thames Tunnel suggested to Brunel the idea of establishing a general principle of construction which could be expected to combine a degree of utility with economy. He began a series of experiments with a range of materials to discover a substitute for those usually employed in the construction of arches, and also to find some means of dispensing with the cumbersome and expensive apparatus essential to centering — the temporary wooden framework needed to support an arch whilst under construction. The first of these experiments was to ascertain the force of cohesion between ties of various characters and dimensions — straw, wood fibres, hemp, reed, laths of fir and birch, and iron in several forms including hoops. These were embedded in cement mixed with sand (two parts of cement to one of sand) and subjected to a variety of tests.

In the yard of the Thames Tunnel at Rotherhithe, Marc decided to construct an experimental arch to prove his theory that centering could be eliminated. In this he was supported by Mr Francis, the principal of the firm of Francis & White, cement manufacturers, who supplied without charge the major part of the materials and labour. No special preparation of the ground was made for the foundation of the pier — ground which was mainly loamy sand mixed with vegetable mould. The original intention was to construct two semi arches each of 25ft to 30ft span, resting on a central pier 4ft by 3ft 4^1/$_2$in. When Brunel started to extend the arches, however,

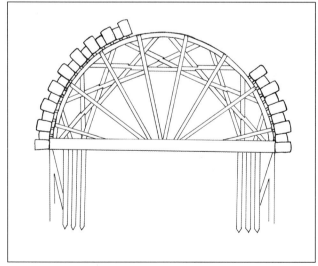

Illustration of arch construction using timber centering. *J. Williams*

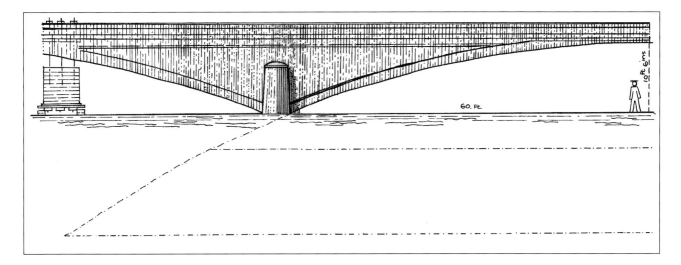

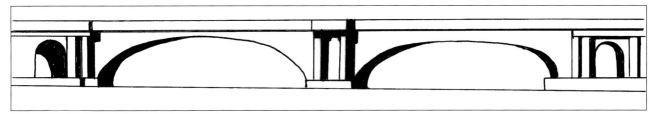

it was found that the space available limited the structure to 40ft. It was therefore decided to extend just one of the arches. To secure equilibrium, weights were suspended from the other.

In the first part of the experiment, fir ties were employed. Later these were substituted with hoop iron. The ultimate length obtained was 60ft with an elevation of only 10ft 6in, using a counterbalance of 57,600lb. The success of this experiment resulted later in the construction of the beautiful railway bridge over the Thames at Maidenhead — one of Isambard's great triumphs but almost entirely the result of his father's innovation.

* * *

In 1833, Marc went to Ireland for six weeks (see Appendix 5). He visited Dublin, Cork, Waterford and Killarney, and was fêted wherever he appeared. As a result of this visit he was invited to submit a design for the delightful little bridge over an arm of the River Lee known as Cork Jail Bridge. Richard Beamish had recently been appointed engineer to County Cork, and it was through his introduction that the authorities engaged Brunel for the work. The bridge was constructed of local limestone with a single flat arch of 50ft span which led to the Physics Department of University College, Cork, behind the old County Jail.

This short holiday, coupled with the success his son was now beginning to enjoy, inspired Marc with new life and vigour. He returned to London convinced that the government would eventually be forced to finance the completion of the Thames Tunnel, even though public interest was now mainly concentrated upon the development of the railways.

The tunnel company shareholders at last discovered their error in supporting a chairman whose views were opposed to those of all the experts, and on 6 March 1832 Smith was deposed. A bill was prepared and introduced to Parliament in the following session. Once more the Thames Tunnel became the main subject of conversation and debate in the coffee houses of London. The Royal Family was no exception, and Brunel was requested to give a full account of all the operations and incidents to King William IV. A Tunnel Club was established, principally from among the Fellows of the Royal Society and on 25 April 1834, Brunel's 65th birthday was celebrated by a magnificent entertainment at the Crown & Anchor inn. When the petition of the tunnel company was supposed to be presented to Parliament in April, it was nowhere to be found. This was very unusual, but even more mysterious is the fact that it was never recovered. This would normally have resulted in the bill being put back once more but Lord Spencer took upon himself the responsibility of granting the necessary loan from the Treasury.

Realising that it would take several weeks before the funds could become available, Brunel decided to pay a visit to his birthplace at Hacqueville. He found it little changed since his childhood. (His brother, who was the Mayor of Hacqueville, had died in April 1833 aged 63.) Marc's fame had spread and he was welcomed by the inhabitants, many of whom had followed his career eagerly, still

Right: The elegant Maidenhead Railway Bridge. Although designed by Isambard Brunel it was based upon experiments conducted earlier by his father at Rotherhithe. *Reg Silk*

Below: The plaque on Maidenhead Bridge. *Reg Silk*

Bottom: An artist's impression of Cork Jail Bridge over the River Lee, designed by Mark Brunel for University College, Cork. *J. Williams*

1838

THE SOUNDING ARCH

I.K. BRUNEL DESIGNED THIS BRIDGE THE BRICK ARCHES ARE THE WIDEST AND FLATTEST IN THE WORLD ~ EACH SPAN IS 128 FEET WITH A RISE OF ONLY 24 FEET

THIS PLAQUE WAS ERECTED IN EUROPEAN ARCHITECTURAL HERITAGE YEAR 1975

remembering the retiring little lad whose family had sent him away at such an early age. A monument was eventually erected to him on a bank beside the church, where it stands to this day. It is interesting to note that among those living in France today, Marc's exploits prior to the Thames Tunnel are almost unheard of.

Waiting for him upon his return was an application from the Viceroy of Egypt to design a secure and permanent passage across the River Nile. It appears from his journal that he spent a considerable amount of time and effort in the British Museum and elsewhere, examining all the records and related information about the Nile. He soon realised, however, that he would need to visit the site in person before being in a position to advise or submit plans. His other engagements did not permit him to make such a journey, so he reluctantly declined the commission. (It was executed some years later by Robert Stephenson, son of George Stephenson, the railway engineer.) During this time of 'enforced idleness' (his words), Marc designed a four-mile-long viaduct to carry 'Isambard's railway' through west London to a proposed terminus near Vauxhall Bridge. Sadly this viaduct was not required when Paddington was chosen instead of Vauxhall for the London terminus of the Great Western Railway. Part of it still exists however near Hanwell, and Brunel University has a section of the 7ft track on longitudinal sleepers which was originally part of a branch line between Uxbridge and West Drayton.

Following an intensive tour of the country calling at places such as Brighton, London, Edinburgh, Glasgow, Newcastle, Sunderland and Wearmouth, Marc ended up at Northampton where he was entertained by Lord Spencer at Althorp. The time was now approaching for his great project under the Thames to be resumed once more.

16.   Tunnelling Resumed

The government had agreed a loan of £246,000 to the Thames Tunnel company and advanced the first instalment of £30,000 in December 1834. This came with the condition that the money should be 'solely applied in carrying on the tunnel itself, and that no advance should be applied to the defraying of any other expense, until that part of the undertaking which is most hazardous shall be secured'.

Isambard Brunel was now too deeply engaged in other projects to permit his personal resumption of the day-to-day management, so Richard Beamish was appointed resident engineer. He had four assistants to work with, chosen by Marc Brunel. They were Lewis Gordon, Joseph Colthurst, Andrew Crawford and Thomas Page.

The stringent conditions imposed by the Treasury resulted in much unnecessary time, money and anxiety being expended, and it was only with great difficulty that Brunel was persuaded out of abandoning the enterprise completely. His scheme would have incurred no additional outlay in sinking the Wapping shaft which ultimately had to be sunk; no expensive preparation would have been needed to set up the new tunnelling shield, and the cost of removing the excavated soil and waste would have been reduced to a minimum.

One of the first requirements was the removal of the old, 80-ton shield so that it could be replaced with a new and improved shield weighing 140 tons. This consisted of 9,000 parts which had to be fitted together underground with absolute precision. It also involved supporting 1,656sq ft of surface during the construction, using a system embracing 300 iron piles, preceded by the formation of a series of drains, without which it would have been impossible to advance. The drainage work was commenced on 19 March and completed on 7 May. By November the old shield had been removed and the extent of the irruption was revealed in all its magnitude. So extensive was the cavity that some of the directors, who had insisted upon inspecting the work for themselves, became so alarmed that they took one look and immediately retired to the safety of the shaft.

Great interest in the tunnel continued to be displayed by the public up and down the country. *The Scotsman* declared in July 1835 that the tunnel was 'a beautiful Archway, brilliantly illuminated with Gas . . . as a work of art, it is unrivalled and the first object that attracts the notice of foreigners who visit London'. The paper's readers were encouraged to go and see 'An exact fac-simile of this wonderful production', which was to be found locally in St David Street. The entrance fee was sixpence, reduced to threepence for working people and children. So even those who lived many miles away from London, and who would doubtless have had little chance of visiting the tunnel in person, had the opportunity to view the intricacies of its workings on their own doorstep. Four years later, the Edinburgh Exhibition of Arts also displayed, among its many specimens, models and drawings of the Thames Tunnel. It is interesting to note that as well as all the items referring to the tunnel, the exhibition also displayed a working model of one of Marc's early inventions — the apparatus for winding cotton balls.

By 1 March 1836, the redesigned shield, manufactured by Rennie, was in place. There had been only one accident, when a man jammed a finger, which Beamish reported 'was the result altogether of carelessness and in spite of warning'. None the less, the health of all was suffering. Joseph Colthurst was compelled to retire and Crawford was appointed in his place.

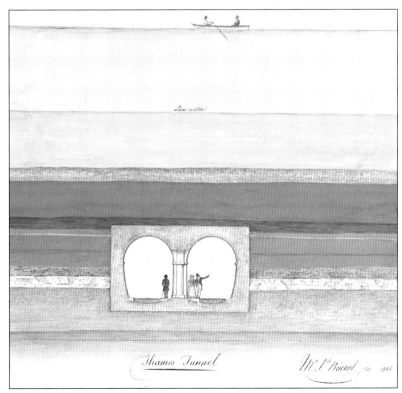

Marc Brunel painted this watercolour as the abandoned tunnel workings were about to be reopened. It shows his idea of how the tunnel would appear, and the boat on the river is being rowed by his future son-in-law with Isambard in the passenger's seat, complete with stove-pipe hat. Lithographs of this painting are available from Brunel Museum, Rotherhithe. *Brunel Museum, Rotherhithe*

By strictly adhering to Brunel's principle that security should be the primary object, very slow progress was made, but as everything was under tight control, all operations were performed with increasing confidence. On 21 June 1836, this was severely tested when, during a high tide, there was a great influx of water into the works and the pumps were overwhelmed. Had the men not stood to their posts and obeyed every order with the utmost promptitude and coolness, the most serious consequences could have resulted. More than 500 gallons of water per minute poured into the top boxes alone. The whole ground in front and above appeared to be in motion, and at one time seemed liable to completely engulf the frames. Relief was obtained with the fall of the tide, and before the next high tide a sufficient quantity of clay and gravel was deposited on the riverbed to secure the work. Contemporary newspapers competed with each other, presenting differing views of the setbacks in the tunnel construction to their readers. Following yet another problem encountered by the tunnel workers, *The Scotsman* commented that 'We have much pleasure in assuring our readers, that this stupendous national undertaking is progressing most satisfactorily, notwithstanding the lugubrious anticipations of the *Times* and other Tory papers, that a complete failure was to follow on the temporary and *anticipated interruption* occasioned by the nature of the soil of the bed of the river . . . Mr Brunel's mode of *stopping the leak* — throwing in a coating of clayey gravel from lighters over this spongy part of the bed — has succeeded perfectly.'

For some days afterwards little progress was made. The ground in front was too loose to permit excavation until Brunel devised a new method to deal with the problem. He abandoned work in the top boxes of the frames, and concentrated on the middle and lower boxes where the ground was firmer. He then had flat iron plates driven upwards from the middle boxes until they engaged with the top staves, thus securing the loose ground above and allowing the top boxes to be excavated in the normal manner.

By the end of August, 655ft of the tunnel had been completed. The top boxes were almost free from water and the future seemed bright, but it was not to be. On 24 August, Richard Beamish was forced to resign through failing health. The following month, Lewis Gordon's health also broke down and he too was compelled to retire. This threw the main supervisory responsibilities upon Thomas Page, who met the ensuing rigorous situations with a courage and ability seldom previously demanded by the enterprise. For some months the work progressed favourably, but as the ground became more treacherous, excavation slowed again. In September 1836, the tunnel advanced nearly 20ft, but in January 1837, this had fallen to just under 7ft. It soon became necessary to apply to Parliament for a supplementary grant of money to cover the great increase in cost caused by the difficulties encountered.

After considering the opinion of Mr James Walker who was employed by the government to report upon the work generally, the Committee of the House of Commons closed its report to the House in the following terms:

'Looking to the importance of a work of this nature, for the first time now undertaken as a means of fixed communication to situations where no other of an equally permanent nature may be available, and also that the sum of 180,000*l.* has been already expended upon the work by the proprietors, and the farther sum of 72,000*l.* by the public, they are of opinion that it will be expedient to authorise the Treasury to continue the advances to the Thames Tunnel Company according to the Act of Parliament.'

Brunel had benefited from the high regard in which he was held by the country as a whole and the importance of the tunnel to the commerce of London in particular. His pleasure was enhanced at this time by the marriage of his youngest daughter Emma to the Reverend Frank Harrison on 31 October. They went to live at Longdon near Tewkesbury. The marriage of his son Isambard to Mary Horsley, a dark-haired, elegant society beauty, and the eldest daughter of William Horsley the composer, had taken place three months earlier on 5 July 1836.

By now Brunel had acquired the habit of going down to the works for an hour or two during the night, to supervise the progress and deal with any problems arising. During the time when he was unwell Sophie had devised a method by which he could be kept informed without actually visiting the tunnel. She arranged for a box to be suspended from the bedroom window in which a sample of the soil from the workface was placed at two-hourly intervals by a messenger. This was then drawn up by Sophie and, after it was examined by Marc, the messenger was given whatever instructions were necessary and conveyed them back to the workings. This arrangement continued for about four years and resulted in the Brunels becoming so accustomed to it that they developed the habit of sleeping in two-hour shifts; a habit they found difficult to break after the tunnel was completed.

It was not only Marc that suffered from bouts of ill health. Marc's concerns for Sophie, his wife, were conveyed to Isambard through various letters of correspondence that have survived. One dated 11 July 1834 urged Isambard to think of the state of his mother's health: 'She has been confined to her bed the last two days,' wrote Marc: 'today she is up, but in so debilitated and nervous a state that it is most distressing to witness it.' Marc then related how his daughter Emma was of little assistance: 'subject as she is to headaches which come suddenly upon her, she is incapable of any kind of exertion in such circumstances'. He ended the letter by entreating Isambard to write often to relieve him. Three years later, another letter to Isambard from Marc, in November 1837, described how his dear mother was very bad: 'She sinks now and then under a bodily depression which alarms me.' Sophie's illnesses must have been a great worry for Marc, but the huge undertaking of building the Thames Tunnel had to continue.

On 23 August the river again broke into the works. There were three other minor irruptions but on 21 March 1838, the river invaded the works for the fifth time in 20 weeks, and in 26ft of tunnelling no fewer than three irruptions of the river occurred.

In spite of all this, by careful adherence to the principle of the shield, only one life was lost and that occurred despite there being sufficient time to escape.

Brunel now directed that no polings should be removed at all. They should instead be forced forward into the loose ground to the required distance by the screws. In this way the ground in front of the frames was compressed and the miners relieved of the difficulty and danger of replacing the poling boards. But the influx of water was impregnated to a large extent with poisonous gases, and the black mud which rolled in from time to time spread its foul, noxious influence throughout the tunnel workings. The assistants Faraday and Taylor together with Babington and Murdock suggested the application of large quantities of disinfectant; but a sudden irruption of 60cu ft of this filthy mud and water immediately neutralised all such attempted remedies. The workforce gradually diminished under such adverse conditions. Inflammation of the eyes, vomiting, debility, and open sores on the body were the most common symptoms. Marc wrote in his journal on 10 May 1839, commenting on the awful conditions: 'air continues very offensive, the assistants are quite exhausted . . . and so far debilitated that they cannot be depended upon for completing their shifts'. One distressing case Marc recalled involved a man called Richard Williams who they had to take to a lunatic asylum 'as being dangerous to be left out of doors'. Brunel's men certainly worked extremely hard; in one eight-hour shift they were allowed only half an hour for refreshment which was brought to them where they worked. However, to compensate for the dreadful conditions, wages were high, amounting to as much as two pounds five shillings a week. This figure, at a time when a skilled artisan's wage was about half that, meant that Marc could employ more highly skilled men to work the shield.

An explosive gas or firedamp also frequently spread dismay among the underground workers. Large bursts of fire would pass across the shield, sometimes as wide as 20 or 25ft. In addressing the members of the Institution of Civil Engineers, Marc commented on how some of the gases which issued forth, ignited very rapidly. The explosions were frequent, often extinguishing the workmen's candles. Marc thought that the gases came from the mud of the river. He added that the reports from Guy's Hospital stated that some of the men had been so injured by breathing these gases that there was small hope of their recovery. Through all these trials, Thomas Page continued his valuable superintendence of the works, a devotion to duty which was acknowledged by the directors in a tribute to him together with his assistants, Francis and Mason.

On 4 April 1840, when the work had advanced within 90ft of the site of the proposed Wapping shaft, the river mud over the shield was seen suddenly to sink at low tide. There was considerable excitement among those of the public who noticed the collapse from the northern shore. People ran in crowds to the wharves and warehouses from which they hoped to get a better view. Those who could obtain a boat took to the water, expecting to view the total destruction of the tunnel and perhaps also the struggles of those

engaged underground in its construction. In an area 30ft in diameter the ground had subsided bodily, leaving a cavity 13ft in depth. According to the report of the superintendent on duty at the frames, a noise like the roaring of thunder was heard, followed by a rush of air which extinguished every light. All the men fled, with the exception of a few of the experienced miners who, worried and bewildered by the unexpected turn of events, still retained their presence of mind and patiently awaited further developments. Until then, water had been the main reason for consternation, and as no water accompanied this unprecedented movement of the ground, they did not panic. Their example quickly reassured the deserters, who began to return to the shield in dribs and drabs as their confidence increased. The few displaced polings were restored and soon an increased sense of security replaced that of consternation and terror. The great cavity formed by the sinking of the ground was filled in the usual manner by the application of bags of clay and gravel, but work at the face could not immediately recommence: time had to be allowed for the repairs to settle and consolidate. In consequence, during the months of April and May, the tunnel advanced only 5ft.

Meanwhile, preparations for sinking the Wapping shaft on the north side of the Thames were in hand. Iron and timber curbs were laid exactly as for the Rotherhithe shaft. However, the company, through lack of funds, was compelled to limit its purchase of ground to a space contiguous to the wharf. It was surrounded by buildings in a dilapidated condition, devoid of proper foundations, with many dependent upon shoring for support. The company was liable for any damage to those premises which were located on the brink of the excavation and one of these was the Irving & Brown Coal Wharf whose building suffered subsidence when the shaft was sunk. They were compensated accordingly. The Ship and Swan public houses, together with some other buildings in the vicinity, were purchased for £8,000 and the occupants were duly given notice to quit.

On 9 October 1840 Brunel began clearing the ground where the curbs of the shaft were to be laid. In so doing, he exposed the remains of a timber floor supported on piles. The piles offered great resistance, while the surrounding ground was soft and yielding, rendering it impossible to obtain a firm level bearing. The area had previously been a ship-breaker's yard, and all kinds of materials and implements common to the business were found in jumbled, disorganised masses. There were wharf pilings of two distinct periods, walings (the side planking of ships), masts, iron ties, bolts, chains and a wide variety of tools. There were also quantities of ships' timber and the wreck of a boat. In spite of all the difficulties, Brunel pressed on with the work and on 27 October the first brickwork was laid at Wapping with all the men present. The sun shone just at the moment of laying the first brick, which was accompanied by great cheers. Progress was interrupted by the severe winter weather of 1840-1: none the less work forged ahead on the shaft, and this amazing structure, 55ft in diameter and 70ft in depth, weighing over 2,000 tons, was completed in 13 months without serious disturbance of the surrounding buildings.

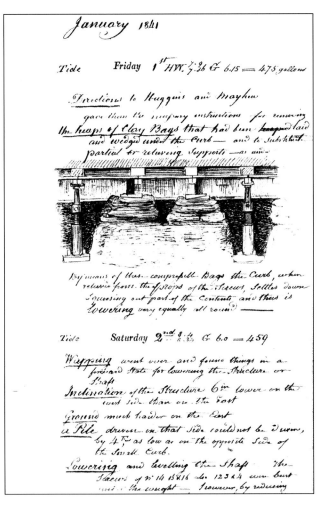

On 24 March 1841 Marc Brunel was knighted by the young Queen Victoria. This was at the suggestion of Prince Albert, who had shown a keen interest in the tunnel project and had always been an admirer of Brunel. Among the many congratulations received, one of the most valued came from his old friend and constant supporter Earl Spencer. He wrote, 'You have fairly earned your title by long continued and able services to the country; and it is a memorial of those services, and consequently highly honourable to you. I therefore most sincerely congratulate you upon receiving it.'

Work on the tunnel resumed in the summer of 1841. Further major irruptions were avoided, although the influx of water was never less than 450 gallons per minute. The extra pressure on the shield as a result of sinking the shaft caused continual fractures in the frames and ultimately the complete collapse of that on the east side, but by then it had served its purpose.

Just before Christmas 1841 the last poling board was removed from No 1 frame. The top staves had come into contact with the brickwork of the shaft, and all that remained was to cut through the side of the shaft and seal the joints. But even at this late juncture difficulties arose. An opening of 930sq ft had to be made in the side of the shaft to allow the tunnel to be completed and sealed, but when only half an inch gap remained, the semi-fluid ground forced its way through. Had this been allowed to continue, it could have resulted in subsidence of the ground beneath the surrounding buildings for which the tunnel company was responsible. Extra workmen were rushed to the scene, but it was not until 7 January 1842 that the work was finally secured. The influx of water, which had been about 450 gallons per minute, was reduced in March to 288, in April to 150, and in May to 70. Bringing the water under control and constructing the spiral staircase occupied the remainder of the year.

The tunnel opened for the first time on the Wapping side of the river on 1 August 1842. Over 500 people passed through the tunnel as far as the shaft on the Rotherhithe shore. The shaft was described by *The Times* as being 90ft in height and surmounted with

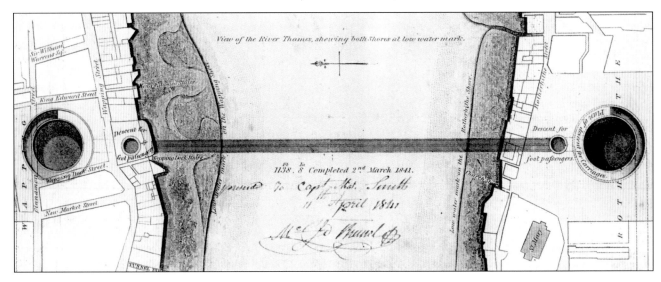

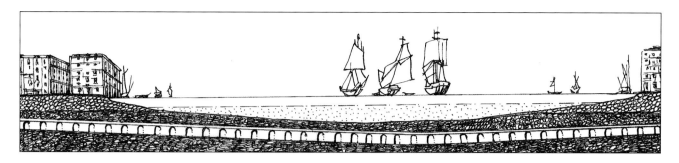

'a handsome dome, which is glazed, and light and air admitted'. According to the paper, an inspection of the visitors' book revealed that 'the names and residences of more than 30 Americans from different parts of the United States were found to be entered, together with those of persons from all parts of Europe, and many from Asia'. The interest shown in the tunnel from foreigners as well as locals, boded well for its future.

On 7 November that year, Marc Brunel suffered a stroke which paralysed his right side. With his normal acceptance of the vicissitudes of life, he immediately began teaching himself to write with his left hand. Fortunately, recovery was almost total and not long delayed, but this says much for the man's character when faced with a situation which could have driven many others to despair.

The date of the grand opening of the Thames Tunnel was 25 March 1843 and, in spite of warnings from his medical advisors, Marc insisted on taking part in the official opening ceremony. Newspaper reports described how two marquees were erected, one for the directors and proprietors and the other for general visitors. Flags were hoisted and bells were rung. Sir Marc was cheered with heartfelt enthusiasm which he courteously acknowledged as he walked through the tunnel. The *Illustrated London News* referred to the tunnel

Top: Section through the Thames Tunnel. *J. Williams*

Above: A contemporary artist's impression of the foreshore at Wapping. It is slightly out of proportion: the shaft was set back further from the riverbank and the coal wharf was closer to the shaft. *J. Williams*

Left and below left: Commemorative medal of the opening of the Thames Tunnel (actual size 1¾inch, diameter). *Author's collection*

as 'this most wondrous of all London's wonders', and judging by the huge influx of visitors to the site, the public appeared to agree. From 6pm on Saturday 26 March to 9pm on Sunday the 27th, in the course of 27 hours, more than 50,000 people passed through the tunnel.

Not everyone was impressed, however. Representatives writing for *The Times* were present at the opening and made the following rather disparaging observations: 'The area of the shaft is too small to give effect to the frontage of the tunnel, while its depth, when viewed from the top of the circular staircase by which passengers descend, looks frightful.' The writers then described how many visitors found the tunnel to be very uncomfortable due to the lack of ventilation, the dampness of the ground, and the flaring of 130 gas lights, which all combined to create 'an atmosphere at once disgustingly heated and fetid'. The paper noted that amongst the majority of visitors there was a perceptible anxiety, and that it was evident that there was a 'lurking chilling fear in the breast of many; and it cannot be denied that the very walls were in a cold sweat'.

Another group of people that disapproved of the tunnel were the watermen. Symbolically they hoisted a black flag on the day of the ceremony itself, to indicate their feelings towards the tunnel — a structure they saw as undermining their own interests. Despite the negative views from some quarters, the tunnel proved to be very popular. Even *The Times* grudgingly conceded that when standing at the bottom of the shaft and looking at the vista which the tunnel presented 'it must be admitted that the view is strikingly agreeable'. The public's confidence in the structure was a reward for the perseverance, the courage and the engineering genius of its architect.

By the end of April 1843 nearly half a million people had visited the tunnel. In the 15 weeks from the day when it was opened to the public, one million visitors were recorded. These included dignitaries from almost every developed nation in the world. News of the tunnel under the Thames had reached far and wide. Even before its completion, a Miss Pardoe visiting Constantinople in 1836 found herself both surprised and pleased to be questioned about the tunnel by an Albanian chief. Likewise in Egypt, the tunnel's progress was eagerly followed by many. In France, a proposal for building a tunnel under the River Charente was eagerly encouraged by the French newspapers, which declared that their government should imitate the English government and construct a work which would be the boast of France as the Thames Tunnel was the boast of England.

The date 26 July 1843 marked another momentous occasion for the tunnel. Queen Victoria and Prince Albert visited the structure. It appears that not much notice beforehand had been given of their visit. Had it been, Brunel would certainly have made sure of being there in person to receive them. Unfortunately he was engaged elsewhere. Instead, Benjamin Hawes and Sir Alexander Crichton, two of the directors who happened to be on site, deputised for him. The satirical magazine *Punch* gave a rather amusing if somewhat tongue-in-cheek account of the royal visit. It described how as soon as the tunnel employees knew of the visit, the chief clerk 'ran helter-skelter to the first shop, and purchased a lot of scarlet baize for the purpose of lining the Tunnel Pier' and the directors 'pulled out their purses to send for "choice exotics" to decorate the staircases'. Unfortunately the messenger, having scoured the whole of Wapping, managed to bring back only a couple of pots of sweet peas and a box of scarlet runners in bloom — the latter having apparently been lent by an inhabitant of Rotherhithe.

Far left: The procession descending the staircase at Rotherhithe for the opening ceremony. *Illustrated London News Picture Library*

Left: The engine house, beside the Rotherhithe shaft, 1843. *J. Williams*

Top: The ceremony of the opening of the tunnel. Sir Marc Brunel follows the band of the Coldstream Guards. *Illustrated London News Picture Library*

Above: Rotherhithe entrance, the Thames Tunnel. *Southwark Local Studies Library*

Captain from walking upon and soiling them'. He was clearly a shrewd businessman, for the hankies were sold during the rest of that day for half-a-guinea each.

Quite a few visitors who had been unceremoniously locked into the tunnel when the Queen arrived, huddled together on one of the landings and then proceeded to sing 'God Save the Queen' as the royal party re-ascended the staircase. The magazine declared that the effect of the national anthem 'sung in different times, different keys, and almost in different tunes, was quite electrical'. *The Scotsman* was slightly more polite in its report, declaring that the anthem was sung 'more loyally than musically'. The Queen expressed her regret at Mr Brunel not being present for her visit and then left the site amid prolonged cheers.

Queen Victoria was not the only royal visitor to the tunnel that year. In October His Imperial Highness the Grand Duke Michael of Russia made an appearance. It was said that during his stay in London he visited Buckingham Palace, the Bank of England, the Tower and the Thames Tunnel. Brunel's engineering masterpiece had made it onto the major list of sightseeing attractions in the capital.

In March 1844 another colourful event occurred at the tunnel. It was visited by the Ojibway Indians. They were five in number including two squaws and a child. They came complete with feathered headdress and painted faces, wearing bearskin clothes and moccasins. Not surprisingly, their arrival excited great curiosity among the people of Wapping, who apparently deserted their shops to take a look. On reaching the double archway, the Indians were said to have expressed surprise with guttural 'ughs!' which apparently showed their satisfaction. It was declared that they thought the tunnel was more wonderful than many of the more showy spectacles that they had already seen on their visit to England.

Above left: Queen Victoria and Prince Albert visit the Thames Tunnel, August 1843. The royal barge is lying off Wapping Stairs, at the Tunnel Pier. The engraving is slightly inaccurate because the entrance to the shaft was further back from the shore. *Illustrated London News Picture Library*

Left: The Rotherhithe entrance to the tunnel. *J. Williams*

Above: Sir Marc Brunel at the opening of the Thames Tunnel. *Illustrated London News Picture Library*

Right: View of the tunnel reproduced from Knight's *London — Pictorially Illustrated*, 1842. In fact, horsedrawn vehicles never entered the tunnel. *J. Bagust*

Once the royal barge drew alongside Tunnel Pier, the party which included Duke Ferdinand, Prince Leopold, Baron de Waggenheim and Captain Zaitsck, was escorted to the shaft. As the royal party proceeded down the shaft, the money-taker 'waved his book of pass-checks with much enthusiasm, and the Queen acknowledged with a smile, this honest outburst of extemporaneous loyalty'. One of the stall keepers in the tunnel, seeing the Queen approach, proceeded to tear down the whole stock of his silk pocket-handkerchiefs in order to lay on the ground for her to walk over. The magazine noted however that the ingenious stallholder, on discovering that Baron de Waggenheim and Captain Zaitsck, who were following, had dirty boots, 'clutched up the handkerchiefs just in time to prevent the Baron and the

Three years after its opening, J. A. Hoy wrote a song in praise of the tunnel in which the visit of Queen Victoria was duly acknowledged:

Have you seen the Tunnel? allow me to inquire,
This most noble structure all the world must admire;
It is well ventilated and lighted with Gas,
And no smoking allowed by any one class.
The Tunnel's constructed like unto a cave,
By stairs you descend its shaft under the wave.

Some thousands of persons to the Tunnel have been,
A visit was paid it by our most Gracious Queen;
At the Tunnel Bazaar, too, stalls there are many,
You may buy what you please; admittance one penny.

The Tunnel's constructed like unto a cave,
A subterranean passage right under the wave.

An extract taken from Knight's volume on London, published in 1842, referred to the costs involved in completing this great project and all the trouble it had caused Brunel. The writer described how 'The expenses of the Tunnel have been, of course, very much greater than were contemplated, and that circumstance has not been one of the least of the engineer's difficulties: in one sense, indeed, it was his greatest, since it did not rest with himself to conquer it.' Nevertheless, income from admissions to the tunnel along with the sale of books reached the figure of £468,250 by 31 December 1844. The total expenditure was agreed at £454,810. So it transpired that in just 20 months from the time the Thames Tunnel opened, the tunnel company was in profit.

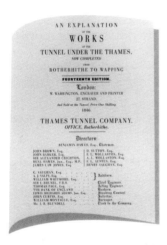

Above: The front page of the brochure sold to visitors to the tunnel from its opening.

Right: Steel plate engraving of the tunnel by H. Winkles from a drawing by Tombleson. It is misleading to some extent because many of the alcoves would have been occupied by stallholders selling their wares and the roadway would have been thronged with people. *Author's collection*

17. Epilogue

By the time the Thames Tunnel was completed, the striking, upright and debonair young Frenchman who had stolen the heart of Sophie Kingdom many years earlier, was almost a broken man. Most of his hair had gone and he was stooped and quite frail, but the spirit which had fought and won so many battles remained as flamboyant as ever.

Although Marc never again accepted any major commission on his own account, he was regularly involved with the numerous works undertaken by his son Isambard. He was constantly suggesting ideas, planning and supplying drawings for many of the projects which now bear the name of Isambard Kingdom Brunel, but which were often in fact joint efforts. The design of Isambard's Box Tunnel on the Great Western Railway was based upon Marc's experiences with the Thames and Chatham Tunnels. His construction of the Royal Albert Bridge at Saltash drew upon Marc's design for the pontoons originally intended for use at St Petersburg, as did the Clifton Bridge construction. The 7ft track gauge used by Isambard when planning the Great Western Railway had already been incorporated by Marc in the Chatham Dockyard reorganisation many years earlier.

Throughout his life Marc had a mischievous sense of humour which never deserted him. When he originally landed in England, he sent for a tailor to make him a coat, impressing upon the man the urgency of the situation. The tailor measured him and went away promising to return the next day with the garment. When he returned it was found that Marc's right shoulder was considerably higher than the left and the coat did not fit. Full of apologies, the tailor went away and altered the garment. Upon his return that evening it was found that he had altered the wrong shoulder for now the left shoulder was the highest. The poor man was distraught. Such a thing had never happened before and he was at a loss to know how he could have made such a mistake. At length Brunel straightened himself and admitted the whole thing had been a joke. He compensated the man and all was well. Marc never entirely lost his French accent, which always charmed the ladies. He was a great favourite with them, yet despite his popularity among many educated society women, there was never any hint of an affair. It appears that he remained true to Sophie throughout his life.

Father and son were totally different although they worked well together. Isambard carried no trace of a French accent; he was born and educated in England until he was sent to France to complete his education. He spoke both French and English fluently, the result of having a French father and an English mother, and was encouraged and supported by doting parents. In stature he was slightly shorter and stockier than his father, but endowed with a strident, forceful nature which did not always endear him to his peers. He never had the slightest doubt that his opinions and conclusions were correct. There were many occasions when his causes could have been better advanced by a little diplomacy, but Isambard was no diplomat. He *knew* he was right and would not be moved — and on most occasions so it proved in the end. He had an in-built, ill-concealed contempt for those whose intellectual powers were inferior to his own and this resulted in many people considering him brash, conceited and bumptious. This does not make him any less of a genius. Many great men have some slight flaw in their character. It just confirms that he was human.

By contrast, Marc was always very sensitive to other people's opinions and could sometimes get depressed when unjustly criticised. On one occasion at Chatham, a Mr Seppings (later Sir Robert), ridiculed Marc's fears about the stability of a large chimney which was being erected (fears which later proved to be well founded). This ridicule continued to irritate him like an open sore until he finally put his feelings in a letter to this gentleman. The contents of the letter are not known but it brought a reply from Seppings in the following terms:

'I trust that you will desist in again addressing me on this subject, it being unconnected with my profession: at the same time I am ready to give you every assistance in my power on this and every other occasion, you still having my best wishes.'

Father and son were both able to draw a perfect circle freehand, and were unusual in the little sleep that they required. Marc seldom slept for more than five hours each night, and Isambard frequently worked for 30 hours or more without rest. Isambard, however, was never known to suffer from absent-mindedness, whereas Marc was notorious for it. On one occasion he stroked the hand of the lady seated next to him at a dinner in the belief that she was his wife. He often mistook the day and hour of an invitation. He was known to go to the wrong address, and on one occasion was refused admission to Earl Spencer's house when he presented another person's visiting card in mistake for his own. It was also not unknown for him to take the wrong coach and to find himself set down far in the country when he should have been in London. One diary entry from May 1823 noted how Marc had written to the innkeeper at Euston, 'for the purpose of tracing my cloak'. Apparently it could have been left in a number of places including Stratford, Warwick or Lincoln. The losing of an umbrella was a regular occurrence. On a particular day when Sophie had impressed upon him the importance of keeping it about his person, he steadfastly kept hold of it throughout a visit to a friend. Upon taking his leave, he noticed an umbrella in the hall very similar to his own and took it to protect himself from the rain during the walk home. It was only when Sophie met him at the door that they discovered he still had his own rolled umbrella tucked under his arm.

In 1829 Isambard, returning after a visit to France, made the

This is the Last Will and Testament of me Sir Marc Isambard Brunel of Park Prospect Westminster in the County of Middlesex Knight. In case my dear Wife Lady Sophia Brunel shall survive me then I nominate and appoint my said Wife sole Executrix of this my Will and I give devise and bequeath all and singular my Estate and Effects whatsoever and wheresoever and whether real or personal unto my said Wife her heirs executors administrators and assigns absolutely for her and their own use and benefit but in case my said Wife shall die in my life time then I nominate and appoint my Son Isambard Kingdom Brunel and Park Nelson of Essex Street Strand in the County of Middlesex Executors of this my Will And I give devise and bequeath my Estate and Effects in manner hereinafter mentioned that is to say I give and bequeath all and singular my Plate, Jewels Pictures, Wearing apparel Furniture and Household Effects unto my three children to be equally divided between them share and share alike. I give devise and bequeath unto my said Executors their heirs executors administrators and assigns all and singular other my said Estate and Effects upon trust as soon as conveniently may be after my decease to collect and get in the same and convert into Money all such parts thereof as shall not consist of Money or be invested in the public funds or on securities for Money of the nature hereinafter mentioned and upon further trust as to one third part of my said Estate and Effects to pay assign or transfer the same unto my said Son Isambard Kingdom Brunel his Executors Administrators and assigns for his and their own use and benefit And as to one other third part thereof to pay assign and transfer the same unto the said Isambard Kingdom Brunel, Park Nelson and The Reverend George Harrison of New Brentford in the County of Middlesex Clerk or the Survivors or Survivor of them or the Executors Administrators &c. assigns of such Survivor upon and for the trusts ends intents and purposes hereinafter declared concerning the same share and to pay assign and transfer the remaining third part of my said Estate and Effects to the said Isambard Kingdom Brunel, Benjamin Hawes of Queen Square, Westminster Esquire, M.P. and Park Nelson or the Survivors or Survivor of them or the Executors Administrators or assigns of such Survivor upon and for the trusts ends intents and purposes hereinafter declared concerning such last mentioned share and I do hereby declare that the

(1.) — M. I. Brunel

---

bearing, and address and even the dress of a French gentleman of the Ancien Regime, for he had kept to a rather antiquated but very becoming costume. I was perfectly charmed with him at this, our first meeting, and from many subsequent ones feel bold enough to say that he was a man of the kindest and most simple heart, and of the purest taste in art, whether architecture, painting, sculpture, or medalling . . . But what I loved in old Brunel was his expansive taste and his love and ardent sympathy for things he did not understand and had not had time to learn. There is no adequate portrait of this very remarkable man. The picture then in the drawing room by Jemmy Northcote, though it presented something like a man of genius and very deep thought, was little more than a map of dear old Brunel's face. It would have required a man of much more fame and genius than Northcote to catch the variety and play of the old engineer's countenance. What I most admired of all was his thorough simplicity and unworldliness of character, his indifference to mere lucre, and his genuine absentmindedness. I had liked the son [Isambard], but at our very first meeting I could not help feeling that his father far excelled him in originality, unworldliness, genius and taste.'

* * *

In his later years Marc took enormous pleasure in being associated with Isambard in some of his greatest successes. On 19 July 1843, he was present at the launch of the *Great Britain* at Bristol, the first large ship to be equipped with screw propulsion. This

acquaintance of Charles Macfarlane, author and traveller, and invited him to dine with the family at Duke Street. Macfarlane's account of the occasion provides an excellent description of Marc's character and personality. He was welcomed in by Sophie:

'I was not ten minutes in her company before I made out that she was devotedly attached to her dear old French husband and enthusiastic of his reputation, and that she doted on her only son, and was proud, as well she might be of his vivacity and abilities . . . I met the head of the house, dear old Brunel, to whom in an instant I flew, and attached myself as a needle to a loadstone . . . The dear old man had, with a great deal more warmth than belonged to that school, the manner,

was a direct result of experiments Marc had made some years earlier with a variety of models; he had recorded in his journal that the screw was one of the means of propulsion investigated.

It was during his excursion to Bristol that Queen Victoria made a surprise visit to see the Thames Tunnel for herself. Not being there to meet her personally was a great disappointment for Marc, so much so that he hastened to arrange a meeting as soon as he returned from Bristol. On 3 August, Marc was granted his wish and met both Prince Albert (who instantly recognised him) and Queen Victoria on the Royal train at Slough. Marc recalled how the Queen spoke to nobody else but him and expressed an interest in visiting the tunnel again.

Having returned to London, the Brunels took up residence in a small house in Park Street, Westminster, fronting St James's Park. Marc lived to witness the happiness and prosperity of his children and thoroughly enjoyed the company of his grandchildren who were frequent visitors to their home.

In 1845, Marc suffered another, more severe stroke, which left him almost totally paralysed down his right side. A full-time nurse was engaged to assist Sophie, who was herself becoming rather frail, but who seldom left his side. His cheerfulness and equanimity never left him and he resumed his attempts to write with his left hand, with some success. Gradually, though, his powers diminished until he was no longer able to speak and seemed to drift in and out of consciousness. On 12 December 1849, at the age of 80, Marc Isambard Brunel died, and on 17 December his remains were interred in Kensal Green cemetery. An obituary in *The Times* generously acknowledged the contribution that Marc had made to his adopted country: 'By birth he was a Frenchman, but his life and genius were almost wholly devoted to the invention and construction of works of great public utility in this country.' The paper noted that Brunel's blockmaking machinery 'is still an object of admiration to all persons interested in mechanics'. The achievement of constructing the Thames Tunnel drew the most praise, despite the rather scathing report by that same paper concerning the tunnel's opening six years earlier. The writer of the obituary declared that the Thames Tunnel, from a scientific point of view 'will always be regarded as displaying the highest professional ability, an amount of energy and perseverance rarely exceeded, and a fertility of invention and resources under what were deemed insurmountable difficulties, which will always secure to Sir I. (sic) Brunel a high place amongst the engineers of this country'. Marc would surely have been satisfied with such an accolade.

* * *

Of all the many and varied projects undertaken by Marc Brunel, the Thames Tunnel undoubtedly stands as the greatest monument to his genius and tenacity. Some idea of the magnitude of the work involved can be obtained by comparison with the more recent Channel Tunnel between England and France. The Channel Tunnel was bored mainly through stable chalk with the aid of modern equipment, whereas the Thames Tunnel was entirely a manual operation, involving the spoil being removed by miners with picks, shovels and wheelbarrows. The various strata encountered in the Thames Tunnel included hard rock, blue clay and soft sand, with a combination at times resembling a fluid with the consistency of thin porridge. Had the geological surveys reported accurately on the condition of the riverbed, it is unlikely that Marc Brunel would ever have attempted the work. The fact that he did and overcame all the unforeseen problems with so little loss of life, is entirely due to the design of the Great Shield and the method of operating it.

The workers in the Channel Tunnel were able to perform their duties in comparative comfort. They were never required to stand up to their waists in water. Nor was there ever much risk of the electric lighting and power failing or of the ventilation system collapsing, all of which were daily hazards for the Thames Tunnel employees. There is no doubt that the Channel Tunnel is an outstanding achievement, but the Thames Tunnel stands as a monument to one of our greatest civil engineers and inventors, whose system of tunnelling was the basis upon which the Channel Tunnel was constructed.

Unfortunately the Thames Tunnel was never opened to wheeled traffic as originally intended. The construction of the approach roads was at first delayed and then abandoned through lack of financial support. However, for some years it was well used by pedestrians, and many of the arches were rented out to stallholders who sold all manner of goods. In March 1844, one newspaper gave an account of a fair which was held under the Thames. Troops of men, women and children apparently besieged both entrances to get inside. Stalls selling jewellery and confectionery were numerous. The more enterprising set up 'Thames Tunnel Grottoes', 'Thames Tunnel Tom Thumbs' and 'Thames Tunnel Boa Constrictors' in the various archways and recesses. A brass band was stationed at either end of the tunnel to create a lively atmosphere which, considering the enclosed space, must have been quite deafening. In 1845, to coincide with the second anniversary of the tunnel's opening, another fair was held under the Thames. The tunnel was brilliantly illuminated and the shafts were decorated with flags, banners and Chinese lanterns. The stall keepers were apparently very keen to hold such an event as they paid a heavy rent to the tunnel company for the privilege of selling their wares in such a location.

Despite these colourful celebrations, it was perhaps as well that Marc did not live to see what became of his great project. As early as 1850, a year after his death, the tunnel was in trouble. The magazine *Punch* wrote that things in the tunnel continued to look black. It noted that there had been a falling off in the tolls over the last year, but that could be partly attributed to cholera. It concluded by saying: 'We should be most happy to offer anything like consolation or encouragement to the proprietors; but truth compels us to say that we utterly despair of ever seeing the concern succeed in keeping itself above water.'

A lifeline came in the form of the East London Railway Company who purchased it in September 1865 for £200,000. Four years later, the first steam train passed through it. On 31 March 1913 the Metropolitan Railway Company introduced electric trains on the East London Line, and under an agreement dated 29 January 1914, the tunnel became part of the London Underground system. Stations were placed at either end of the tunnel, which are still in use today. The shaft at the Rotherhithe end is sealed, but the Wapping shaft is in daily use as the means of access to the platforms of Wapping station.

If you go down the steps to the platform at Wapping instead of taking the lift, you can observe the construction of the shaft much as it was in Brunel's day. In the last few years a certain amount of renovation has taken place in the tunnel but the basic structure remains unaltered, thanks to the activities of the Heritage Trust which has supervised all the work undertaken by the railway authorities.

The original engine house which stands beside the sealed Rotherhithe shaft is now a museum run by a trust called The Brunel Engine House Rotherhithe. Just a stone's throw from the engine house is the Mayflower public house, formerly the Shippe Inn, and it was from the river steps here that the Pilgrim Fathers set sail for America in the 1600s. The pub was almost completely destroyed during the World War Two Blitz on London, but the present building has been restored in similar style to the original. The Mayflower is situated almost directly above Brunel's tunnel, now surrounded by modern buildings, but there are still excellent views of London's river to be seen from its windows.

Plans are afoot to refurbish the Rotherhithe shaft and to create a two-storey museum above it. This will replace the one presently located in rather cramped conditions in the old engine house. It has been awarded a Heritage Lottery Grant of £50,000, which should go a long way to making the necessary improvements. On 18 May 2004 the museum unveiled a blue plaque to commemorate the banquet held under the Thames 177 years earlier. The scene was captured in oils at the time by the artist George Jones, and his painting was given on loan to the museum at this ceremony. Ironically, but perhaps not unexpectedly, the blue plaque unveiled that day was in honour of Isambard Kingdom Brunel, not his father. To mark the event, the poet John Hegley wrote the following verse:

Isambard Kingdom dug down
To Wapping from Rotherhithe Town
He tunnelled the Thames
and this unearthly gem's
A jewel in Isambard's crown . . .

How ironic that once more the son is receiving all the plaudits for a project that was entirely his father's idea. It is about time that Marc's achievements are given their true recognition and I hope that this book has gone some way to rectifying the imbalance, highlighting what an extraordinary life Marc led. Barely acknowledged today in both his native country and his adopted one, Marc Brunel is a forgotten genius but remains one of the greatest civil engineers of the 19th century.

The Brunel family grave at Kensal Green cemetery. *C. Bagust*

Certificate of Citizenship — United States of America

Particulars supplied by the US Department of Justice Immigration and Naturalization Service

Be it remembered, that at a stated District Court of the United States, held for the District of New York, at the City of New York, in the said District of New York, on Tuesday the second day of August, in the year of our Lord one thousand seven hundred and ninety-six, Marc Isambard Brunel came into the said Court and applied to the said Court to be admitted to become a Citizen of the United States of America, pursuant to the directions of the Act of the Congress of the said United States, entitled, 'An Act to establish an uniform rule of naturalisation, and to repel the Act heretofore passed on that subject;' and the said Marc Isambard Brunel having thereupon produced to the said Court such evidence, and made such declaration and renunciation as by the said Act is required, it was considered by the said Court that the said Marc Isambard Brunel be admitted, and he was accordingly admitted by the said Court to be a Citizen of the United States of America.

In testimony whereof the seal of the said Court is hereunto affixed.

Witness, John Laurence, Esquire, Judge of the said District, at the City of New York, in the said District of New York, this second day of August, in the year of our Lord one thousand seven hundred and ninety-six, and of the Independence of the said United States, the twenty-first.

Letter to the Navy Board, 2 May 1817

Transcribed from the original by permission of the University of Bristol Library

Honourable Sirs,

Although your Honourable Board's last communication, which reached me today, may be considered to have brought my late correspondence relative to the Trial of Towing Ships of War to a close; still, as the transaction does not appear to be completely understood, I cannot, in justice to my own character, allow one day to elapse without offering such explanation as the nature of the case demands.

After the opinion given on the present state of the Regent Steam Packet, your Honourable Board must be at a loss how (to) reconcile my statements with the ostensible facts; and how I would have justified the application of a sum of 60 pounds, to which I had estimated the expenses likely to be incurred in the Trial, when it is reported that the whole of the materials and work, seen on this occasion, does not exceed £13.00. In order therefore to remove any unfavourable interpretation that might, under such circumstances, be attached to the whole of my proceedings, as far as they are known to your Honourable Board, I beg leave to observe that if the Trial I had proposed, had been confined to the mere action of Towing a Body of such magnitude as a Ship, I am aware that any other Steam vessel would have answered as well as the Regent, and that too without the least alteration; but, if your H. Board will take into consideration the various communications I have made upon this subject, it will then be obvious that I did not intend to limit my experiments to the single service of Towing a Ship, but that I had contemplated it farther, as connected with more important duties.

Whatever pursuit I direct my attention to I devote to it all the resources of my mind, and I may add that in this instance, I have been through a series of very extensive experiments the results of which, convince me of the practicability of accomplishing what I have mentioned to My Lord Melville and to some other Lords of the Admiralty, that is, of attaching to a vessel purposely constructed, such Machinery, as to render it capable of riding very heavy seas and a gale of wind, and fit to carry cables and anchors to the assistance of Ships in distress, or for any service equally important: in a more comprehensive language, such a vessel, which, if at Spithead would under circumstances wherein all attempts from other vessels would be unavailing, make its way and be able to steer through the shortest direction practicable by the report of guns of distress. I might enlarge upon this interesting subject; but shall not trespass upon your Honourable Board's time farther than what may be necessary to explain my proceedings. I am obliged to add, however, that since the month of July or August last, I have been at considerable expence in obtaining self-navigating models, which have been made at my own cost, at my own works here, and under my own immediate direction for the particular purpose of exhibiting in a satisfactory and comprehensive manner, what I am now advancing.

On the few additional points I expected to have ascertained in the course of the Trial in question, were likely to have occasioned an expence of about 50 or 60 pounds including what I expected to have had to pay for alterations to the vessel, I trust it will be found that I have acted consistently; and that with respect to the terms for hiring the Vessel, I have done for the best as the proprietors were most adverse to allow her to be engaged upon that service on any account.

I beg to be understood that I am not soliciting any compensation for what I may have had done or provided through Mr Maudslay or any other on this occasion. If the decided and acknowledged superiority of the Regent as a Steam Vessel has already brought one application of some magnitude, I trust I shall at the end and through other channels, reap the fruits of my labour and perseverance for the further improvements to which I have alluded, although they are better calculated for a specific than for a general service.

I hope I shall not be charged with a want of that respect which is due to your Honourable Board by observing that I conceive myself to be very ill used in the course of this transaction, or rather at the termination of it. *When my suggestions were considered to be worthy of attention they were acknowledged and encouraged*, but from the moment that circumstances have operated in a contrary direction, then, in return I experience a treatment I was not prepared to have met from your Honourable Board.

I have the honour to be

Yours etc

M. I. Brunel.

The Footway Bridge at Liverpool

Letter from Marc Brunel to the Liverpool Docks Committee dated 13 June 1822

Transcribed from the original by kind permission of the Institution of Mechanical Engineers

Sir,

In answer to your favour of the 10th requesting me to furnish the Committee of the Liverpool Docks with a plan and estimate of a Footway Bridge to be placed over the entrance to the old and two other Docks: and specifying at the same time, the various points that are to be attented to, namely That it should be *as light* as may be considered consistent with security because of its having to be frequently and hastily opened and shut: and also that it should possess a considerable degree of strength in order to enable it to resist concussions of Vessels which, in tempestuous weather or by accident, may unavoidably be forced against it.

With regard to the strength to resist concussions of that nature, I beg leave to observe that the only provision that can be made to prevent the consequences of concussions from bodies of such magnitude, is to make the Bridge in two parts *opening indiscriminately either way* and meeting in the middle as represented at C.

Thus, a great force cannot do any harm, and the *lighter the structure*, the less the resistance in yielding to such force, whatever its magnitude may be.

Under this point of view, *lightness* becomes a necessary condition, which, by that arrangement, qualifies the bridge for all its subsequent duties.

Now with respect to its services as a thoroughfare, it is equally important to make the necessary provision for the safety of the passengers: and to that effect, I propose to divide the Bridge into two narrow paths by a stout railing in the middle — it is well known that a file of men will easily pass at the rate upwards 150 persons in one minute and an equal number in opposite directions if there are no obstructions. This precaution has the additional advantage of clearing the Bridge from all those who, having once passed the middle, must run as fast as they can, to get on shore before the parts are too widely thrown open. By this

disposition, those who remain, finding themselves between two rails, are under no apprehension of danger.

Being requested to report what sort of bridge I should think most eligible, and what sort of materials I think best for the construction, I take the liberty of stating that, on the ground I have assumed, the double bridge, *swinging either way*, is incontestably the best: and as to the materials, Sheet Iron is the chief substance I should employ; and to give you an idea of the manner I should constitute this substance into a stiff structure, I beg to refer you to the longest tubes that are daily before your eyes, namely the chimneys of Steam Vessels, some of which are 3 feet in diameter and 40 feet in height. If, instead of the round form, you were to shape the same substance into a triangular one, you would obtain a spacious floor for two files of foot passengers; and with a modification of the principle, you may have the lightest structure that can be put together for the service required.

If these suggestions meet with the approbation of the Committee, I will, on receiving further instructions from you, proceed in making the plans, estimates and, as early as possible you will have the goodness to decide whether you have room for the full sweep on both shores for the reception of the parts of the bridge.

Allow me to observe that whether the Committee should reject my plans, or not execute them, I should make such charge as the nature of the object and the time I may devote to it, may reasonable entitle me to.

I am, Sir

Your obedient humble servant

M. I. Brunel (signed)

Chelsea, June the 13th 1822

A = 2 ins. Deals applied to support the middle Railings and the platform.

BB = Foot boarding to protect the sheet iron from too rapid wear.

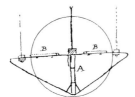

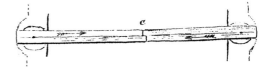

Sir Marc Brunel's sketches from his letter above to the Liverpool Docks Authority illustrating points concerning the construction of a footbridge.
Reproduced by courtesy of the Institution of Civil Engineers

Letter to the Thames Tunnel Directors dated 22 January 1828

Transcribed from the original by permission of the University of Bristol Library

Gentlemen,

Since my last report of the 13th instant, I have been endeavouring to ascertain the extent of the accident, the means of remedying it and the preservation of the work.

I now report a detailed account of my proceedings since the accident.

On the morning of the accident, in consequence of the intensity of the Fog, no observation could be made, nor could any soundings be taken. The Diving Bell was sent for and as soon as it could be removed from the West India Docks, was towed to Rotherhithe.

It could not however be used at the night tide. On Sunday the 13th Mr Gravatt descended first, but the chain, which is fit only for the service of the West India Docks, was found too short for our situation. Nothing could be done during the very short time of slack water but sounding, which is very unsatisfactory at best; this was a very great disappointment as the West Side of the Shield was still open, that is, free from the Ground which the current has subsequently washed in. In the mean time all hands were employed on shore and in the Barges in tempering Clay and filling Bags. It being Sunday a proper chain could not be procured, a Rope was accordingly tried for the Diving Bell, but it broke; having succeeded in procuring a chain of sufficient length, Mr Gravatt went down on the 14th in the morning and reported that the Brickwork appeared to be sound and undisturbed.

The frame No.1 in place and Top Staves level, but at No.2, two of the Top Staves appeared to be deranged. I was preparing to go down but the air pipe which had been lengthened from an ordinary (hose) not proving air tight was obliged to come up again before I could reach the bottom. The weather was all the time most tempestuous, cold and rainy, the tides very violent both up and down, and our Barges were exposed to be fouled at every instant by Crafts and Shipping. At 7 p.m. of the 14th I went down, Mr Gravatt with me: our Sextant having fallen overboard we were not very sure of our position; the signals not being well understood above, we were carried against the west side of the Cavity and exposed to danger of being upset; obliged again to come up.

Tuesday 15th at 6 a.m. having procured a Sextant and found that we had been driven off near 20 feet; but having removed our position, I was able to stay under water as long as the chain would enable us so to do. I found No.1 covered with about 2ft 6ins to 3 feet of ground, which the tide had washed down upon it. I was not able to ascertain with sufficient degree of accuracy the state of No.2 where Mr Gravatt had made his observations before. He had however brought up a Brick, a Wedge and a Rope which must have come from No.1 from the Side Brickwork. The ground rises on the East, which indicates that the frames on that side have not been disturbed.

Wednesday the 16th. I went down again and remained upwards of $2^1/_2$ hours having then about 42 feet of water, but could obtain no additional information. I therefore ordered the men to throw 2 Barge loads of Bags of Clay, and after they had thrown in the way we had done before, I descended to examine how they laid: finding them as well as could be expected from the State of the Tide, I ordered on the 17th to proceed in throwing large bundles of Bags through an opening made in the Raft. All hands were employed in tempering Clay, filling the Bags, and in throwing in, when the Tide was low enough. I thus proceeded as we have done before. Eighteen continued the operation of filling Bags of Clay; and when the Tide was low I availed myself of that time to have soundings taken which indicated that the Bags were taking their right direction.

Thursday the 17th. Having thrown a sufficient quantity of large Bundles of Bags, I ordered to throw small Bags with Iron Rods, instead of Hazel Rods over No.2.

Sunday 20th. Continued throwing Bags of Clay and Gravel.

Monday 21st. All hands employed, as before, some on Shore to prepare Clay Bags, others afloat. The Pumps being tried for the purpose of ascertaining whether they had been disturbed by the sudden irruption of the water, were found in good order.

I have thus reported the proceedings I have adopted since the accident, and the means used for securing the works, but the boisterous weather, the velocity of unusually high Tides, the shortness of the days, the insufficiency of the Diving Bell's apparatus for our purpose and the absence of my Son, have prevented my investigations being as satisfactory as they were immediately after the last accident, which occurred at the most favourable period of the year for out door work and on the River.

I think it however proper to add that from my inspection of the ground over the Shield, and from the report made to me from one of my assistants which I have before referred to, there does not appear to me any reason to doubt but that the means which I have formerly adopted for remedying the first accident, may be successfully applied to remedy the present, and with the experience of the First, I trust at less expense.

That the Directors may be perfectly convinced of this, I have, I beg to add, discharged a great number of men, and reduced the wages of some that are necessarily retained, as the TimeKeeper's return will show.

I shall continue my utmost exertions to restore the Works, and I shall take the earliest opportunity of making a further report.

M. I. Brunel (signed)

Letter to Mr Coxson on the Advantages of Steam when Crossing to Ireland

Transcribed from the original by permission of the University of Bristol Library

Dear Sir

I have found so much to do on my reaching home that it has not been in my power to put my pen to paper sooner on any subject but that of business. Being now somewhat relieved I will not go one day longer without expressing how much I was disappointed at being prevented, when I reached Bath, calling on your son as I had proposed. Out of sight, out of mind is frequently exemplified; but I can assure you that it was not so with me as you will convince yourself by the following Syllabus of my movements in the course of a week. It is not now as of old, for I have, in the course of one week, seen Killarney, Cork, Bristol, Bath, London, Ramsgate and London again; and all that without much bodily exertion, and being some portion of my time with my family.

Yesterday being outside of a Coach, and accounting thus for my weeks's doings, a friend of mine said that in 1809 he had travelled from Killarney to Cork, a memorable event he observed by the almost unsurmountable difficulties in getting through and along places designated Roads *(high roads they were)*. There, detained at Cork for some days by the wind — he embarked on a tuesday, as I did, and reached Bristol on the wednesday at night as I have done likewise; but, with this difference, that two weeks had elapsed between his wednesday and mine. Sick all the way, whereas with us, the Ladies walked the Deck the whole time.

Then the Roads were so excessively bad from Bristol to Bath and the whole way to London that he swore he would never revisit his own Country again. However, the farther from sight the nearer to mind with him. Steam is a Leveller of the greatest difficulties. Trusting therefore to the encouraging accounts of modern performances, he ventured with confidence and reached Bristol in 14 hours, just in time to get in a steamer; the next morning after his arrival he was in Cork, somewhat amazed, still somewhat disappointed too, that his trunk was not yet come for he would have leaped into a Coach just starting for Killarney. He had the gratification of seeing a little more of Cork and, admitted that the progress of improvements had been wonderfully rapid.

Thus my week's performance is outdone; however, whatever gratification he might have derived, he has not had to record so delightful a week as I have had at Killarney under the hospitable roof of Mr & Mrs Coxson, to whom I feel indebted for one of the most interesting weeks I have ever enjoyed; I must not omit the little companions whom I have left without even taking leave of them to my great regret, had they even been *Sans culottes.*

The great Enterprise of the Rail-way may some day or other bring one in contact with them and will afford me an opportunity of making an apology and renewing acquaintance.

Now I must close with that hope, most particularly as may put me in your way again; and that of the opportunity I may have of receiving you under my Roof or at the bottom of the River.

With my most respectful compliments to Mrs Coxson and love to the little ones, I am, my dear Sir, very Sincerely yours

M. I. Brunel (signed)
23 Parliament Street. 1st October 1833.

Letter Concerning Postal Services and Roads

Transcribed from the original by kind permission of the University of Bristol Library

My Dear Sir,

I am very much obliged to you for this most valuable Document of yours which will furnish me with a splendid illustration of those great monuments which this country possesses — not merely in one Department but in many. But, that under your immediate inspection and direction may be said to take a precedence as the most prolific agent of civilisation and improvement.

38,000 Letters daily and 43,000 in return!

36,000 news papers which in point of matter, therein contained, exceed 40,000 volumes of Mr Babbage's Book, for instance!

If we couple the active service of the Post office on the public Roads, with that of the public coaches in general, we find that in England and Wales, they altogether perform upwards of 1,200,000 miles weekly, consequently nearly 50 times the circumference of the Globe (per week).

We have in this little Island 120,000 miles of Roads, that is Turnpike and By Roads, all kept by the public; and in whatever place you are you may convey a letter or receive one from the general office or any office.

These Roads must now be macadamised (a new word in the vocabulary) and kept to such a degree as to efface from our records the word *Rut*; could it be imagined that, to cover that extent of Road, with road materials, the average quantity for yearly exigencies is I compute equal to the volume of two Pyramids, that is, the largest Pyramids of Egypt: all to be broken up of the size of an Egg.

The public Roads are great monuments, that is, considering their state their extent and their use — I think that, from Documents I have — the number of miles performed on them weekly by all that sort of vehicles, is little short of two millions and half of miles!

The number of Stages that enter and go out of London weekly amount to 52,000! that is the number of journies performed per week by each coach.

This subject is too extensive to proceed with I must therefore close with the expression of my thanks for your valuable contribution.

Believe me my dear Sir most sincerely

Yours,

M. I. Brunel (signed)
14th March 1834

Robespierre

Maximilien Marie Isidore Robespierre, the most fanatical of the leaders of the French Revolution, was born in Arras on 6 May 1758. His family was of Irish descent, having emigrated from Ireland at the time of the Reformation for religious reasons. His grandfather had settled in Arras and established himself as an advocate. His father followed the same profession. Maximilien studied law in Paris and was admitted an advocate in 1781, after which he returned to Arras to set up his own practice. He was appointed criminal judge in the diocese of Arras in March 1782, but resigned soon afterwards to avoid having to pronounce a sentence of death. His fame as a careful and painstaking advocate soon became widespread and some of his arguments were translated into other languages including English. He read widely and became a supporter of the theories of Rousseau, which he tended to accept quite literally and later tried to put into practice, with disastrous results for France. The Commune which was set up following the storming of the Bastille consisted of a dozen or so opposed and opposing groups, each determined to advance its own ideas; consequently the Commune was at war with itself right from the beginning. Robespierre was the leader of one of these groups and as a respected lawyer, gathered a large number to his cause, especially when, at the trial of Louis XVI, he declaimed on 3 December as follows:

'This is no trial; Louis is not a prisoner at the bar; you are not judges; you are — you cannot but be statesmen, and the representatives of the Nation. You have not to pass sentence for or against a single man, but you have to take a resolution on a question of the public safety, and to decide a question of national foresight. It is with regret that I pronounce the fatal truth; Louis ought to perish rather than a hundred thousand virtuous citizens; Louis must die, that the country may live.'

Following the execution of the king, a Committee of Public Safety was convened which included members of all the opposing factions, many of whom were atheists and sworn enemies of Robespierre. It was shortly after this time that the 'Terror' was instigated but it is not clear how far Robespierre was initially implicated in its enforcement although it certainly met with his approval. It gave him the opportunity he needed to eliminate many of his enemies, particularly since France was then in turmoil, seeing itself in grave danger of being invaded by England who, it was believed, was intent upon restoring the monarchy.

Meanwhile Robespierre was determined to increase the pressure of the Terror especially in Paris and on 10 June brought in a new law by which even the appearance of justice was abandoned; since no witnesses were to be called or examined, courts became simply places of condemnation and sentencing. As a result of this law, between 12 June and 28 July 1794 a total of 1,285 victims perished on the guillotine in Paris alone. The total number of people guillotined during the French Revolution has always been disputed; according to the source consulted it can vary between 10,000 and 60,000, but D Greer's book the *Incidence of the Terror in the French Revolution. A statistical interpretation*, published in 1935, suggests a figure of between 35,000 and 40,000 and this is now accepted by most authorities.

The remaining members of the Committee were not particularly displeased at the intensification of the Terror for it labelled Robespierre as the chief instigator and increased the general public's fear of him, making any move to eradicate him more likely to receive the approval of the masses. On 17 June Vadier, one of the more powerful of the remaining opposition, made fun of Robespierre in the Convention by repeating a report of his collusion with Catherin Théot, a mad woman who had claimed that Robespierre was a divinity sent by God to cleanse France of all evil. For a senior French statesman to be ridiculed at that time was fatal to his position and power, and Robespierre was forced to retaliate.

On 27 July he moved a motion in the Committee for the ending of the Terror and for those members who had been guilty of inciting atrocities to be brought to justice. His speech was violently interrupted with cries of 'Down with the tyrant', and at 5pm he and his remaining supporters were arrested. However, he was quickly rescued from his prison with the other deputies by troops of the Commune and taken to the Hotel de Ville where he was surrounded and protected by faithful followers.

The decision to arrest him was almost certainly a defensive move by his opponents in the Committee who had a fair idea who was to be next in line for the guillotine, but the day was past when the Commune could overrule the Convention. They denounced the members of the Commune as outlaws and sent the National Guard to the Hôtel de Ville to arrest them. It was claimed that Robespierre was shot in the jaw by a young gendarme named Meda whilst signing an appeal to one of the sections of Paris to take up arms on his behalf, but others contended that the wound was self-inflicted in an attempt at suicide.

Following a night spent in agony, Robespierre was taken before the tribunal where his identity as an outlaw was proved. Without further trial or delay he and 21 faithful acolytes were guillotined on 28 July 1794.

The period between 12 June and 28 July, when the Terror was at its height under the direct influence of Robespierre, lasted only 45 days, but it is that period of the French Revolution which has made the greatest impression on the rest of the world and probably reversed any move towards Socialism in England by at least a hundred years. We still suffer today through the influence of the French Revolution, for the metric system of measurement was introduced at the same time as it altered the calendar and imposed the 10-day week.

Taylors of Southampton

The Taylor Workshops and Factories

The steady development of the Taylors' business may be assessed by a survey of their workshops and mills. On the death of Messer *c*1755, the Taylors acquired the business and workshop in Westgate Street, Southampton, next to the medieval tin cellar, where they carried on the trade of blockmaking as a livelihood until about 1770. The family also owned property in nearby Bugle Street which had been purchased by the first Walter Taylor in 1713. This property which included dwelling houses and workshops was faced on the east by Bugle Street and was bounded on the west by the ancient wall of the town, washed at that time by the sea.

After the adoption of Taylor's blocks by the Navy Board, a horse was used to power the machinery. This was described by David Steel in his *Elements of Rigging and Seamanship* published in 1794. As the business flourished, especially with the increased demands from the Navy following the fire at Portsmouth in 1770, expansion became necessary. Walter Taylor III therefore purchased a piece of land adjoining Weston Lane in the Mayfield Park area of Southampton where he built a water mill to power the machinery on the stream which fed into the Itchen estuary.

The mill house has been demolished but Miller's Pond, a local landmark which originally supplied water for the mill, is still in existence. The tiny stream which fed the pond dried up for several months of the year and Taylor was soon forced to look for a more reliable source of water. He moved further up the Itchen to the site of an ancient mill on the estate of Hans Sloane at South Stoneham. Here he built a new mill of timber from which the factory derived the name of Woodmill, a name which it retains to this day. Unfortunately the original Woodmill burned down in about 1820 and the present brick structure beside the river in Woodmill Lane was built later on the site.

Taylors' factories and workshops were not the 'satanic mills' of William Blake's poem *Jerusalem*. The Bugle Street and Westgate Street workshops were small but they were close to the sea, so the workers were able to enjoy the waterside pleasures in any spare time available to them, and the transfer to Weston brought them into more rural surroundings of great beauty. The view down Southampton Water at that time was picturesque. As William Cobbett wrote: 'To those who like water scenes (as nineteen-twentieths of people do) it is the prettiest spot, I believe, in all England.' Today's tower-block monstrosities which scar the landscape, with the Fawley refinery defiling the opposite shore, cause the view to be seen in a somewhat different light although it is still a pleasant walk beside the Solent on a sunny day when the tide is in.

Woodmill was equally rural although larger. Over a hundred workmen were employed there and they formed an important little community in the Southampton suburbs of Portswood and Swaythling. Taylor was a model employer. There is a picture by Maria Spilsbury in the Tudor House Museum in Southampton which portrays a tea party for the children of the workpeople which was held in a schoolroom built by Walter Taylor as an extension to his home, Portswood Lodge.

Walter Taylor III was a Nonconformist and Elder of the Above Bar Church in Southampton which, until its destruction by enemy action in 1940, was located close to where Woolworths is today within a few yards of the medieval Bargate. He was also a great friend of John Newton (1725-1807) the clergyman and writer of the *Olney Hymns* at a time when Newton's daughter, who was suffering from some malady, had been prescribed sea bathing by her doctor. The Newtons often stayed with the Taylors at Portswood, visiting the beach at the end of Bugle or Westgate streets where a line of bathing huts was located just outside the city walls. During these visits John Newton would often preach in the schoolroom attached to Portswood Lodge which resulted in the locals calling it 'Newton's Chapel'. Although the sea once lapped against the walls of the city, today it is over a hundred yards distant, with a main road and car parks in between. By the end of the 18th century the Taylors also had factories at Deptford and Walton-on-Thames, but no details of these works now exist.

There is considerable confusion regarding the name of Walter Taylor's residence in Portswood. In some journals it is referred to as Portswood Lodge, in others as Portswood House. The explanation is that in the early 1770s a house was built for General Giles Stibbert, a retired officer of the East India Company, who owned the Portswood estate which included the old St Denys Priory. This was the original Portswood House situated at the southern end of the estate close to the present Lawn Road before the coming of the railways. When Walter Taylor III built his house at the opposite end of Portswood, he named it Portswood Lodge to avoid confusion with Portswood House. It remained so until 1875 when it was acquired by Walter Perkins, a Southampton estate agent, who renamed it Portswood House, the original Portswood House having been demolished in 1852. Walter Taylor's house was demolished in 1924.

The gravelled drive from the Broadway to the house commenced roughly at the place where the NatWest Bank is today, curved up to the front of the house, then continued back to the Broadway, emerging almost opposite present-day Westridge Road (a study of the roof line over the shops at that point clearly shows where the drive emerged). At that time Portswood Road (now called The Broadway) was little more than a rutted country lane with the northern side almost continuously hedged. The grounds of Portswood Lodge included all the area enclosed by Brookvale Road, The Broadway and Highfield Lane, with the exception of a farm which followed the line of Highfield Lane and was later associated with Southern Counties Dairies. The whole of the Portswood and Swaythling area was quite rural and mainly agricultural. The chief industries of Southampton were located down by the waterfront — development of the suburbs was a feature of the Victorian era much later.

The Taylors as Millwrights

The Taylors' horse mill used a friction drive to transfer the power of the vertical shaft to a countershaft from which the machinery was driven via leather belts. Several such countershafts could be taken from one source and the countershafts, carried on a hinged frame, could be disconnected from the main drive by a lifting lever.

The belt pulleys were provided with a curved surface for belt-centring, the belts being about 3in in width. The shaft was of steel and ran in what was described as 'a square iron box enclosing patent rollers on which the axis of the friction wheel works'. This was almost certainly a type of roller bearing.

Taylor and the Circular Saw

Southampton City Council has a portrait of Walter Taylor II which affords valuable evidence to support the claim of Taylor III and his father to have been the inventors of the circular saw as we know it today. The portrait was purchased by the Corporation from descendants of Walter Taylor whose family had held it continuously until its sale. The portrait is attributed to Gainsborough Dupont, the nephew and pupil of Thomas Gainsborough, who lived between 1755 and 1797, and it is thought the picture was painted about 1784 because it shows Taylor II as a man of around 50 years of age.

The saw held by Taylor in the portrait is almost identical in form to a modern circular saw of 7in diameter, with gulleted teeth of approximately 1in pitch — the type used for ripping soft woods. It is not known who made these saws for the Taylors; it is possible that they employed the services of a local blacksmith, but it is equally possible that they solved the problems of forming, hardening, tempering and pre-stressing their own saws in their dark cellar at Westgate.

The claim often put forward that Taylor was the inventor of the circular saw, emphasised as it is by its representation in his portrait in the Tudor House Museum, Southampton, warrants a digression to examine the claim closely. Since the 14th century, reciprocating saws had been known, operated at first by water and later by steam power. It must have occurred to several minds that the saw blade could be made circular to work continuously by rotative movement, and the milling cutter used in the machine for cutting the teeth of clock and watch wheels, invented by Robert Hooke, c1670, is supposed by many to have afforded some hints. However, the difficulties of making and tempering a thin disc of metal of a diameter of 18in or more, so as to project above the table, must have presented almost impossible obstacles to the earlier inventors. There was also the additional problem of designing suitable bearings to run at high speeds, because existing pivot bearings could not stand up to the friction, mainly because at that time lubrication was almost entirely by whale oil.

The claim that the saw originated in the Netherlands in the 17th century has been made frequently, but no definite evidence has been forthcoming. A patent was granted to one Claes Cornelisz Jonge Calff in June 1645, for an invention which is described in vague terms, but nothing more was heard of it so we may assume that the inventor had been unable to solve all the problems detailed above.

It is thought that Taylor introduced the saw in 1781 when the business moved to Woodmill, but the first positive reference to a circular saw occurs in the patent specification of Sir Samuel Bentham (No 1951, 23 April 1793) for 'improved methods and means of working wood, metal and other materials'. On page 18 of the specification he states that 'Working by a rotative motion has already been used, as I understand, in a few instances such as cutting timber into boards or in cutting logs for firewood, cutting mortices for ships' blocks and, where the diameter of the saw is considerable, the saw instead of being in one piece may be more advantageously composed of annular segments, fastened on the face of the flaunch.'

Whilst this is only circumstantial evidence it certainly pre-dates other claims such as that of L. C. A. Albert in France (who patented in 1799 [No 335] a *scie sans fin* [endless saw] which consisted of a disc to which were attached segments with cutting teeth) and Marc Brunel's English patent (7 May 1805, No 2844). There was also a patent granted to George Smart (19 June 1800, No 2415) for a circular saw for cutting the staves of the wooden canteens then carried by every private soldier.

It is reasonable to credit Taylor with having mastered the inherent difficulties of making the circular saw into a workshop tool and of inaugurating a new era in woodworking, whether or not he actually invented the saw.

Boring Machines

Ships' blocks frequently failed through the sheave rubbing on the mortise of the shell. This was usually due to the pin, on which the sheave ran, being slightly out of square with the mortise. In 1762, Taylor patented a boring machine which ensured that the pinhole was square with the faces of the block. It is not clear whether the same boring machine made the holes to form the mortise ends, as does the Brunel-Maudslay machine of 1810, but there is certainly a similarity. The Taylor blocks were rough-slotted with the circular saw and finished by hand; the Brunel blocks were mechanically slotted in a mortising machine with a chisel and needed no finishing.

In addition to the small machines used for making ships' blocks, Walter Taylor III designed larger machines for boring wood and metal with which pumps and water pipes were bored. A wooden pipe 24ft long and of 5in bore, made for the Southampton water supply, may be seen in the Tudor House Museum in Southampton.

Sawing Machinery

Being carpenters by trade, the Taylors' machines were, naturally, framed in wood, but wearing surfaces were faced with metal. Their first sawing machines are described in the patent of 1760, granted to Elizabeth Taylor following her husband's death. They consist of frame saws for cutting parallel slices of hardwood from the log for shaping into sheaves or blocks. Following the invention of the circular saw, Walter Taylor III designed machines for the later stages of block manufacture. A letter to William Kingdom included the following: 'It is then removed to a round saw where the piece cut off is completely shaped, and only requiring to be turned under the saw.'

Ships' Pumps

Pumps are an essential part of the equipment of any ship, but in the days of sail, with the risk of tempest, battle or collision ever present, the efficient working of the ship's pumps could be a matter of life or death to the mariners.

Before Taylor's time, pumps for ships were of three kinds: the chain pump, the bucket pump or baling machine, and the piston suction pump. All were unreliable and subject to frequent failure. The chain pump worked until the chain broke, which often happened; the bucket pump was, as its name implied, a simple bucket with a flap valve in the base, which was liable to choke with gravel or flotsam, as were the valves of the suction pump. All types required a great deal of manpower and maintenance and would work only with low heads of water.

Taylor first improved the baling machine mainly by better workmanship. This type of pump continued to be manufactured throughout the life of the firm, probably to meet the requirements of his more conservative customers, whilst Taylor experimented with more advanced methods using a variety of ideas, including glass cylinders to enable him to watch the pumps in operation.

The barrel of the last Taylor pump was made of copper, the pump rod having a leather fitted bucket. The barrel was enclosed in a wooden casing bored from a log, the pump suction being an extension of this wooden tube. This type of pump with a 7in bore would raise one ton per minute a height of 24ft, with a crew of 10 men hauling five-a-side alternately. Pumps up to five times this capacity were eventually made.

Taylor-Brunel Links

Sir Samuel Bentham, who became Inspector-General of Naval Works, took a great interest in woodworking machinery, and from c1770 produced a series of inventions for woodworking machines including the planing machine. In 1791, Bentham's assistant Goodrich visited most of the manufactories in England, but found no efficient blockmaking machinery except that of the Taylors at Southampton and one other at Plymouth.

With the assistance of his famous brother Jeremy, the founder of University College London, Samuel Bentham constructed a series of prototype machines for general woodworking at No 19 York Street, Westminster. It was intended to install these machines in the Navy's shipbuilding yard at Redbridge, Southampton, but the steam engine intended for the power source was not ready in time and so the complete equipment was stored at Portsmouth Dockyard. Henry Maudslay, the foremost engineer of his time, was well known to Bentham and may well have been responsible for producing some of the machinery.

Brunel had already (in 1801) offered, through his brother-in-law Mr Kingdom, his designs for shaping ships' blocks to Samuel Silver Taylor, who was managing the Woodmill works on behalf of his ailing father. Samuel Taylor declined to take up Brunel's designs in a letter describing Taylors' methods which the writer felt could not be surpassed, as detailed in Chapter Three.

Samuel Taylor has often been criticised for his lack of foresight, but at that time Brunel had only designed the 'shaping machine', which was a type of lathe for turning the block to its finished oval form. Furthermore, to change over to Brunel's new methods and machinery would have meant closing the works for several months whilst the new equipment was being installed, at a time when they were fully stretched to meet Admiralty requirements. The Taylors never opposed new ideas without due consideration, but in this case they really had no option but to refuse Brunel's approaches.

Decline of the House of Taylor

In 1803 a commission was set up to enquire into corruption in some branches of the Navy. Partly as a result of this commission and partly in anticipation of the fruition of Sir Samuel Bentham's scheme for the Navy to manufacture its own blocks by Brunel's invention, the Taylor contract was cancelled. The original 21-year contract contained break clauses at seven and 14 years, 'if it was found to be unfavourable to the public', but the Admiralty thought fit to cancel after nine years. Walter Taylor III disputed the cancellation on the grounds that it was illegal and sought a public enquiry, as it 'might be a matter of doubt how far it was safe to enter into contracts (with the Government) when they were liable to be dissolved without explanation'.

Taylor's case was taken up by Sir H. Mildmay who made an appeal in the House of Commons for an inquiry, at the same time praising the work and dedication which the Taylor family had shown over the years. However, by the time this tribute to him was made in Parliament, Walter Taylor was dead. He died on 23 April 1803, his widow dying seven weeks later.

The death of Walter Taylor III, followed closely by the construction of the Brunel Portsmouth blockmaking workshops, left his descendants with equipment and expertise but no government orders. No doubt some sections of the business could have been continued with other buyers, but the core of the enterprise had been destroyed. In 1810 an unwritten agreement of partnership was dissolved between four of Taylor's sons: Samuel (by his first wife) and William, Thomas Ebenezer and John (by his third wife). The workshops at Walton-on-Thames and Woodmill had already been closed and the one at Deptford was wound up.

It is often assumed that Taylor's blocks were crude and badly formed in comparison to those made by Brunel's methods. This is quite untrue. Brunel's blocks were little better than Taylor's; their great advantage was that they could be mass produced by unskilled labour at a 10th of the man-hours previously required.

In March 1798, Walter Taylor raised, at his own expense, a corps of Infantry (The Portswood Green Volunteers) which in 1799 took part in the expedition to Holland. The principal achievement of this expedition, the capture of the Dutch Fleet, was celebrated a few weeks later with great enthusiasm. The volunteers, after parading, were regaled with many loyal and constitutional toasts and dinners provided mainly by Captain Taylor at various venues in the Portswood and Swaythling areas.

* * *

Part of the above account is reproduced with the permission of The Institution of Mechanical Engineers. It is condensed from an address given at the Summer Meeting of the Institution at Southampton on Tuesday, 5 July 1955 by J. P. M. Pannell, MBE, MICE, entitled *The Taylors of Southampton: Pioneers in Mechanical Engineering*.

Also included with permission are extracts from an address read at the Iron & Steel Institute London, on 12 January 1955 by H. W. Dickinson, DEng (Hon), MIMechE, Past President, entitled *The Taylors of Southampton: Their Ships' Blocks, Circular Saw, and Ships' Pumps*.

The Dockyard and Town of Portsmouth

By 1800 Portsmouth Dockyard was the largest and most efficient in the world. This was partly as a result of its location in the era of sailing ships. It was sheltered to a large extent by the Isle of Wight and easily accessible from the open sea without enduring the tortuous and dangerous approaches to London via the Thames or to Southampton, 10 miles or so up the narrow Southampton Water. Portsmouth Dockyard was equipped with more dry docks than all the other naval dockyards put together, and these were linked by a complicated underground drainage system coupled with a series of pumps which could empty the docks irrespective of tidal conditions.

In addition to its many other functions, it included workshops for the copper-bottoming of ships. Vessels built elsewhere were routinely sent to Portsmouth for copper lining and for repairs to the copper shell.

In 1800 it was estimated that 97% of the population of Portsmouth was employed in or derived its living from the dockyard or its employees. In times of war, when its products were in great demand, Portsmouth flourished; peace brought hardship and bankruptcies. Without a regular income the inhabitants could starve and small traders such as butchers, grocers and greengrocers fail almost overnight as debtors defaulted *en masse*. And because Portsmouth Dockyard is located on what is virtually an island, the nearest opportunities for alternative employment were at least six miles distant on the mainland — a daunting prospect for pedestrians whose journey would have taken them through copses and woods inhabited by footpads and assassins bent upon robbing anyone so unprotected.

The situation was aggravated by the government, through the Navy Board, paying all salaries on a quarterly basis. This meant that most employees had to obtain credit for their day-to-day requirements on the promise that they would settle the accounts when payment was received from the Board at the end of the quarter. This was a fairly satisfactory arrangement whilst a war with France appeared to be imminent, but as soon as the threat lessened, the system collapsed into chaos as the Navy Board cancelled contracts, often without prior notice.

Payment was in coin, and at the appointed time a coach loaded with cash departed from London accompanied by a company of Dragoons, to make the dangerous journey to Portsmouth via Guildford, Godalming, Liphook and Petersfield, a route considered to be one of the most hazardous in England. Its appeal to highwaymen lay in the six-mile-long climb between Godalming and the Devil's Punchbowl, where all coaches were forced to proceed at a slow walking pace. The gibbets erected at the crest of the hill, from which the bodies of convicted highwaymen were suspended, apparently did little to discourage other members of the fraternity. (One of these gibbets survived near the Devil's Punchbowl until *c*1940.)

The inhabitants of Portsmouth town had a reputation for thieving, swindling, robbing and whoring, and were particularly attentive to visitors and crews arriving on the Merchantmen who used the port in addition to the Naval vessels. The port area was kept reasonably clean, but the town was filthy. The narrow streets were lined with small shops and tenements, from all of which rubbish and excrement was thrown into the street, lying there until dispersed by heavy rain. Horse-drawn vehicles added to the problem, and in hot weather the wind whipped up the dried horse dung into dust clouds which it was impossible to avoid breathing.

Corruption was endemic in both the dockyard and the town. Anyone appointed to a position of authority expected to supplement his income by bribery of some kind. Even a minor chargehand expected his charges to reward him for preferential treatment in the workplace, and a higher official could expect to double his salary by unrecorded additional emoluments. In the town corruption began with the Mayor and descended through every department of local government down to the lowliest crossing-sweeper who, although an employee of the town council, would demand payment before clearing the way for a lady or gentlemen who gave the appearance of being wealthy.

This was the position when Marc Brunel came to Portsmouth. He could have had no idea of the den of intrigue and iniquity into which he was entering. The appointment of Burr to manage the blockmills was no accident: the Navy Board knew he was one of its most reliable underlings. Not only would he watch Brunel (a French suspect) and report upon his progress, but he was in a position to counteract any rapid increase in production which would jeopardise the Board's income from rival enterprises. The fact that he encouraged visitors into the mills upon payment of an admission fee was probably unknown to the Board, but had it been aware of him lining his own pockets, it is unlikely this would have been regarded as anything unusual.

There were several ways in which production could be restricted, and although there are those who will come to Burr's defence, to an unbiased observer the results of his efforts are fairly obvious. Brunel was constantly kept short of supplies; even when all the machinery had been installed and was ready for full production, no coaks were available so the units could not be assembled. The wood from which the sheaves were made, lignum vitae, was always in short supply and when delivered was often of unsuitable quality. But the shortage of labour was the most serious of all the problems with which Brunel was beset. It is often assumed that the Brunel machines were automatic, but this was not so. Almost all required the services of a trained operator. He did not need to be a skilled craftsman, but training was essential if the machinery was to operate efficiently without damage. Each machine produced just one component part of a block, after which the unit was transferred to the next machine where it was processed one stage further. Brunel was continually requesting more hands to train, but his requests fell mostly upon deaf ears.

When Brigadier-General Samuel Bentham gave his approval to Marc Brunel's proposals for reorganising block production at Portsmouth, it was inevitable that he would be in conflict with the Navy Board. Bentham was an honest man, one of the few persons involved with government contracts not engaged in some kind of chicanery. As such he was anathema to the Navy Board which did its best to unseat him and confound all his efforts. Eventually it succeeded in getting him sent on a mission to Russia where he was powerless to influence the Board's plans.

This was the situation into which Brunel was precipitated when he accepted the appointment and began work on the installation at Portsmouth. He may have known that bribery and corruption were accepted world-wide at that time, but he could not have been prepared for the intrigue and conniving endemic in the Portsmouth Dockyard.

Maudslay and Bramah

Nicknamed 'The Woolwich Powder Boy', Henry Maudslay was born on 22 August 1771 at Woolwich, London, in a house standing almost opposite the gates of the Royal Arsenal. His father, William Maudslay, was a native of Bolton in Lancashire and came from an ancient family of the same name, the head of which resided at Mawdsley Hall, near Ormskirk, at the beginning of the 17th century. William Maudslay was a skilled joiner, but left the Bolton area after an alleged indiscreet liaison which he always denied. He enlisted in the Royal Artillery as a sergeant-wheelwright and was posted to the West Indies. He was engaged in battle on several occasions, and during his last action was hit by a musketball in the throat. The leather stock he was wearing diverted the ball and, although severely wounded, he survived. He brought the stock home and preserved it as a relic, afterwards leaving it to his son. Henry would often point to the stock, hung against a wall and say, 'But for that bit of leather there would have been no Henry Maudslay'. William Maudslay married Margaret Laundy at Woolwich Church in July 1763. They had seven children, of whom Henry was the fifth. Henry also had seven children and his eldest, Thomas Henry, also had seven.

William Maudslay was invalided out of the Army and returned to Woolwich, the headquarters of his corps, where he was discharged shortly afterwards. He obtained employment at the Royal Arsenal and later became a storeman in the Dockyard. It was during his period at the Arsenal that his son Henry was born. As was usual in those days, Henry was put to work at the age of 12 and, probably at his father's request, was taken on at the Arsenal filling cartridges at first, but afterwards being apprenticed to a blacksmith in the metalwork shop. By the time he was 18 his skill as a craftsman was renowned.

* * *

Joseph Bramah took out the first patent for a lock in 1784 but ran into considerable trouble owing to the impossibility of machining parts to the precision required. Tolerances in those days were measured in eighths of an inch or a little less. When Bramah was introduced to Henry Maudslay the latter was only 18 years of age and still indentured to the Royal Arsenal, but Bramah was so impressed with the young man that he immediately offered him a position in his workshop.

At the completion of his apprenticeship Maudslay joined the staff of Joseph Bramah, the pioneer of hydraulics and the inventor of the Bramah press and lock, where he was eventually appointed manager. Bramah, originally a cabinetmaker, was intrigued by the possibility of producing a pick-proof lock, and in 1784 patented and exhibited a lock invented by himself, and offered a reward of two hundred guineas for anyone who could pick it. Despite attempts by many individuals it defied all efforts for 67 years. It was finally opened by A.C. Hobbs, an American locksmith, after 51 hours of concentrated effort.

This very complex piece of machinery required a whole range of well-designed and precisely manufactured engineering tools made to a tolerance unknown at that time, so Henry Maudslay was in his element. The prototype machines designed and produced by Bramah and Maudslay laid the foundations of the British machine-tool industry, and Bramah's experiments with the new science of hydraulics were to prove invaluable to the Brunels later. Bramah was responsible for several other inventions, including a water-closet, a numbering machine for banknotes and a wood-planing machine.

Bramah always admitted that he owed a major part of his success to Maudslay's skill and ingenuity, but even after promoting him to manager of the works, he refused to pay him more than 30 shillings a week. As a result, Maudslay left to establish his own business at Wells Street off Oxford Street in London, where he developed a method of cutting screw-threads on a lathe. Until then, large threads had been forged before being finished by hand-filing, and small threads were all cut by hand by the most skilled of craftsmen. Maudslay's new screw-cutting lathe gave such precision as to permit previously unknown interchangeability of nuts and bolts and the standardisation of screw-threads. It also enabled him to produce sets of taps and dies.

Using his new device, he cut a long screw with 50 threads per inch (about 20 per centimetre) and made this the basis for the world's first micrometer, which came into daily use in his workshop as a means of checking the standard of work produced. In an engineering world where a tolerance of about one sixteenth of an inch was considered to be the utmost limit of precision, this was a revolution. Maudslay's tolerance was down to one thousandth of an inch.

Maudslay's firm, with him as designer and chief working craftsman, went on to design and produce marine steam engines. The first was a 13kW/17hp model; later engines of 42kW/56hp were built, and in 1838 after Maudslay's death, the company went on to build the engines for the first successful transatlantic steamship, I. K. Brunel's *Great Western*, which developed 500kW/750hp. In the early 20th century the company went into motor manufacturing, mainly commercial vehicles, until it was absorbed into AEC which, in turn, eventually became part of British Leyland. At about the same time one of Maudslay's descendents became a joint founder of the Standard Motor Company.

Early in 1831 Maudslay caught a chill on his return to England after a trip to France, and died on 15 February that year. He was buried in a cast-iron tomb of his own design in Woolwich churchyard. It is generally accepted that he was an honest, upright, hardworking and intelligent Englishman without whose genius the industrial revolution in England would not have progressed so rapidly. When he took Joshua Field into partnership, it was to enable him to devote all his attention to engineering, whilst Field took over the administration and accounts side of the business.

Several of Britain's outstanding engineers learned their trade in Maudslay's workshops including James Nasmyth, Sir Joseph Whitworth and Joseph Clement.

1814 Report on Prisons

In 1814 a report was presented to Parliament by a committee appointed to enquire into the state of the King's Bench, Fleet and Marshalsea prisons, and to suggest 'any improvements which may be practicable therein'. Although seven years had elapsed between the publication of the Report and Marc Brunel's committal to the King's Bench prison, very little improvement had taken place, and conditions endured by inmates were similar to those described in the report. The following is a condensed version of the report as submitted by the committee. It is reproduced by kind permission of the Corporation of London's Guildhall Library.

Your Committee find the King's Bench to be under the jurisdiction of the Court of King's Bench, and that it is a national and not a county prison. Persons arrested for debt, or confined under the sentence, or for contempt of the Court, are imprisoned therein.

The Marshal is appointed by the King, and *William Jones, Esq.* the present Marshal, was nominated by sign manual on the 19th March 1791, and holds his office 'for so long time as he shall behave himself well, and shall be resident within the walls or rules of the prison, and no longer'.

There is no salary attached to the office of Marshal of the King's Bench; his income is derived from fees on commitments and discharges, and from other sources, such as the rent of rooms in the prison, profits on the sale of porter, ale and wine, the rent of the coffee-house, etc., etc., but his principal emoluments arise from granting the rules, or the liberty of living without the walls of the prison, within a certain area. The average gross amount of revenue derived from these sources, may be estimated annually, for the last three years, at £5000.15s.8d. according to an account delivered in by the Marshal; the expenses and outgoings, including the payment of clerks, turnkeys, watchmen, taxes, etc., etc. are £1730.9s.6d. leaving a net income to the Marshal of £3270.6s.2d. To this must be added the fees taken on bails and judgments, which are calculated at £320 per annum; making on the whole a sum of not less than £3590. Amongst these emoluments, that on beer sold in the prison amounts to £872, and that on the rules to £2823 per annum.

The outgoings consist principally in salaries, taxes, and law expenses, and the small sum of £75 for repairs; though the Committee observe, that 'the Marshal shall at his own cost and charges, by and out of the fees and profits incident to the said office, well and sufficiently repair, and keep in good repair, the said prison, and all the buildings and appurtenances thereunto belonging'.

The officers of the prison are all in the appointment of the Marshal, and consist as follows:

The Deputy Marshal is *Mr. Josiah Boydell*. This gentleman has nothing to do with the management of the prison, and has no salary, but his emoluments arise from fees taken under the rules and orders of the Court, which amount from £350-£400 per annum,

and would, it is stated, be considerably increased, if the fees were not voluntarily relinquished in all cases of poverty and distress. [The duties of this officer are to attend the Chief Justice in Court, and to accompany him to the royal levees.]

The situation of Clerk of the Papers was held by a nephew of *Mr Jones*, who is dead. No new appointment has taken place, but the Marshal receives the profits of the office, which, arising from fees, may be estimated at from £600 to £700 per annum.

The office of Clerk of the Rules is held by the Marshal himself; who pays also three clerks fixed salaries. There are in addition, three turnkeys and four watchmen, whose incomes are derived partly from salaries and partly from fees and emoluments arising out of the prison.

The Chaplain is the Rev. *William Evans*, who is appointed by the Marshal: he has no salary, but he receives a fee upon the commitment of each prisoner, which is paid to the Clerks of the Judges at their chambers.

The prison contains within the walls about 200 rooms, eight of which are called state rooms, and are let for 2/6 each per week, unfurnished; the remaining 192 are (or ought to be) occupied by the prisoners who are compelled to pay weekly 1/- for a single room, unfurnished: if two persons live in the same room 6d each, if three 4d. But the Marshal states, that he never demands any rent from those who are unable to pay. On a prisoner's arrival at the gate, he is called upon to pay his commitment fees, which amount to 10s.2d. Your Committee have been assured that, whether the fees are paid or not, he receives on demand a chum ticket (as it is called) which is a ticket of admission to some room in the prison. Your Committee however, observing that the Marshal seems to consider he has a right to refuse the ticket if the fees be not paid, think it is essential to remark, that a question as to the legality of such a refusal was put to Mr Templer, the visitant in 1791, and his answer was, 'That any prisoner, immediately on going into prison, has a right to the possession of a room or part of one, if any shall be found capable of receiving him, although at such time he should be unable to pay any fees'. The principle upon which this chummage takes place may be thus explained: Supposing the 192 rooms in the prison are occupied by one prisoner each, and there is an arrival of fresh persons, which in term time often occurs to the number of 20 or 30 of a night, and chum tickets are demanded from the chum-master; if the prisoner so requiring a ticket is of decent appearance, and has the air of good circumstances, one is given him upon a room already occupied by a person of his station in life; but if the applicant be poor, he receives his ticket upon a room held by one who is enabled to pay him out, that is to say, to give him so much per week, which generally amounts to 5/- whereby he yields to the existing occupier the whole right to his room, and pays for his lodgings with persons of his own class and situation: so that it is not uncommon to find 6 or 8 persons of the poorer classes sleeping two in a bed, or on the floor, in rooms of the dimensions of 16 feet by 13; some also of these sleep at the tap on benches and tables, and as many as 48 have slept there at one time. The choice then of the chummage is thus perfectly optional with the chum-master, who is one of the turnkeys, and has the sole management of the business as far as the ordinary rooms are concerned; but those of a better description, from their situation, are considered as being at the disposal of *Mr Brooshooft*, the first clerk to the Marshal, who has in point of fact the direction and management of the whole prison. The prisoner who has sold his share of his room is considered as entitled to re-enter whenever he chuses (*sic*) to break the bargain, it lasting only for one week; but it appears in evidence that this right has been denied or evaded, and that persons who have interest with the officers of the prison may either keep a room free from chummage, or prevent those who are chummed upon them from returning to their rooms, if the payment of 5/- per week be regularly made: in this latter case, the person insisting upon his right to return is shifted from his own room, and chummed on another.

No care seems to be taken to acquaint the prisoners, on their first entrance to the prison, that a chum ticket is to be obtained upon application. Some have been several days within the walls, paying a heavy rent for their lodgings, before they learnt from their fellow prisoners that they had a legal right to a share of a room. The ordinary proceeding is for one of the turnkeys to take the prisoner on his arrival to the coffee-house, the master of which provides a room at the cost of about 3/- a night, or a lodging is engaged from some one of that numerous class of persons who, having been long in the prison, gain their livelihoods by letting out their own rooms, or their share of a room, to new comers. Eight or ten shillings a night have been given for a bed; but the usual price is from fourteen shillings to one guinea per week. *Mr Brooshooft* says, that some delay necessarily takes place in the delivery of chum tickets, and that it is sometimes difficult to provide situations on the emergency of the moment, fitted to the station of life of the claimants, and who are therefore inclined either to look out for themselves, and find a lodging in the prison, or to wait, in the expectation of someone going out, when they can succeed to the vacant room.

The rule of chummage is, that the person who has been longest in prison keeps his room free from having another prisoner chummed on it, till all the rooms held by those of a junior date to himself have each a prisoner chummed on them. The system purports to be one of rotation; and if the prisoner be poor, and wishes to be bought out, he is chummed upon one who can afford to pay him; if he wish to remain, he is placed in the room of a person who will keep him, and he has accordingly a chum ticket upon the youngest prisoner in one or other of these classes.

Your Committee sat several days within the prison, and have endeavoured to understand the manner of delivering of the chum tickets; but though there be a rule stated to exist, by which this delivery is regulated, yet it appears that so many exceptions are made to that rule, that the whole system seems to be one of favouritism and partiality, and liable to great abuse. It is in evidence that, contrary to the orders, and, what is more extraordinary, without the knowledge of the Marshal of the King's Bench, the turnkeys, criers, waiter at the coffee-house, etc. hold rooms within the prison, which they furnish and let to prisoners, the criers and the waiter at the coffee-house having no wages or salary, but deriving a portion of their income from this resource. It was also admitted by *Mr Morris* the chum-master, and confirmed by *Mr Brooshooft*, and by some of the parties themselves, that a person by the name of *Gore* was a prisoner within the walls of the prison; that as such he was chummed on a room; that he paid out his companion; that he furnished the room and let it; and that, though he be now out of the prison, living with his wife within the rules, being also an assistant tipstaff, he was permitted to keep possession of the room, and to let it to prisoners at the rate of 20/- per week. It appears also that *Mr Morris* and *Mr Brooshooft* were both acquainted with all the circumstances of the case, and voluntarily winked at the abuse.

Your Committee have also to observe, that the whole management of this branch of the economy of the prison seems most defective. Though the Marshal has published an order, that extravagent demands shall not be made by one prisoner on another, and that no alterations shall be attempted in any of the rooms, and that a system of antient date for the successor to a room to buy the fixtures, and pay for the alterations of his predecessor, should be broken through, and no longer be permitted to exist, yet the practice is most prevalent, and several witnesses have declared that they have paid considerable sums of money for what were called fixtures. Though no evidence went directly to prove the fact, as having recently occurred, yet Your Committee feel themselves warranted in declaring it to be their opinion, that there appears to be an understanding in the prison on the part of some of the officers, to chum those persons only in particular rooms who are enabled to be at the expense of making these purchases. Great abuses prevail in the whole system, and the orders and regulations of the Marshal are nugatory. Of the 192 rooms in the prison, though there were 440 prisoners within the walls, 80 were occupied singly; many of the other rooms were occupied by 6 persons, or more, of the poorer classes. *Mr Burnet*, a prisoner, and by profession a surgeon, and who gratuitously attends the sick, informed Your Committee, that such was the crowded state of some of the rooms, and the consequent offensive smell, that, if he had not been used to an hospital, he could not have borne it; and that a corpulent patient under his care was getting into a bad way, from the crowded state of his apartment.

Mr Brooshooft says too, that there is no limit imposed by the rules of the prison to any number of prisoners as shall so think fit, crowding together in the same room, the effect of rapacity on one side, and necessity on the other; and that if ten persons were to lie together, no one would interfere. In some of the rooms, when Your Committee visited them, there were 5 or 6 persons; not including women and children, of whom there were within the walls, on the 13th of March, 180.

Besides the 440 persons confined within the walls of the prison, there were 220, on the 13th of March, who enjoyed the benefit of the rules; these rules extend in circumference about two miles and a half. But it is ordered by a rule of the court of King's Bench, that all taverns, alehouses, and places licensed for public entertainment, should be excluded out of and deemed no part of the said rules. With these exceptions, and with the permission of the Marshal, any prisoner for debt may live in any house within the precincts of the above-mentioned district. The Marshal, on an application for this permission, takes security from the applicants by way of indemnity for the debt for which they stand charged; and the purchase is made on the following terms:- If the prisoner takes the rules for a debt of £100 or more, 8 guineas is demanded for the bond, stamp, and other fees, and 4 guineas for every succeeding £100; if the sum be large, and the security good, a fee is sometimes taken in a smaller proportion; if the debt be under £100, 6 or 5 guineas are demanded. But in many instances the rules are given without expense; and some persons are now within the rules, who have paid nothing for that permission: the Marshal indeed grants these rules as a mode of clearing the prison, and in 1813, a great number of persons were admitted to them by a public notice circulated in the prison, which gave that liberty upon the claimant finding proper security, and paying only for the stamp. The profits arising from the rules are very considerable: *Mr Jones* states, that he once received £300 from one person for granting them; and that the annual average value of this source of his emolument amounts to £2,600.

There are also day-rules, which any prisoner can obtain during term time, by permission of the Marshal, upon sending a petition to the clerk of the day-rules, which petition is presented to the court of King's Bench, and the liberty is so granted; the fee paid on this by each prisoner, is 4s.2d. which is divided in the manner set forth in the table of fees. The safe return of the prisoners within the rules, is considered as sufficiently secured by their recognizances: and those within the walls, give a bond to the tipstaff, renewable every term, the expense of which is 3 guineas. When the sum for which a prisoner detained is small, and under £600 the tipstaff can of his own authority, and at his own risk, consent to his applying for a day-rule; when the debt amounts to a larger sum, a special application must be made to *Mr Jones*, who requires security before he will consent to the rule; but in case of a prisoner not being able to find such security, he may avail himself of a day-rule, by giving the tipstaff one guinea, who furnishes a man to take care that the person does not escape, on whom the expense of maintaining this guard falls; the prisoner may be out of the prison from nine in the morning to nine at night. The profits to the Marshal from the day-rules, amount to £223 per annum.

Besides these, there is a manner by which a prisoner has been permitted to go out of prison, which is known by the name of a 'run on the keys'. It is a liberty granted by some of the officers of the prison to any of the prisoners, to go into the rules for the day. The Marshal, being examined upon this practice, stated, that he had discovered that such permission had during the last year been given, that he censured his officers for acting so improperly, and that if he detected them in the repitition *(sic)* of the act, they should

certainly be discharged from their situations. Your Committee have not only discovered that the practice, though diminished in its extent, still continues, but that money is taken for that permission; one of the prisoners stated to Your Committee that he paid, and *Mr Morris* acknowledged that he received, a one pound note which was to have been divided among the turnkeys, who considered themselves, if the prisoner escaped, as liable for the debt with which he stood charged. *Mr Brooshooft* however thinks differently from the Marshal upon this subject; he has thought proper to tell the Committee, that though the Marshal has forbidden it, he should have no difficulty, if a gentleman came to him on the morrow, and intended to go out either to his attorney or to buy any thing of which he stood in need, in granting him that indulgence, 'and that he should think it extremely odd if he had not a right to act for himself in many cases'. The nature of the situation of *Mr Brooshooft*, beyond that of private clerk to *Mr Jones*, Your Committee have not been enabled to learn: he appears to have the sole direction and management of the prisoners, and is known to the prisoners by the name of the Deputy Marshal, and described as such: but his conduct in this respect, to say the least of it, is in direct opposition to the orders of his superiors, and Your Committee think it their duty to mark their sense of the impropriety of such a procedure.

The receipt of money on the part of the turnkeys opens a wide field for every species of abuse; and the mere circumstance, of their considering themselves as liable to the payment of the debt, neither lessens the impropriety of disobedience to the commands of a superior, nor changes the real nature of the transaction, which is neither more nor less than the taking of a bribe to commit a breach of their duty.

The present system of the rules is of considerable long standing, but Your Committee are of opinion that it ought to be subjected to some regulations. It would perhaps be too severe an aggravation of the law of arrest, to deprive prisoners of the liberty of the rules altogether; but, as the practice now stands, the fraudulent, no less than the honest and unfortunate debtor, can, upon the payment of a sum of money, place himself in a situation that can hardly be called imprisonment, by which he eludes the sentence of the law, and may dissipate that fortune which, in justice, is no longer his own. It appears to Your Committee, that the Marshal ought not to be permitted to exact such large sums from the persons who apply for the liberty of the rules. As a mode of relief to the crowded state of the prison, he occasionally grants that permission, upon the parties finding the security for the debt and the payment of the law expenses; and there therefore seems to be no reason why the practice should ever be otherwise, except that of enabling the Marshal to derive a large income from those sources. The sum of £2,600 per annum is surely much too large to be drawn from the pockets of debtors; and it is ever to be remembered, that whatever is paid by that unfortunate, though sometimes criminal class men, is not paid by them, but by their creditors, to whom the property they are worth legally belongs.

Your Committee are also of the opinion, that the practice of exacting from the applicant for the day-rule a term bond, which bond is renewable every term, at the cost of £3 besides a fee of 4s.6d. daily, is most objectionable, and which, though warranted by antient custom, ought to cease, for the same reasons which have been urged against the heavy payment demanded by the officers of the prison on the granting the liberty of the rules.

By the Act 53, Geo.III, all persons . . . . . . can, upon swearing that they are not worth £10, receive 3s.6d. per week in weekly payments. The whole amount of the money paid by the different counties in England is £700 per annum: and if there be any balance remaining in the hands of the Treasurer, it is to be paid to some of the public charities.

The Marshal of the King's Bench has no control over the disposition of this fund, it is left to a magistrate: and Mr Hope, a person in the office of the Clerk of the Peace, attends weekly to pay the allowance.

Persons committed for misdemeanors have no claim to relief under the Act, and if poor, have no means of obtaining assistance, except from the charity of the Marshal; who states, that last year three poor men were committed under the Excise laws; they were without money, and were in the King's Bench for some time; he allowed them 1s.6d. per day out of his own pocket. Though these cases do not often occur, it surely cannot be right to leave any one in a prison without the means of subsistence, and to be kept from starving by the charity of individuals.

In addition to the allowance of £700 per annum under Mr Thornton's Act, the former class of debtors become entitled to receive a share of all the charities and benefactions paid for the use of the common-side prisoners, on taking an oath that they cannot command the sum of five pounds, and that they cannot subsist without the above charities. A list of these has been laid before the Committee, and is also hung up in the lobby of the prison, and their annual amount may be estimated at £40 per annum. Very few persons, however, entitle themselves to a share of these charities; there are seldom more than two or three, who form a kind of corporation, consisting of a major, clerk and assistant, who regulate all its concerns. Those who claim these charities are besides compelled to hold alternately the begging-box at the grate; and many who are in great need relinquish this relief, rather than submit to that which is deemed a degradation, and which seems to Your Committee to be a condition that ought not to be enforced. The proceeds of this box are divided every night by the major. *Mr Ledwell*, who holds that office at present, values its annual receipts at £15; it is however most probable that there is a concealment of the profits so derived, as the assistant, *Owen Owen*, acknowledged that, one day with another, he receives from sixpence to a shilling.

When Your Committee first visited the prison, there were two persons on the list, viz. the major and his assistant: since that period the assistant has been superseded, and the major, *Ledwell*, only remains. This man has been imprisoned in the King's Bench nearly seven years, originally for debt; on his discharge he was committed for a contempt, for not answering before the commissioners under the Bankrupt laws; and though now he could without difficulty obtain his discharge, yet he remains in confinement, for the sake of the subsistence which he draws from the charities and allowances of the prison.

Your Committee called for the book in which an account of the monies received is purported to be kept; and having examined it, there appears to be no entry since the 6th of April, 1813; and though the major affirmed that he kept an account of all the charities, yet he afterwards acknowledged that he could only just write his name, 'and that he made sometimes a loose copy on a scrap of paper, and sometimes he forgot it'.

It appears, too, that of late years the entries have been most irregular, there being none whatever in the year 1812; the book is also much mutilated and defaced; it contains the rules and regulations made by the court of prisoners, for their own government, which in general are just and equitable.

There are no other allowances in the King's Bench; no coals, nor bedding, nor even blankets, except from the occasional charity of individuals. *Mr Brooshooft* states, that he has general orders from the Marshal, never to withhold assistance from any person who is sick and in poverty, and that expenses so incurred are charged in his account, and invariably paid; yet he adds, that few demands are made for blankets, etc. unless it is known that the benevolence of private individuals has provided them, 'for the applicant could not otherwise expect to be successful'. It is thus clear, that though the distress arising from the want of these articles may be great, the relief is scantily, if at all, afforded.

In the rules and orders for the King's Bench prison (made in 1760) . . . . there is a charge of so much a night for bed and bedding, to be paid to the Marshal, for the first night 6d. and for the succeeding $1^1/_2$d. That practice has long ceased. But as it is in evidence that 10 shillings a night have been paid by poor prisoners for the use of a bed, it is the decided opinion of Your Committee, that some regulations of a similar nature to those which heretofore existed, providing for the supply of bed and bedding, should take place.

There is no infirmary within the walls of the prison, and no medical attendance whatever. It appears, however, by the rules and regulations signed by Lord Raymond in 1729, that the Marshal was directed to take particular care of all prisoners on the common side who shall happen to be sick, and that all proper necessaries shall be provided for them by the steward and assistant, who shall be reimbursed out of the county money; also, that two rooms under the dining room be reserved for the use of such prisoners as shall be afflicted with any disease or any infirmity that may require such an accommodation. This order, though the situation of the rooms be changed, is confirmed by the regulations signed by Lord Mansfield in 1760. Though these rules were framed for the government of the prison in its antient state, yet the principle of affording medical relief, and establishing an infirmary, are acknowledged; and it is the opinion of Your Committee that these rules, being still in force, ought to be acted upon.

The Marshal informed your Committee that the prison is uncommonly healthy: and that during the 24 years he has filled his present office, he has never known a contagious fever. He added, moreover, that at present only one person was ill in the prison.

On inspection, however, the latter statement was found to be by no means correct; many prisoners were discovered to be afflicted by sickness, and, if it had not been for the charitable assistance of *Mr Burnet*, a surgeon, and then a prisoner, would have been in absolute want of all medical relief.

Mr Burnet has been a prisoner for four or five months. He informed your Committee, that he gratuitously attends the sick, and that he has occasionally furnished money to purchase medicines for those that require them. On the 16th of March there were under his own attendance about 30 sick prisoners; and he said that he had been professionally employed since 6 o'clock in the morning. He adds also, that within his own knowledge the distresses are very great, for want of medical relief; and that it is his opinion, that the dirt and filth suffered to accumulate at the back of the prison are greatly prejudicial to the health of the prisoners. *Mr Burnet* never heard of medicines, or soup, or slops, being given at the gate to sick prisoners; but upon being asked if he had ever known of an application being made and refused, replies, 'I am sure they give nothing. There is a poor Dutchman, unwell of a jaundice, and dying of a liver complaint; he is well known; he is sitting on the cold stones: if I had any place to put him in, I would do it at my own expense.' Several members of your Committee observed this miserable Being, who was certainly a fit object for an hospital; and in the room where he lived, there were five men, two women, and two children.

Mr Audley, one of the prisoners, when examined by your Committee, affirms he has seen the same man leaning against the state house, close to the gate, apparently very ill. He complained he had no friends, nor no money, only a few pence in his pocket; and that on Sunday he had nothing to eat but a piece of dry bread. The same evidence informed your Committee, that a prisoner of the name of *Rogerson* had been ill a long time of the rheumatic gout; that he is in indigent circumstances, lodging in a room with 5 or 6 other persons, that he had nothing but what was furnished by the charity of individuals: he had received an old rug from the officers of the prison, and lately some relief-money from the same quarter.

Upon these cases Your Committee make no comment; they feel it quite unnecessary to prove, by reasonings, that medical attendance and relief under the calamity of sickness and disease, are claims which the poor prisoner has to make from the laws of his country which incarcerate him. Surely there is no situation more deplorable than that of a man shut up in a crowded prison, suffering under diseases, perhaps caused by want, and the unwholesome air he is condemned to breathe, and unable from poverty and distress to provide himself with medical advice or the medicines essential to his cure.

Your Committee would feel they were not doing their duty, if they did not recommend in the strongest manner the establishment of an infirmary, and the constant attendance of a physician or surgeon, to be paid out of the county allowance, which, if not sufficient for that purpose, ought to be increased.

Your Committee wish to bear the testimony of their praise to the humanity and charity shown by *Mr Burnet*, who appears in the evidence, though a prisoner, and poor, to give gratuitously his time and skill to the sick and needy, and, as far as his own means allow him, provides medicine for those whose distress prevent them from so furnishing themselves.

Your Committee have called for a return of the number of deaths within the prison: they amount, in 10 years to 96. *Mr Audley* informed them, that a prisoner, by the name of *Samuel Roysen*, who was chummed in the same room as himself, was taken ill on the 2nd of February, and languished, confined to his bed, to the 17th, when he died. *Mr Audley* mentioned repeatedly the illness of this man, but no notice was taken; no one came to see him, but the surgeon *Mr Burnet*, who gratuitously attended him. The witness repeatedly pressed to be removed out of the room, but without avail. During the period of the illness, a third person was chummed on the room, though at that time but a small part of the prison had three persons in a room. After much remonstrance with *Morris*, the chum-master, the new chum, sleeping in the room with the sick man for two nights, was removed. The witness slept in the same apartment all the time, from the 2nd to the 17th. No medical relief, or nourishment, or attention of any description, was afforded; 'nor no more notice taken of the sick man than if he was a dog'. As soon as *Roysen* died, which was on the Friday evening, information was given at the gate; but it was not till Monday that the coroner viewed the body. The witness slept one night in the room with the dead body, and must have continued so to do until it was removed for interment, if a fellow-prisoner had not given him permission to shift his bed into the adjoining room. During the illness of *Roysen* the room became so offensive, that myrrh and frankincense were burnt, to remove the smell.

Mr Brooshooft being examined upon this subject, said, that persons whose timidity might make them desirous to be removed out of a room where a dead body lay, might be so enabled, upon application. He added too, that the body is always left in the room until it is removed for interment; and that though he cannot say that the person chummed in the room has continued to lie there during that period, yet he can say they were not chummed out by him.

Your Committee find, in a Report of a Committee of this House that sat in 1791, cases of a similar nature. That Committee forbore to comment upon them, having received assurance from the then Secretary of State, the late Lord Melville, 'that provision was made for the purpose of preventing such miseries in future'. From 1791 to 1814 these miseries have continued; and Your Committee, in detailing with these transactions, confidently hope, that those, whose duty it is to put a stop to a practice so disgraceful, and who have under their care the health and comfort of so many of the King's subjects, will lose no time in remedying an evil that so loudly calls for redress.

The discipline of the prison, as connected with its morals as well as the comfort and security of the prisoners confined, comes next under the consideration of the Committee. The first day they visited the King's Bench, the escape of *Lord Cochrane* was announced to one of their members. By what means *Lord Cochrane* escaped from the prison, has not yet been ascertained; but he got out on the

Sunday or Monday morning, and on Thursday, when the Committee met at the King's Bench, the Marshal and his officers were unacquainted with the event. It seems then evident, that a great neglect must exist somewhere. That a prisoner should escape at all, implies inattention or neglect of duty in the officers who held him in charge; but that any one, of the rank and quality of *Lord Cochrane*, should be out of the prison for three days and nights, or more, and the Marshal of the King's Bench owe his information on the subject to a Committee of the House of Commons sitting accidentally within the walls of that prison, is so extraordinary, that, if it had not happened, the possibility of such an occurrence would have been disbelieved by everyone. The Marshal and his officers were questioned upon the subject of escapes; they declared them to be extremely rare; that no record was kept by them, but all that was done was the payment of the debt by the Marshal. To what extent the number of escapes has really amounted, or what is the general standard of remissness on the part of the officers of the prison, Your Committee have no means of judging; all that they remark is, that two have taken place within a month, and that in both cases the event was communicated to them before it was known to the Marshal. The 2nd case to which they allude, is that of a prisoner by name of *Underwood*, who is said to have volunteered on board the tender; a person was procured, who swore a debt against him; upon this he was arrested and brought to the King's Bench, from whence he escaped, supposed in the disguise of a sailor. *Underwood's* attorney had got his discharge, but did not use it, thinking he might as well lie in the prison as on board the tender. No search was made in the prison after him, and no communication to the Marshal of his escape, though he is liable under an Act of Parliament, in the penalty of £100, for the safe delivery of the person to His Majesty's officer, on the discharge of the debt. This case carries with it much suspicion. The party was under age. *Mr Brooshooft* did not think the escape an event of sufficient importance to report it to the Marshal, as there was no debt to be paid, the action being friendly and fraudulent, and relying on the remissness or forbearance of the Government not suing for the penalty of £100.

By what means escapes are prevented or discovered, it is not easy to determine; it is in evidence that no examination takes place of the prison, or of the prisoners; nor is there any roll-call or muster, daily, weekly, or monthly. On the Monday of every week, the chum-master goes round to receive the rents; but if he be paid by another prisoner, or in the court-yard, or on the staircase, he does not think it necessary to enter each room; and upon inspection of the book, it appears in a great variety of instances that the chamber rent has not been paid for many weeks together. So that as far as the officers of the prison are concerned, a prisoner might be absent a month, or for a year, or longer, before they found it out.

In respect of the morals of the prison, drunkenness is most common, and there is a constant sale of spirituous liquors. *Mr Jones* says, he has information of ten places where they are sold; and another witness states, that on the staircase where he resides, there is a sort of public-house kept, where spirits are not only sold to be carried away, but that people assemble there of an evening to drink

them; that he is disturbed by the riotous intoxication in his vicinity, and that, from the strong smell of gin, if the officers visited the room, they must discover the practice.

Of the extent to which gambling is carried, the Committee have not learnt; for they observe, that though complaints are readily made on matters where the rules or want of rules of the prison affect the comfort of the prisoners; yet where their own indulgences are concerned, they are not so communicative.

They do not complain of extravagance of living, or criminal dissipation, or those habits of self-indulgence by which creditors are defrauded, properly dissipated, and morals tainted. There is however a practice, which, though it would be injudicious altogether to prevent, might still be put under regulation. The practice alluded to arises out of the racket-grounds. For a long period, a portion of the court has been divided into four racket-grounds, which are held by six masters; the successor to each ground pays something to the first occupant, on his quitting the prison, partly for the purchase of the rackets, and partly for the goodwill. One of the present racket-masters paid as large a sum as 6 guineas. A small fee is paid by the players: and though it be said that anyone may play there who chuses (*sic*), whether he pays or not, yet it is also said that this circumstance hardly ever happens, there being a point of honor (*sic*) felt by the prisoners upon this subject. The objection seems to be, the permission given to strangers to play there; it is an open place, where any one may play, and as such it may become the resort of the idle and the dissolute. As an exercise and amusement to the prisoners it is unobjectionable; as a place of resort, from the opportunity it may furnish to gambling, it is fitting that some regulation should be adopted to limit the practice as at present existing.

Your Committee see, by a rule of Court in 1781, that the Marshal is ordered not to suffer the wives and children of any of the prisoners to lodge in the prison, under any pretence whatsoever. On the 13th of March, 1815, there were 180 women and children, not prisoners, residing within the walls of the King's Bench. Your Committee, while they deeply feel for the condition of those whom our laws sentence to imprisonment for debt, cannot but remark, that in a prison often too crowded by the number of prisoners who are lodged therein, the addition of 180 persons must augment every inconvenience to a most alarming degree. Independent of the bad consequence thus arising from the over-crowding of the prison, the result of this indiscriminate admission of women and children of all ages and descriptions, there are moral evils connected with the subject that strongly weigh on the minds of Your Committee. It is in evidence, in the Report of a former Committee of the House, that a system of early prostitution was the effect of the introduction of female children within the walls, thus exposing them to the contaminating manners of a prison. The inquiries of Your Committee have satisfied them, that, as it is at present administered, the King's Bench exhibits scenes of vice and debauchery, the contemplation of which must aggravate these evils. Women of all descriptions are here freely and without inquiry admitted; and men, women, and children, sleep indiscriminately in the same room. Though the separation of a man from his family may fall hard upon some individuals, yet in balancing the good against the evil, the advantages against the disadvantages, Your Committee are decidedly of opinion that the positive rule of the Court ought not to have been disobeyed, but on the contrary it ought to be acted upon. For certain hours each day, the longest time that can be given, the friends or family of a debtor ought freely to be admitted to him; but no stranger ought to be permitted on any account to sleep within the walls, except on the special leave of the Marshal, the permission, and the reason for giving it, being entered in a book and shown to the Visitors. The prison is ill lighted, and, even when the Committee saw it, extremely dirty; though they were informed, that for some months it had not been so clean. It smells not only from the sewers, but from the piles of dirt heaped up behind the prison, which are offensive and unwholesome. The scavengers are paid by the Marshal, but they make a profit by the dirt, and of course only take it away when the quantity repays the labor of removal. They sell also the urine, which is collected in tubs, the smell of which is generally complained of. Your Committee recommend the prison to be better lighted, at least one lamp to be in each staircase; that the scavengers should be made to do their duty, and remove all the dirt out of the prison weekly; and that the urine should be carried off by an underground drain. At present no attention is paid to cleanliness in the prison; each individual is as cleanly or as dirty as to him seems meet, and the apartments are kept according to the habits of the occupier; there is no attention paid by the officers of the prison to these particulars.

Among the fees levied in the prison, is a tax of one shilling a week taken from the butchers selling meat within the prison, by the scavengers; of this the Marshal knew nothing. It is an extortion; the establishment of the butchers in the prison is for the general benefit of the prisoners, and the individuals should not be taxed at the discretion of the servants of the Marshal.

The prison is managed and governed by certain rules laid down by the Judges of the King's Bench, etc. which are called Rules of Court . . . The Marshal is a magistrate, but acts only as such within the district of his own prison. He has the power of committing riotous and disorderly prisoners to Horsemongerlane, which however is but seldom exercised. The places of confinement within the walls of the prison, are called the Strong-rooms: there are two of them, and are both of the same size, one floored with stone, the other with wood; but both without fireplaces, and without glass to the windows, which open into a small court about 6 feet wide. In these miserable places persons have been shut up for months. Those who have beds are permitted to bring them, but to those who have none, straw is furnished. It is in this place that *Lord Cochrane* has been confined for above three weeks; the windows were however sashed and glazed on the third day after he came into it, and the room carpeted; yet the cold and damp and offensive smells so affected his health, as to render his removal necessary.

Your Committee observe in the examination of *Mr Brooshooft*, and *Morris* the chum-master, that it is no uncommon circumstance for persons to be confined in this room for weeks and months: one individual was so shut up for 42 weeks; and though the witnesses seemed to speak most unwillingly upon this transaction, yet enough has been extracted from them to satisfy Your Committee that the privations and sufferings endured by that person were most considerable. *Mr Jones* informed Your Committee, that no one was ever confined there, except for a few days. They cannot suppose he wilfully concealed from them the fact above mentioned; and they must therefore consider this circumstance as furnishing another instance of his entire ignorance of all that is transacted within the walls of the prison.

Your Committee are of opinion that these strong rooms are not fit places for the confinement of debtors, or indeed of any other class of prisoners, for any length of time. They strongly recommend the selection of some other apartment in the prison, to be used for the purpose of temporary confinement, against which the same objection cannot be urged.

Your Committee would suggest, that whenever the necessity for punishing any prisoner in this or any other way should occur, it would be advisable that the Marshal should immediately report the reasons for such to the Visitors; and that the period of confinement should not be extended beyond a week, without their special authority.

The prison is visited once or twice a year by the Master of the Crown Office, and some others connected with that department. Your Committee called before them *Mr Barlow*, who had visited the prison many years in company with Mr Templer, the late Master.

No record is made of such visitation, and no minute entered in any book kept for that purpose. If any complaints are brought before the Visitors, a report is made to the puisne judge of the King's Bench. The duty of these Visitors is only to see that the rules and regulations are observed: and in respect of complaints, *Mr Barlow* does not remember that any have been made since the time of Lord Mansfield.

You Committee examined also *Mr Lushington*, who, as Master of the Crown Office, visited the prison for the first time last year, he having been recently appointed. He made a report (but of which he had no copy) and delivered it to Mr Justice Dampier. The principal objects of that report were to complain of a contravention of the order of 1729, which forbids persons committed for any criminal charge, having any receipt of the charities, or of being concerned in the distribution, as well as to recommend a book being kept of the charities, and the amount. Nothing has yet been done upon those subjects; the same man *Ledwell*, has the receipt and the distribution of these charities, and no account is kept of them. *Mr Lushington* also recommended that the rules and orders of 1729 should be hung up in the prison.

Your Committee are of opinion, that a mode of visitation going much further than that which has hitherto been practised, should be considered as the duty of the Visitors.

They recommend, that they should visit occasionally without previous notice, as well as at stated times; that a book should be kept in the prison for the purpose of their writing the observations as to the condition in which it then was; the number of prisoners; the cleanliness or dirt of the prison; the observance or breach of the rules and orders; the state of the discipline; and lastly, that these visitations should at least take place once a quarter, and the report of the same be made in writing to the Chief Justice of the King's Bench.

Your Committee having inspected the prison, and sat several days within its walls, and having heard evidence from all those who were desirous of being called before them, cannot conclude this part of their Report, without offering some further observations upon those arrangements within the prison, and the mode by which it is governed and administered, which appear to them to call for the interference of the Legislature and the Government.

The King's Bench prison cannot conveniently hold more than 400 persons: *Mr Jones* indeed says it is capable of containing 500; but, considering the size of the rooms, and the necessity of having some single apartments for the accommodation of persons under sentence of the courts of justice, Your Committee conceive that two prisoners in each room are as many as ought to be so placed. Some further accommodation is therefore wanting, as during the last year, nearly 600 persons have at one time been confined within the walls of the prison.

Your Committee having observed, that it is not unfrequent for persons to prefer remaining in the King's Bench after they have become entitled to their discharge, are of opinion, that the Marshal should take such measures, as soon as he could conveniently so do, for removing those individuals, as their stay must increase the many inconveniences which arise from the want of room within the prison.

They are also of opinion, that a separation ought to be made between the male and female debtors. That separation takes place in all the well regulated gaols in the kingdom, and Your Committee believe that the corruption of manners prevailing in the King's Bench is much augmented by the system which now exists in that prison.

The mode by which the Marshal is remunerated, appears to be most objectionable. The sum itself is great; amounting, on an average of the three last years, to a net income of £3,590 per ann. A minute detail of the profits and disbursements is contained in the Appendix to this Report. It is the decided opinion of Your Committee, that the mode of remuneration arising out of fees paid by the prisoner, and by emoluments made at the expense of those who are confided to his charge, is the most objectionable means by which a salary is given to the keeper of a prison. His interests are thus set invariably against his duty; and his profits are made at the expense of those whom the law supposes to be pennyless (*sic*), and whose property belongs not to themselves, but to their creditors. The Marshal of the King's Bench and his officers are public servants, and as such they ought to be paid at the public expense. The allowing a gaoler to make a profit from coffee-houses and taverns, and on the sale of beer and wine, within the prison, is also most objectionable. He thus becomes interested in the promotion of drinking; and his emoluments increase in the degree that sobriety and good conduct cease to exist among those whom the law places under his care. An Act of Parliament declares, that no keeper of a prison shall be a publican; why then

is there to be an exception made in favour of the Marshal of the King's Bench? Your Committee observe, that during the last year, 540 butts of porter and 65 barrels of ale, were consumed within the prison; the profit on which gave the Marshal a net emolument of £872.

Your Committee are however aware that the marshal holds his office and emoluments for life, as long as he shall behave well; they therefore recommend that some arrangement should take place, by which a just compensation should be made to him; that the fees exacted from the prisoners should be put under some regulation, or abolished altogether; and that the mode by which the gaol is supplied with malt liquor should no longer continue. The Marshal should have a control on every thing, a personal profit on nothing.

The Marshal, *Mr Jones*, from all that appeared to Your Committee, is not disposed to press upon the poverty of the class of persons whom he has under his charge; the evidence before them however warrants Your Committee in stating it to be their opinion, that he is little acquainted with what occurs in his prison; he avows, himself, that he seldom or never enters within its walls; and the numerous contradictions given to his evidence by others, who are in the constant habit of seeing and hearing what is really going on, would have left no doubt of that being the case, even if he had not acknowledged it. A keeper of a prison, receiving from it a net annual income of £3,590 and not daily inspecting and visiting the prison, and not being personally acquainted with all that is transacted therein, seems to Your Committee to prove the existence of a state of things that ought not to be suffered to continue.

Mr Brooshooft, though only the private clerk of the Marshal, is in fact the keeper of the prison; he directs and manages every thing, though he has no legal appointment, and consequently is vested with no legal authority.

Mr Morris, the chum-master, seems to be a most improper person to have the management of any part of the prison. It is proved before your Committee, that he took money (which was divided among the turnkeys, all of whom are therefore equally criminal) for the purpose of obtaining from *Mr Brooshooft* permission to do that, which all parties knew the Marshall had forbidden to be done. His whole evidence is shuffling and prevaricating; and your Committee would have felt themselves justified in reporting him to the House, if they had not attributed his conduct as much to arise from his ignorance as from wilful and criminal prevarication. There is enough in the evidence to justify the opinion, that to ensure the due execution of the duties of the officers of the prison, the attendance of the Marshal should be permanent, steady, and uniform; and that the management of the prison should not continue any longer in the present state.

It appears therefore on the whole to your Committee most advisable, that in case the Judges of the King's Bench should not have time to enter into a minute investigation upon the subject, that a commission should issue from the Crown to form rules and regulations for the better government of the prison of the King's Bench; that the plan should be submitted to parliament, and that some legislative enactment should take place upon the subject.

Patents Filed by Marc Brunel

	Specification	**Number and date**	
(a)	Machine for writing and drawing ('Polygraph')	**2305**	11.4.1799
(b)	Ships' blocks	**2478**	10.2.1801
(c)	Trimmings and borders for muslins, lawns & cambrics	**2663**	25.11.1802
(d)	Saws and machinery for timber sawing	**2844**	7.5.1805
(e)	Cutting veneers	**2968**	23.9.1806
(f)	Circular saws	**3116**	14.3.1808
(g)	Boots and Shoes	**3369**	2.8.1810
(h)	Sawmills	**3643**	26.1.1813
(i)	Rendering leather durable	**3791**	12.3.1814
(j)	Forming drifts and tunnels underground	**4204**	20.1.1818
(k)	Tinfoil manufacture	**4301**	5.12.1818
(l)	Stereotype printing plates	**4434**	25.1.1820
(m)	Copying presses	**4522**	22.12.1820
(n)	Marine steam engines	**4683**	26.6.1822
(o)	Gas engines	**5212**	16.7.1825

In many cases the sketches that accompany the patent specification are too large to be included in this volume and a section has been substituted. The complete full size drawings are held at the Patent Office in Newport.

A.D. 1799 . . . . . . N° 2305.

Machine for Writing and Drawing.

BRUNEL'S SPECIFICATION.

TO ALL TO WHOM THESE PRESENTS SHALL COME, I, MARC
ISAMBARD BRUNEL, late of the Parish of Saint Mary, Newington, in the County
of Surrey, but now of Canterbury Place, in the Paris of Saint Mary, Lambeth,
in the said County, Gentleman, send greeting.

5 WHEREAS His present Majesty King George the Third did, by His
Royal Letters Patent under the Great Seal of Great Britain, bearing date at
Westminster, the Eleventh day of April, in the thirty-ninth year of His reign,
for Himself, His heirs and successors, give and grant unto me, the said Marc
Isambard Brunel, my exŏrs, admŏrs, and assigns, His especial licence, full
10 power, sole privilege and authority, that I, the said Marc Isambard Brunel,
my exŏrs, admŏrs, and assigns, from time to time and at all times hereafter, during the term of years therein expressed, should and lawfully might
make, use, exercise, and vend, my Invention of "A CERTAIN NEW AND USEFUL
WRITING AND DRAWING MACHINE, BY WHICH TWO OR MORE WRITINGS OR DRAW-
15 INGS, RESEMBLING EACH OTHER, MAY BE MADE BY THE SAME PERSON AT THE SAME
TIME;" within that part of His Majesty's Kingdom of Great Britain called
England, His Dominion of Wales, and His Town of Berwick-upon-Tweed;
in which said Letters Patent is, among other things, contained a certain
proviso, that if I, the said Marc Isambard Brunel, should not particularly
20 describe and ascertain the nature of my said Invention, and in what manner
the same is to be performed, by an instrument in writing under my hand
and seal, and cause the same to be inrolled in His Majesty's High Court of

---

Chancery within one calendar month next and immediately after the date of
the said Letters Patent, then the said Letters Patent, and all liberties and
advantages thereby granted, should absolutely cease, determine, and become
void, any thing therein-before contained to the contrary thereof in any wise
notwithstanding, as in and by the said Letters Patent will more fully and at 5
large appear, in relation being thereunto had.

NOW KNOW YE, that in compliance with and obedience to the said
proviso, I, the said Marc Isambard Brunel, do, by this instrument in writing
under my hand and seal, and which I shall forthwith cause to be inrolled in
His Majesty's High Court of Chancery, declare that my said Invention is 10
particularly delineated, described, and ascertained by the drawings and sections
thereof hereunto annexed, and by the letters and figures thereon respectively
marked, and the description thereof following (that is to say) :—

Figure 1 represents the machine as intended for making two writings or
drawings by the same person at the same time, in any situation where it can 15
be placed at a proper and convenient height for the person using it, under
which Figure the letter A denotes a desk with folding hinges, which, when the
parts fixed upon it are removed, may be folded together like a common portable
writing desk. B denotes a sliding board, divided in the middle so as to admit
its folding, and covered with rough leather or cloth, to prevent the papers 20
placed thereon for writing or drawing on from being too easily moved from
their places, upon which leather or cloth the lines as shewn in this Figure are
drawn, that the paper for drawing or writing on may be placed with greater
facility, so as to correspond properly with the situation of the pens intended to
write upon them. The intent of the sliding board is for the moving of both 25
papers written or drawn upon at the same time, and in an equal proportion, as
the person writing or drawing upon them wishes. Underneath the sliding
board is a hollow space, where the machinery above, when folded up, may be
placed, and the whole, when the desk is folded, locked up and secured. C, C,
denotes two ink bottles, about an inch and a half deep, let into the desk in such 30
manner as to slide up and down, in proportion to the height at which the pens
are required to be used, according to the thickness of the books or papers
intended to be written or drawn on; b, b, represents the one a book, and the
other a paper, to be written or drawn on. D, D, an oblong square frame, fixed
with hinges to and supported upon two uprights E, E, which rest upon the 35
desk, and have two pins or tenons proceeding from the bottom of each, and
let into the desk to render them steady, but admitting of their being taken
off and put on at pleasure. F denotes a thin brass plate, sunk edgeways into
one of the upper sides of the frame, and even with the surface thereof, for the

---

wheels 1 and 2 to run on the upper edge of the said plate, there being a groove
made in the frame on each side thereof, to admit of their running upon the
edge of the said plate without touching the frame. G, G, represents two
stoppers, which turn, and when elevated, prevent that part of the machine
5 which runs on the frame from running too far and falling off. That part of
the machine which runs on the frame, and is called the autograph, consists of
the following parts, which are denoted as follows, namely :—J denotes an
oblong exterior frame, to which are fixed four metal wheels, 1, 2, 3, 4, (see
also Figure 2,) the edges of 1 and 2 being grooved like the sheaves of pullies
10 and running on the edge of the brass plate F, which makes the autograph
move in the same right line when drawn from one end of the frame D to the
other, 3 and 4 run on the opposite side of the frame. K denotes an oblong
interior frame, fixed within the exterior one J by two pivots (for which pivots
see L, L, Figure 2), at the end thereof, upon which pivots the interior frame
15 librates freely. M, M, M, M, denote two metal pieces called beams, as the
motion of each, in some degree, resembles that of a beam for scales; the said
beams are fixed within and to the inner part of the interior frame K, by four
pivots a, a, and two others opposite to them (see also Figure 2), upon which
they move or librate freely; the inner ends of the said beams are connected
20 together by a pin N passing through them, the one having a tongue and the
other a slit to receive it, are put the one into the other; the hole made to the
tongue, and through which the pin N passes, should be oblong, in order to
admit a free motion when the beams librate. O, O, denote two metal pieces,
depending from the interior sides of and near the outer ends of the beams
25 M, M, M, M, and fixed thereto, as seen in Figure 2 at e, e, e, e, by pivots or
screws, upon which they swing freely in some measure like a pendulum. P, P,
represent two flat metal sliding barrs, with two metal racks f, f, screwed to the
outer sides thereof; g, g, denote two cocks screwed to the pins o, o, serving to
keep the pieces P, P, close to the pieces O, O, and supporting two pinions, (see
30 also S, S, Figure 2,) by which, by turning two milled heads R, R, fixed to them,
the racks are moved up and down. h, h, two metal clamps fixed to the lower
extremity of the pieces O, O, and enclosing part of the sliding bars P, P, on each
side, to keep them close to the pieces O, O. Q, Q, denote two metal springs
fixed to the outer side of the pieces O, O, pressing against the barrs P, P, to
35 stiffen their motion as they slide up and down. These springs extend in length
a little below and within the upper edge of the clamps h, h, to a little above
and within the lower edge of the cocks g, g, the pieces O, O, having their
exterior sides cut in sufficiently to receive the said springs Q, Q. T 1 and T 2
represent two sockets, having projecting parts with a slit in each to receive the

---

lower end of the racks, which are fixed therein, with a screw passing through,
upon which they turn stiffly, in order to set the pens or pencils at any inclination required. U, U, denote two pen cases, into which the pens, pencils, or
any other instruments for drawing or writing are put tight. The cases slide
up and down freely in the sockets, and have a small pin near the head which 5
is received into a slit in such socket at the top thereof to prevent their turning
round and also falling through when the sockets are raised. V, a rod extending from one of the sockets to the other, and fixed with pins to the above-
mentioned projecting parts of the sockets, on which pins the rods have a free
motion. m, m, m, m, (see also m, m, Figure 2,) denotes two metal rests, fixed 10
by their vertical sides to the inner sides of the interior frame K, to support the
springs W, W, which are screwed upon the horizontal sides of the rests under
the beams M, M, M, M; these springs support, by their extremities, the ends
of the beams; their power is just sufficient to counteract the weight of the
apparatus, depending from the exterior end of each beam, and prevent the 15
hand which holds the hand X from feeling any part, or, at least, but a very
small part, of such weight, when making the various movements with the
handle, which are requisite to writing or drawing with the machine. x, x, x, x,
in Figure 2, denote small wheels, which cannot be seen in Figure 1, near
the extremity of the beams, the ends of the springs bearing upon the edge 20
of these wheels, move with much less friction when set in motion. To
the under part of the socket T 1 is screwed a small brass arm k, (represented
of its full size in Figure 8,) in which the lower end of the handle X
which terminates in a ball is inserted, and forms with it a ball in a socket,
so as to enable it to play freely in all directions, the intent of which is 25
to give greater freedom of motion to the hand that holds the handle X. Z, a
wooden rod fixed to the exterior frame J of the autograph in order to move it
from one end to the other of the frame D, D, as the person writing or drawing
proceeds from one side of his paper to the other. In case the machine should
be made use of on a table without a desk, and not require to be often taken off, 30
then, instead of using of the rod Z to move it from one end of the frame to
the other, D, D, two strings fastened to the sides of the exterior frame J,
and passing from thence upon pullies fixed near to the stoppers G, G, and in
the direction of the dotted lines y, y, corresponding with and being fastened to
a pedal or treader under the table, which pedal or treader is put in motion by 35
the foot of the person using the machine, will communicate to the autograph
the same motion as the rod Z.

The second Figure represents the upright E, E, and the autograph as folded
up in order to be put into the desk; the springs W, W, being shortened, in

Brunel's Duplicate Writing and Drawing Machine.

order to shew parts of the machine which they would otherwise have concealed. In this Figure the letters, which are also to be found in Figure 1, denote corresponding parts.

Figure 3 represents in what manner the autograph might be used for writing or drawing without the assistance of the handle X, the person making use of it holding the case U 1 instead of the handle X. O, O, P, P, *f*, *f*, Q, Q, *h*, *h*, denote similar parts with those denoted by the corresponding letters in Figure 1. *a*, *a*, semicircular pieces of metal (see also *a*, Figure 5,) fixed to the lower part of the rack in the same manner as the projecting parts of the sockets T 1 and T 2, Figure 1. The pen case U 1 moves freely within the semicircular piece *a* 1 upon two pivots *b*, *b*; (see also Figure 4) the second case U 2 is fixed in a socket K, K, (see also *k*, *k*, *k*, Figure 4,) which moves very freely within the semicircular piece *a* 2, upon two pivots denoted by *b* 1 and *b* 2 in Figure 5. V, a rod extending from the piece *a* 1 to the piece *a* 2, and fixed to one of the outer sides of each piece by a pivot, as seen at *b* 1, Figure 5, upon which it turns freely. *g* denotes a second rod extending from the right side of the pen case U 1, to the right side of the socket *k*, *k*, *k*, of the second pen case U 2; one end of this rod is fixed by a pin to the pen case U 1, and the other end to a screw or pin fixed into the side of the socket *k*, *k*, *k* (see also *d*, Figure 4); the pen case U 2 slides freely up and down in the socket *k*, *k*, *k*, and has a similar pin near the head and for the same purpose as mentioned in describing the pen case U, U, Figure 1.

Figure 4 represents the front part of the socket *k*, *k*, *k*, together with the case U 2.

Figure 5 represents one of the semicircular pieces *a* 1 and *a* 2, and the manner in which the rod V is fixed to them, the other parts in the Figures 4 and 5, corresponding with those exhibited in Figure 3, are denoted by corresponding letters.

Figure 6 represents as much of the autograph as is necessary to shew what alterations and additions are made to it when more than two pens or other instruments for writing or drawing are fixed to it and in what manner they are applied, the letters which correspond with those in Figure 1 denoting corresponding parts. For this purpose the rod V is fixed near the lower end of the pieces O, O, in the same manner as it is fixed to the projecting parts of the sockets T 1 and T 2, Figure 1; then a second rod *q* is fixed in the same manner as the one V, and above it, near the middle of the pieces O, O. The rods V and *q* in this Figure are longer than the rod V in Figure 1, as three papers to write on, if of the same size, will necessarily require a greater length than two, in which case the other part of the machine, were necessary, must be

Brunel's Duplicate Writing and Drawing Machine.

lengthened in proportion, but these rods may be made longer or shorter, and the rest of the machine larger or smaller according to the uses it is intended to be applied to when made. O 3 denotes a metal piece of the size, length, and form of that part of either of the pieces O, O, which extends from the lower end thereof to the upper side of the second rod *q*, which piece O 3 is fixed on the two rods V 5 and *q* by means of sliding sockets *t*, *t*, one of which is exhibited in its full size in Figure 7; the rods V and *q* pass through these sockets, which slide stifly on them, in order to set the third pin to such a situation as may be most convenient for drawing or writing with it; the piece O 3 has fixed to it a sliding bar P 3, and rack *f*, *g*, like those denoted in Figure 1, with a third socket and 10 pen case like those denoted by T 2 and U, U, Figure 1, and is itself fixed to the sockets *t*, *t*, by pins passing through it and one side of each socket so as to admit of its truly turning upon such pins. A third ink bottle of the same form as those described at C, C, is set in a situation corresponding with the third pen in the same manner as the two before mentioned; if a greater number of pens or instruments for writing or drawing at the same time are required, an apparatus similar to that described in Figure C is fixed to the rods V and *q* for each.

Figure 9 represents a stand, independant of a table or desk, in order to make use of the autograph on any table or desk to which there is no sliding board. 20 The uprights E, E, which support the frame D, and with it constitute a stand for the autograph, are fixed the one upon a wheel, and the other upon a board which contains the ink-bottles C, C, under which board are fixed two other wheels, and by means of these three wheels the whole is moved backward and forward as required. The uprights E, E, are kept steady at their upper part 25 to the frame D, either by pins or screws, admitting however their being taken out at pleasure.

In order to use the machine the person who is to write or draw with it should first fix the pens or other instruments he intends to write or draw with tight in the pen cases U, U, see Figure 1; then place the books 30 or papers he means to write or draw on as flat as possible upon the sliding board B by the assistance of the lines drawn thereon, so as to make the parts of each which are intended to be written or drawn on correspond with the pens or instruments so fixed, as above mentioned, which are meant to write or draw upon them. He should then, by turning the milled heads belonging 35 to the pinions R, R, raise or lower such instruments according to the thickness of such books or papers respectively, so as to make each press moderately upon the book or paper under it. The ink-bottles, having a sufficient quantity of ink in them, should be next raised and lowered to the height of the pens opposite

Brunel's Duplicate Writing and Drawing Machine.

to them, or moved to the distance required to correspond with the situation of the pens, so as to enable them conveniently to take ink. The person meaning to write or draw should then take the handle X, Figure 1, in his hand in the same manner as a common writing pen, and direct with it the pen or instrument next to him, writing or drawing as he thinks fit, when the other or others will make similar movements at the same time, and execute similar writings or drawings, as, on moving one of them in any direction, the other or others will move in similar direction, and when ink is taken with one, the other or others will likewise take ink at the same time. In proportion as the person writing or drawing proceeds from one side to the other of his paper or book he slides along the autograph upon the frame D, D, by gently touching as often as he finds it requisite the rod Z, which moves all the pens or instruments equally, and as he proceeds from the top of his paper to the bottom he slides from him the sliding board B (or draws to him the stand, if the machine is used as described in Figure 9) as often as he finds it expedient, the sliding board carries all the papers equally along with it. If the machine is fixed to a table there should, of course, be a sliding board in such table. When the machine is meant to be folded up and put into the desk, the rods V and Z are taken off, and the sliding bar P 1 Figure 1, with the apparatus fixed to it, is taken from the piece O, to which it belongs, and the remainder of the autograph being taken from the frame D, D, is folded up, as seen in Figure 2, the uprights E, E, are then taken from the desk and folded by means of a hinge to each, one of which is denoted by *v*, Figure 2, and the autograph placed within it, as represented in the same Figure, after which the whole may be put into the desk, which folds and is locked like a common portable writing desk.

In witness whereof, I, the said Marc Isambard Brunel, have hereunto set my hand and seal this Tenth day of May, in the year of our Lord One thousand seven hundred and ninety-nine, and in the thirty-ninth year of the reign of our Sovereign Lord George the Third, by the grace of God of Great Britain, France, and Ireland King, Defender of the Faith, and so forth.

Mᶜ I. BRUNEL.   (L.S.)

Sealed and delivered (being first duly stampt),
in the presence of
Robᵀ Bayly,
Gᴱᴼ Nelson, Mid. Temple.

Brunel's Duplicate Writing and Drawing Machine.

AND BE IT REMEMBERED, that on the Tenth day of May, in the year of our Lord 1799, the aforesaid Marc Isambard Brunel came before our Lord the King in His Chancery, and acknowledged the Specification aforesaid, and all and every thing therein contained and specified, in form above written. And also the Specification aforesaid was stampt according to the tenor of 5 the Statute made for that purpose.

Inrolled the Tenth day of May, in the year of our Lord One thousand seven hundred and ninety-nine.

LONDON:
Printed by George Edward Eyre and William Spottiswoode,
Printers to the Queen's most Excellent Majesty.

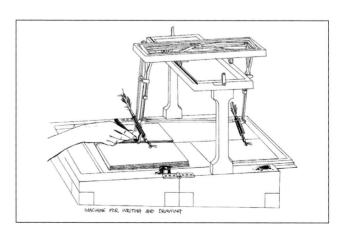

MACHINE FOR WRITING AND DRAWING.

A.D. 1801 . . . . . . N° 2478.

Ships' Blocks.

BRUNEL'S SPECIFICATION.

TO ALL TO WHOM THESE PRESENTS SHALL COME, I, MARC ISAMBARD BRUNEL, of Bedford Street, Bedford Square, in the County of Middlesex, Gentleman, send greeting.

WHEREAS His Majesty King George the Third did, by His Royal Letters 5 Patent under the Great Seal of the United Kingdom of Great Britain and Ireland, bearing date at Westminster, the Tenth day of February, in the forty-first year of His reign, for Himself, His heirs and successors, give and grant unto me, the said Marc Isambard Brunel, my exŏrs, admŏrs, and assigns, His especial licence, full power, sole priviledge and authority, that I, the said 10 Marc Isambard Brunel, my exŏrs, admŏrs, and assigns, from time to time and at all times thereafter during the term of years therein expressed, should and lawfully might make, use, and exercise, and vend, my Invention of " A NEW AND USEFUL MACHINE FOR CUTTING ONE OR MORTICES FORMING THE SIDES OF AND CUTTING THE PIN-HOLE OF THE SHELLS OF BLOCKS, AND FOR TURNING AND BORING 15 THE SHIVERS AND FITTING AND FIXING THE COAK THEREIN," within that part of Great Britain called England, His Dominion of Wales, and His Town of Berwick-upon-Tweed ; in which Letters Patent is, among other things, contained a certain proviso, that if I, the said Marc Isambard Brunel, should not particularly describe and ascertain the nature of my said Invention, and 20 in what manner the same is to be performed, by an instrument in writing under my hand and seal, and cause the same to be inrolled in His Majesty's High Court of Chancery within one calendar month next and immediately after the

---

Brunel's Machine for Cutting Mortices, and Forming and Cutting Blocks, &c.

date of the said Letters Patent, then the said Letters Patent, and all liberties and advantages thereby granted, should absolutely cease, determine, and become void, any thing therein-before contained to the contrary in anywise notwithstanding, as in and by the said Letters Patent will more fully and at large appear, on relation being thereunto had. 5

NOW KNOW YE, that in compliance with and obedience to the said proviso, I, the said Marc Isambard Brunel, do by this instrument in writing under my hand and seal, and which I shall forthwith cause to be inrolled in His Majesty's High Court of Chancery, declare that my said Invention is particularly described and ascertained by the Drawings and sections thereof 10 hereunto annexed, and by the letters and figures thereon respectively marked, and the description thereof following, that is to say :—

The whole machine is divided into four different parts :—The first represents an engine for cutting one or more mortices at the same time, and is exhibited in different views under the Figures 1, 2, 3, 4. Figure 1 repre- 15 sents the front elevation of the engine ; the second, its lateral section and elevation ; the third, the plan of the box or carriage in which the piece of wood intended to be cut is laid ; and the fourth, the shape of the cutters, upon a a scale of double the dimensions of that according to which the other parts are represented. Observation.—The piece of wood intended for making the shell 20 of a block ought to be well seasoned, that the exactness of the mortices may not be injured after they are cut by the warping of the wood. It should be cut to the length required for the size of the shell and squared nearly to its dimension ; it should be afterwards fixed in the box A, Figure 1, 2, 3, and a round hole for each mortice bored through it towards one end of it, either 25 before or after it is fixed. A, Figure 1, 2, 3, represents a carriage or box without a bottom, sliding horizontally and lengthwise between the two uprights B, B, Figure 1, and resting on the platform C, Figures 1 and 2. At the front extremity, and on one side of this carriage, are two screws F, F, Figure 3, to steady the piece of wood W in the carriage ; the same, when much smaller 30 than the carriage, being steadied by pieces of wood T, T, Figure 3, placed in the carriage on each side of it. When the wood is set in the carriage, which is drawn back for the purpose of cutting the mortices, so as to bring the holes already bored in the wood exactly under the cutters E, E, Figure 1, and E, Figure 2. These cutters are set into the square sockets F, F, Figure 1, and 35 F, Figure 2, and fixed firmly therein by screws a, a, Figure 1, a piece of wood b, Figure 1, is intended to keep the sockets at the required distance from each other ; they are fixed to the plate G, Figure 1, by the clamps e, e, and are adjusted as required by the screws d, d, d, d, which make them more horizon-

---

Brunel's Machine for Cutting Mortices, and Forming and Cutting Blocks, &c.

tally to the right or left. The plate G slides up and down between the two uprights B, B, in the brass grooves H, H, Figures 1 and 2 ; this plate is set in motion by the crank J, Figures 1 and 2, by means of the fork g, which is connected with bolts. This engine is intended to be moved by turning the 5 handle K, Figure 1 and 2 ; and in order to render the power applied for this purpose more equal, a fly-wheel of about five feet in diameter, and about two hundred and fifty pounds weight is fixed to the extremity of the spindle L. The cutters, when in motion, playing up and down begin to cut at first from the inner side of the holes previously bored, cutting at each stroke quite 10 through the wood from the upper to the under surface, the wood after each stroke being, by the forward motion of the carriage, made to slide lengthwise and horizontally an equal space in order to receive the next stroke, till the mortices are cut of the length required ; they are represented in Figure 3, as when cut half way. The forward motion of the carriage is given by a 15 double thread screw M, Figure 2, which is propelled by the rocket wheel N, Figure 1 and 2, which acts only when the cutters are raised above the surface of the wood. Behind the sliding plate G there is a pin O, see Figure 2, which sets in motion the lever P, Figures 1 and 2, which is connected to the spindle Q, Figure 2, with the lever R, Figures 1 and 2 ; at its lower extremity 20 is a pall S, Figures 1 and 2, which moves the rocket wheel N, Figures 1 and 2. The carriage advances an equal space every time the cutters rise, which space may be made greater or lesser by setting the pall S higher or lower on the lever R, according to the size of the blocks. The cutters are enabled by this motion of the carriage to cut at each stroke from $\frac{1}{30}$ to $\frac{1}{100}$th of an inch, 25 more or less, as the nature of the work may require. When the mortices are entirely cut, the piece of wood is taken off to make room for another, which when fixed and screwed in the carriage, the latter is drawn back by the assistance of the same screw which propels it, by turning the handle U, Figure 2, towards the left, after having taken off a stopper V, Figure 2, which 30 prevents the screw from turning when the rocket wheel acts. B, Figure 4, represents the front, and C the side view of our cutter out of the socket. This cutter is a kind of mortice chissel, round on the bevel side, of the size in proportion to the breadth of the mortice required to be made, the upper part of it is square and let into the hollow part of the socket. At 35 the bevel of the chissel there is a pin A, which clears the chips out of the mortice at every stroke. This pin, which goes into a hole in the chissel, is knocked out when the chissel is to be ground.

Second part represents the engine for sawing the sides or edges F, F, Figure 2, so as to give them the circular form required. Figure 1 represents

---

Brunel's Machine for Cutting Mortices, and Forming and Cutting Blocks, &c.

the front elevation of the engine ; second, the plate of the table W and the shell A, one side of which is nearly shaped ; third, the lateral section of the engine ; fourth, the rack and wheels by which the shell is moved in a circular direction ; and fifth, the spindle explained under the letter S, Figures 1 and 2. The mortices being cut, the shell A, Figures 1 and 2, is fixed within the two 5 arms B and C, Figures 1, 2, and 3, by the means of the pieces described hereafter and horisontal rail h, Figure 2, fixed on the upper extremity of two uprights t, t, Figure 2, forms a rest n, Figure 1, fixed to the left-hand side of the table W, W ; another rest m, Figures 1, 2, and 3, with a single upright is fixed to the right hand. These rests may be raised or lowered when required. 10 Before the fixing of the shell the rest H is turned towards the left as much as required to be at right angle with the horisontal rail h, Figure 2 ; a board n, Figure 1, being passed through one of the mortices, and the end of the shell, where the mortices are formed square, turned towards the plate a, Figures 1 and 2 ; the screw b is turned in order to keep the shell steady in an horizontal 15 direction between the arms B and C, Figure 1. The shell being fixed, the rest H is turned towards the saw O, which is then set in motion. At the upper extremity of the arm B is a screw b, Figures 1 and 2, to fix the shell ; and at the upper extremity of the arm C is a plate a, with several sharp points, which are also intended to keep the shell steady when pressed by the screw b, the 20 two arms B and C, are either contracted or extended to receive different sizes of shells, and held fast by the pin E passing through the clamp d, Figures 1 and 3, and the sliding barrs D, D, Figure 1, which are capable of being drawn backwards and forwards against each other, as the distance between the arms B and C is required to be increased or diminished, there being several holes 25 through the barrs at equal distance. The clamp d, Figure 1 and 3, is screwed upon a round plate F, Figures 1 and 3, and keeps the barrs fast and close together ; the plate turns upon the extremity G of the rest H, Figures 1, 2, and 3, to which it is fixed by a center pin screwed underneath, in order to change the sides of the edges of the shell when required. The plate F is pre- 30 vented from turning when the saw O, Figures 1 and 3, acts by a small pin which passes through it and the head G at some distance from its center, the small pin is taken out when the sides of the shell are to be changed. The rest H, Figures 1, 2, and 3, is fixed and screwed on the upper extremity of a spindle J, Figure 3, on which the rest, by means of an oblong hole through 35 it, which is in part represented at g, Figure 2, may be sled horizontally, in order to move the head G to or from the saw O. The plate X, Figures 1, 2, and 3, to which the spindle J is fixed, may be sled horizontally between the two side pieces Y, Y, Figure 2, when the spindle is to be set nearer to or farther from the saw O. On the spindle J, Figure 3, is an horizontal cog wheel K, Figures 3 and 4, which is set in motion by an horizontal rack L, Figures 3 and 4; this rack is moved by a pinion m, Figures 3 and four, which is connected by a spindle Q, Figure 4, with a vertical rocket wheel N, Figures 3 and 4. The saw O, Figures 1 and 3, is set in a frame P, P, P, Figure 3, which moves up and down between the two uprights Q, Q, Figure 1, which motion may be given to it either by a crank or a lever. The motion of the saw frame P, P, P, Figure 3, upwards and downwards, moves the arm R, Figures 3 and 5, which is connected by the spindle S, Figures 1, 2, and 5, to the arm T, Figures 1 and 5; the spindle S turns upon its ends. At the upper extremity of the arm T, Figures 1 and 5, is a pall V, Figures 1 and 3, which pushes the rocket wheel N, Figures 3 and 4, and, of course, gives the shell a circular motion proportionally to the cutting of the saw, when one side or edge of the shell is cut, as partly seen in A, Figure 2; the arms B and C, Figure 1, must be turned previous to the cutting of the other side; both sides being done, the shell is taken to the third part of the machine.

Third part.—The engine for the third operation consists of a common stout lathe, and the additional machinery herein-after described. Figure 1st represents the lateral elevation of the engine; Figure 2nd, its plan; Figure 3rd, its front elevation; Figure 4th, the eliptical pieces which shape the shell and the manner in which the shell is fixed when the angles of it are to be cut; and Figure 5th, the instrument for boring the pin-hole, and the manner of using it. A, Figure 1, a chuck in the head of the mandrill of the lathe by the center screw of the piece K, Figures 1 and 5. The shell, shaped as it is seen in a, Figure 3, is set between the two arms of the chuck a, and fastened at its extremitys by the pin B, Figure 1, fixed at the inner end of the pin B, and by the screw C, Figure 1. On the plate D, Figures 1 and 3, are several sharp points which are intended to keep the shell steady; and in order to set it true in the chuck, these sharp points are to be placed in the holes already made by the fixing of the shell within the arms B and C, Figure 1, Part the Second. The pin B is set in a socket F, Figures 1 and 3, in which it may be sled in or out according to the length of the shell, and to which it is fastened by a pin passing through both the pin B and socket F, there being several holes through the pin B at equal distance; this socket turns in the part H of the chuck A, Figure 1, in order to change the sides of the shell when required; G, Figures 1, 2, 3, and 4, a round plate, with several notches to serve as a dividing plate, fastened to the socket F; a stopper J, Figures 2 and 4, catching in either of those notches to hold the plate in the situation it is when one of the sides of the shell is turned towards

---

the cutter O, Figure 1, to be shaped at the inner extremity of screw. C, Figure 1, is a sharp center pin to hold the shell with steadiness, when the shell already shaped on two sides h, h, Figure 3, is fixed within the arms of the chuck A, to be cut to the form required. The first operation is boring of the pin hole Z, Figure 3; Figure 5 represents the borer r fixed upon the forks b, b, of the rest M, M, Figures 1, 2, 3, and 5; it must be observed that as the pin-hole is nearer to the end, Figure 5, than to the opposite end of it, the chuck is set as much from its center by the assistance of the adjusting screw N, Figures 1 and 5. The chuck is fastened on the plate K, Figures 1 and 5, by two bolts L, L, passing through holes both in the chuck and the plate, the holes through the chuck being oblong to admit of its sliding on the plate K by the assistance of the screw N, which turning in a collar k, Figures 1 and 5, fastened to the chuck, and passing through the nut m fixed to the plate K, moves the chuck to or from its center, as required, when the shell had been put off its center in proportion to its size. Then the center bit r, Figure 5, is set upon the forks b, b, of the rest M, M, Figures 1 and 3, and being held by the handle is used as in the common way for boring on a lathe, Figure 5, by pushing the borer towards the shell. The pin-hole being bored, and the chuck replaced to its center by the adjusting screw N, the sides V 1 and V 2 of the shell are then cut with a gouge or cutter O, Figures 1 and 2, which is set, and by a screw d kept firm, in the hollow part of the head e of the slider P, Figures 1 and 2. This slider slides lengthwise and horizontally within the rests M, M, Figures 1 and 2, which is screwed on the plate R, Figures 1, 2, 3. This plate R slides horizontally and crosswise on the frame Q, Q, Figure 1 and 2, by the power of a double thread screw F, Figure 3, fixed underneath and fastened to one extremity of the plate R by two coupling brasses g, Figure 3. The nut through which the screw works being fastened to the frame Q, Q, the turning of the handle S, fixed to the screw f, draws out or in the plate R, which carries with it the rests M, M, and the centre O; the iron frame Q, Q, Figure 1, slides lengthwise and horizontally on the frame of the lathe by the assistance of an adjusting screw h, Figures 1 and 2, fixed on the same principle as the one f, Figure 3; the coupling brasses being fixed on the frame o the lathe and the nut screwed to the frame Q, Q. The sliding of that frame is intended for the adjustment of the cutter O, according to the size of the shell. The sliding of the plate R in a direction horizontal and crosswise to the lathe is intended for moving the cutter or gouge O, fixed to the slider P, when the workman is forming either one of the sides of the shell. The slider P, Figures 1 and 2, is guided in its forward and eliptical motion to give the flat sides V 1 and V 2 of the shell the requisite form by the eliptical piece T 1, T 2

---

is used when the angles n, n, Figure 4, are to be cut. This eliptical pieces T 1 and T 2 are fastened on the frame Q, and may be slid horizontally in the direction of the plate R, as required for several sizes of shells. Through the socket V, at the extremity of the slider P, is a square barr having at its lower end a small roller W, Figures 1 and 4, and a knob q, Figure 1 and 2, at its upper end. This square barr is intended, when raised, to cause the roller W to bear on the outter edge of the eliptical piece T 1, Figures 1 and 4, and when lowered to bear on the edge of the piece T 2, Figures 1 and 4. The workman proceeds in the following manner to shape the sides of the shell:—The side V, Figure 1, is supposed to be the first formed. The workman turning slowly the handle S, Figures 1, 2, 3, and at the same time pushing the slider P towards the shell, in order to keep the roller W close to the edge of the eliptical piece T 1, moves the gouge or cutter O, which in its horizontal motion cuts and forms the surface V 1 of that side to the shape given by the eliptical edge of the piece T 1. The cutter O having operated as far as the size of the shell requires, it its drawn back to its place in order to repeat the same operation, forming the side V 2, which is previously turned towards the cutter O, Figures 1 and 2. Both sides V 1 and V 2 being thus formed, the angles are next to be cut, see Figure 4, which represents the shell a in a vertical position, and one of its angles turned towards the cutter O. Before these angles are begun to be cut, the roller W is caused to bear on the edge of the eliptical piece T 2, Figures 1 and 4, by pushing the knob q downwards till it touches the surface of the slider P. The angles are then cut in the same manner as the sides V 1 and V 2, observing, after each angle is formed to turn the next towards the cutter till they are all cut. The shell being then finished as far as it is intended to be by that part of the machine, is taken from the chuck, and another may be put in the place.

The fourth part represents the engine for making the shivers, and is particularly intended for coaked ones. Figures 1 and 2 represent the elevation of part of the lathe to which the machinery herein-after described is added; Figure 1 represents a lateral elevation, and Figure 2 a front elevation of it. Figure 3 represents the rest on which the tools are set, and the chuck seen from above. Figure 4 represents the front of the chuck; the two last Figures are upon a scale of double the dimension of that according to which the other parts are represented. Figure 5 represents the form of one end of one of the cutters, upon a scale four times larger than Figures 1 and 2. Figure 6 represents the one of the pieces H, H, H, Figure 4. Figure 7, a cutter, seen on two sides, intended for turning parts of the shivers; these two last Figures are upon the same dimensions as the

---

Figures 3 and 4. A and B, Figure 1, 3, and 4, represent the chuck on which the piece of wood or lignum vitæ L is set to be formed. A, Figure 1, 3, and 4 round and flat rim turning upon two centers C, C, Figures 1, 3, and 4, which are fixed by screws to one side of the rim at D, D, Figure 1 and 4. These centres, which are opposite each other, divide the rim into two equal parts, and are intended for turning it, and presenting either side towards the cutters, as required. B, B, Figure 1, a semicircular piece, fixed at its middle part to the mandrill of the lathe. The centers C, C, pass through and turn in the extremities E, E, Figure 1 and 3. F, a stopper, fixed at one end on the side, and at the middle of the semicircular piece B, B, to keep the rim A in its proper place, and to prevent it from turning upon the centers C, C; this stopper F has two notches at its other end to receive the pins G, G, Figures 3 and 4, which are let in the edge of the rim A. On one side of the rim and at equal distances, are three pieces of brass H, H, H, Figures 4 and 6, which are fastened at one of their ends by a center screw J, J, J, Figures 3, 4, and 6, passing through both the rim and brass pieces. The screws J, J, J are the centers on which the pieces of brass H, H, H turn, and the screws T, T, T, Figure 4, are intended to keep them close and fast to the rim; an oblong hole U, Figure 6, enables them to turn on their centers as much as may be required. These pieces of brass are intended to receive and to hold the lignum vitæ L. At the extremity of each piece H is a screw and a nut K, K, K, Figure 3, 4, and 6, the latter of which, being screwed upon the piece of lignum vitæ, fixes it tightly between the nut and pieces H, H, H. It must be observed that the piece under the letters Q, Q, R, S, f, g, and h are the same, and nearly identical for the same purpose, as those already specified under as the same letters in the third part; Q, Q, therefore is a frame which slides lengthwise by the assistance of a screw h, Figure 1 and 2, part the 3ᵃ, at the extremity of which a handle e, Figure 1 and 2, is fixed. R represents a sliding plate, moved by the assistance of the handle S and of the pieces f, g; the rest M, Figure 1, 2, and 3, is fixed upon the sliding plate R at the same place and in the same manner as the rest M, M, Figure 1. Part the 3ᵈ of this rest, Figure 1, represents the lateral elevation; Figure 2 the front elevation, and Figure 3 a view of it, seen from above. N, Figure 1 and 3, a spindle fixed and turning on the rest M, and having at its end m a hole to receive the cutter O, Figure 3, the extremity of which is formed as at O, O, Figure 5. This cutter is intended to cut the notches a, a, a, Figure 4, for the coak, the spindle N is also intended to receive a borer or drill to bore the holes through the coaks and the wood for the rivets which fasten them together. When this spindle is used, the check is stopped and kept steady by a piece of iron b, Figure 1,

fixed, at one of its ends to the frame of the lathe, and having at the other a pin *k*, which when the piece *b* is raised meet with either of the holes *l, l, l*, Figure 4, bored through the rim A.   The lignum vitæ L, Figure 1, 3, and 4, being fixed in the chuck, the first operation is the boring of the centre hole X,
5 Figure 4.   The cutter *d*, represented in Figure 7, being set and screwed on the part U, U, of the rest M, Figure 2 and 3, and its end *n* turned towards the lignum vitæ L, the workmen having at first set the chuck in motion, and, by the assistance of the handle S, slid the rest to the place required for the size of the center hole X, turns slowly the handle *e*, which turning the screw *h*,
10 gives to the cutter *d* an horizontal and lengthwise motion, which forces the cutter through the wood.   The center hole being thus bored, the notches *a, a, a*, Figure 4, are next cut, and for that purpose the chuck is stopped and steadied by the piece *b*, Figure 1.   The cutter O, Figure 3, being fixed to the spindle N, and set exactly opposite the center hole already bored, and
15 regulated according to the depth of the notches by the handle E, the workman proceeds in the following manner:—The wheel P, Figure 1 and 2, is set in motion in order to communicate it to the spindle N by the assistance of the string *t, t, t*, which turns round the wheel P, Figure 1, 2, and 3, fixed to the spindle N, passing over the two wheels *r*, Figure 1 and 2; the workman turns
20 the handle S, which moves the rest M, and of course the cutter O, in a crossways direction; the cutter then cuts sideways from the lignum vitæ as much as it is required for the depth and breadth of either of the notches *a, a, a*; their length is determined by the turning of the handle S.   One notch being cut, and the rest slid back to its place, the pin *k* is taken off from the rim, and the chuck
25 turned, till the next hole *l* meets with and receives the pin *k*; the second notch, or any other, is cut in the same manner as the first.   The three notches being cut on one side, and the stopper F, Figure 1 and 3, being taken off from the pin G, the rim is turned to present the other side of the lignum vitæ towards the cutter and the stopper F replaced.   The three notches being cut on this
30 side in the manner above described, the coaks are next put in place, and the cutter O being taken off, a borer is put to its place.   The holes being bored through, the coaks and the wood pins are put into the holes and rivetted in the common way; the chuck is again set in motion in order to finish the shiver, and the cutter *d*, Figure 7, used in the manner already described, in order to
35 cut the center hole through the coaks quite true; this hole being made, the faces of the lignum vitæ are turned one after the other by sliding the cutter *d*, adjusted at first according to the thickness required for the shiver, from the centre to the extremity; which being done, as far as the diameter of the shiver requires it, the cutter *d* is pushed forward by the handle *e* in order to cut the

lignum vitæ nearly through.   When this operation is done, the end *q* of the cutter is turned towards the lignum vitæ, and let in the groove already cut by the other end of the same cutter; then the workman sliding gently the rest towards the center cut the groove of the shiver to the exact form of the end *q* of the
5 cutter.   The shiver is then cut quite off by a sharp pointed cutter, turning previously the other side of the lignum vitæ towards the cutter.   The shiver being thus finished, a new piece of lignum vitæ may be put in its place; some graduations cut on the sliding plate R might assist the workman in finding the exact proportions of all parts and sizes of the shivers.

In witness whereof, I have hereunto set my hand and seal, the Ninth 10 day of March, in the year of our Lord One thousand eight hundred and one.

(L.S.)   MARC ISAMBARD BRUNEL.

Sealed and delivered, being first duly stamped,
in the presence of   15
GEO. NELSON, Palsgrave Place, Temple.
A. M^cDONALD, his Clk.

AND BE IT REMEMBERED, that on the Ninth day of March, in the year of our Lord 1801, the aforesaid Marc Isambard Brunel came before 20 our Lord the King in His Chancery, and acknowledged the Specification aforesaid, and all and every thing therein contained, in form above written. And also the Specification aforesaid was stampt according to the tenor of the Statutes made for that purpose.

Inrolled the Ninth day of March, in the year of our Lord One thousand eight hundred and one.   25

LONDON:
Printed by GEORGE EDWARD EYRE and WILLIAM SPOTTISWOODE
Printers to the Queen's most Excellent Majesty.

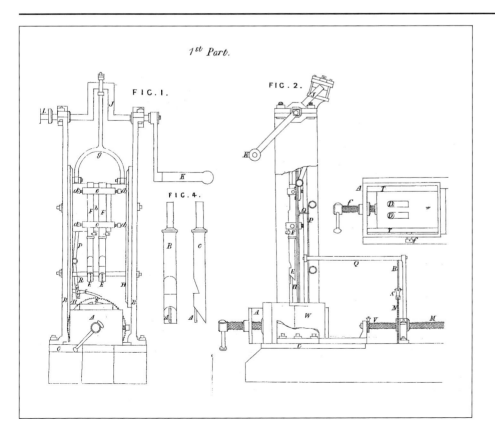

1st Part.

FIC. 1.

FIC. 2.

FIC. 4.

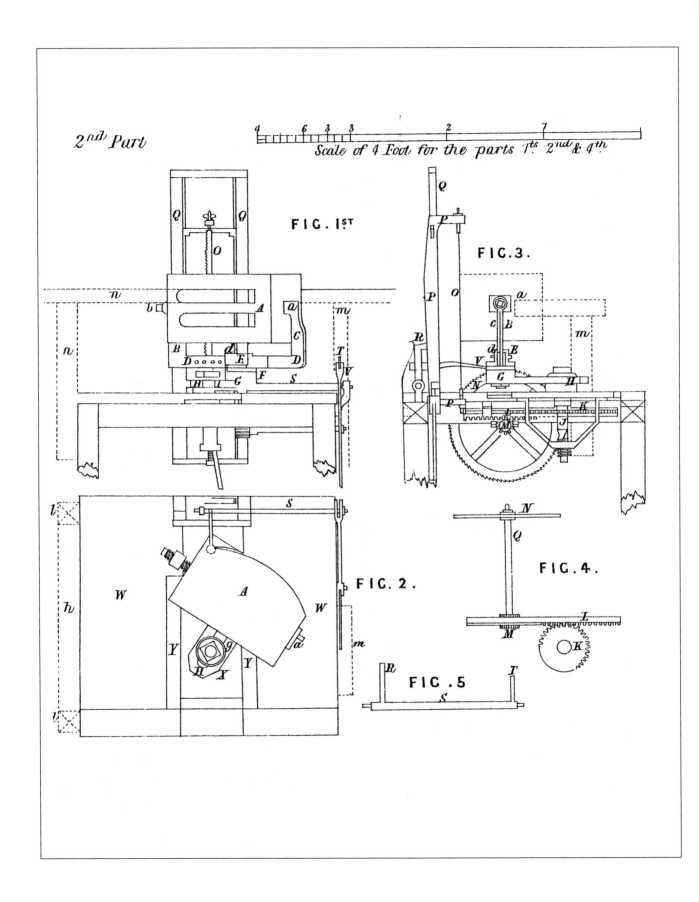

2nd Part

Scale of 4 Foot for the parts 1^{ts} 2nd & 4th

FIC. 1ST

FIC. 3.

FIC. 2.

FIC. 4.

FIC. 5.

A.D. 1802 . . . . N° 2663.

Trimmings and Borders for Muslins, Lawns, and Cambric.

BRUNEL'S SPECIFICATION.

TO ALL TO WHOM THESE PRESENTS SHALL COME, I, Marc Isambard Brunel, late of Gerrard Street, Soho, but now of Queen Square Place, Westminster, in the County of Middlesex, Gentleman, send greeting.

WHEREAS His most Excellent Majesty King George the Third did, by
5 His Letters Patent under the Great Seal of the United Kingdom of Great Britain and Ireland, bearing date at Westminster, the Twenty-fifth day of November, in the forty-third year of His reign, give and grant unto me, the said Marc Isambard Brunel, His especial licence, that I, the said Marc Isambard Brunel, during the term of years therein mentioned, should and
10 lawfully might use, exercise, and vend within England, Wales, and the Town of Berwick-upon-Tweed, my Invention of "NEW TRIMMINGS AND BORDERS OF MUSLIN, LAWN, AND CAMBRIC;" in which said Letters Patent there is contained a proviso, obliging me, the said Marc Isambard Brunel, by an instrument in writing under my hand and seal to cause a particular description of the nature
15 of my said Invention, and in what manner the same is to be performed, to be inrolled in His Majesty's Court of Chancery within six calendar months, after the date of the said recited Letters Patent, as in and by the same, relation being thereunto had, will more fully and at large appear.

NOW KNOW YE, that I, the said Marc Isambard Brunel, in compliance
20 with the said proviso, do hereby declare, that my Invention consists in weaving narrow webs of muslins, of cambric muslins, of lawns, or of cambrics with a

---

2   A.D. 1802.—N° 2663.

Brunel's New Trimmings and Borders of Muslins, Lawns, and Cambrics.

proper selvage at each edge, varied according to the purpose for which the article is wanted, but in all cases adapted to prevent its ravelling out in washing and of any breadth not exceeding ten inches, so as that these narrow webbs shall without hemming, whipping, or otherwise securing the edge by needle work, be ready to be used for the different kinds of trimmings required 5 to be made of the materials above specified. By trimmings to be made of the materials above specified, I mean frills, borders, trimmings for bonnets, caps, cloaks, or for any other article of dress, or even of furniture, provided the trimmings are muslin, cambric muslin, lawn, or cambric, and not exceeding ten inches in width. The advantage of this Invention, beside that the trim- 10 mings produced are in many cases neater than the trimmings of which the edges are secured from ravelling by needle work, is that the operations of hemming, whipping, or otherwise securing from ravelling the edges of trimmings cut in narrow slips out of broader webs, as they have unavoidably been hitherto, are by this Invention altogether saved. In cases where trimmings 15 of the materials above specified are required to be gathered, or to be whipped, for the purpose of being drawn up full, I introduce into the warp a strong or double thread or threads of cotton, flax, or silk, on which the trimming may be drawn or gathered up, without the need of employing any needle work for this purpose. 20

The trimmings may be woven either in looms, similar to those in use for weaving muslins, cambric muslins, lawns, or cambrics, according to the kind of trimming, with the only difference that the looms for these trimmings must be proportioned to the narrowness of the web required; or these 25 trimmings may be woven in riband looms, and very advantageously by the engine looms such as are used by riband weavers with the precaution of using such sizing, starching, and other dressing of the yarn, as is proper for weaving flaxen or cotton webs, as the case may be, of the usual breadth and of the same quality respectively. These trimmings may be woven with plain selvages, so as 30 not to shew any appearance of hem, roll, or turning in, or they may be woven with a thick stripe at one or both edges, so as to give the appearance of a hem, as well as to give additional strength to the edge or edges. The edge or edges of these trimmings may be woven double, so as to imitate a hem more exactly, and so that a quill, bodkin, or wire might be drawn through the 35 hollow part, as is sometimes customary in regard to the broad hems made to trimmings worn in mourning, and trimmings, in which this double part is required, may be woven in looms of a construction similar to those in which ribands with hollow edges are woven, as is the case in some of those woven for hat-bands, but when it is required that each of these folds of the double part

---

A.D. 1802.—N° 2663.   3

Brunel's New Trimmings and Borders of Muslins, Lawns, and Cambrics.

should be of the same texture with the single part of the web, it is necessary, that the weft should be woven four times through the warp of the edge (that is twice for each fold) for every twice that the weft is woven through the warp of the single part of the web. In order that these trimmings may wash
5 and wear equally well with trimmings cut crossways out of broader webs of the same quality, yarn similar to the weft of the broad webs may be used for the warp of the web for trimmings, and yarn similar to the warp of the broad webs may be used for the wefts of the web for trimmings. These trimmings may be woven either plain, striped, checked, or figured in white, or in colours,
10 by means similar to those practiced in the weaving figured ribands, and they may be finished by singeing, dressing, bleaching, printing, calendering, clear starching, or by any other of the operations used in finishing webs of the usual breadth according to the nature of the materials and the purpose to which the trimmings are to be applied.
15   In witness whereof, I, the said Marc Isambard Brunel, have hereunto set my hand and seal, this Twenty-fifth day of May, in the year of our Lord One thousand eight hundred and three.

M° I. BRUNEL. (L.S.)

AND BE IT REMEMBERED, that on the Twenty-fifth day of May, in
20 the forty-third year of the reign of His Majesty King George the Third, the said Marc Isambard Brunel came before our said Lord the King in His Chancery, and acknowledged the instrument aforesaid, and all and every thing therein contained and specified, in form above written. And also the instrument aforesaid was stamped according to the tenor of the Statutes made in
25 the sixth year of the reign of the late King and Queen William and Mary of England, and so forth, and in the seventeenth, twenty-third, and thirty-seventh years of the reign of His Majesty King George the Third.

Inrolled the Twenty-fifth day of May, in the year of our Lord One thousand eight hundred and three.

LONDON:
Printed by GEORGE EDWARD EYRE and WILLIAM SPOTTISWOODE,
Printers to the Queen's most Excellent Majesty.

A.D. 1805 . . . . . . N° 2844.

Saws and Machinery for Sawing Timber.

BRUNEL'S SPECIFICATION.

TO ALL TO WHOM THESE PRESENTS SHALL COME, I, MARC ISAMBARD BRUNEL, of Portsea, in the County of Hants, Gentleman, send greeting.

WHEREAS His most Excellent Majesty King George the Third did, by His Letters Patent under the Great Seal of the United Kingdom of Great Britain and Ireland, bearing date at Westminster, the Seventh day of May, in the forty-fifth year of His reign, give and grant unto me, the said Marc Isambard Brunel, His especial license that I, the said Marc Isambard Brunel during the term of years therein mentioned, should and lawfully might use, exercise, and vend, within England, Wales, and the Town of Berwick-upon-Tweed, my Invention of " SAWS AND MACHINERY UPON AN IMPROVED CONSTRUCTION FOR SAWING TIMBER IN AN EASY AND EXPEDITIOUS MANNER ;" in which said Letters Patent there is contained a proviso obliging me, the said Marc Isambard Brunel, by an instrument in writing under my hand and seal, to cause a particular description of the nature of my said Invention, and in what manner the same is to be performed, to be inrolled in His Majesty's High Court of Chancery within one calendar month after the date of the said recited Letters Patent, as in and by the same (relation being thereunto had) may more fully and at large appear.

NOW KNOW YE, that in compliance with the said proviso, I, the said Marc Isambard Brunel, do hereby declare that my said Invention is described by the drawings and explanations thereof hereunto annexed, and the following description thereof, viz:—

The saws are intended to be of a circular form, and, in order to obtain a great diameter, made of two or more pieces of sheet steel, properly adjusted and fixed together : see Figures 1, 2, 3, 4, and 7. The saw, Figure 1, is made of eight pieces. The pieces A, A, A, having been cut to the proper shape, and

---

fitted together at the edges, as represented at B, Figure 4, are screwed (see 1, 1, 1, Figure 1) against a flanch A, Figure 2, which has been previously turned very flat. The holes through which the screws pass are cut in an oblong form in order to admit of adjustment. When the plates have been thus fastened to the flanch A another flanch C, Figure 1 and 2, is laid upon 5 the plates, and in order to make it fit and bear with equal power upon each plate several thicknesses of. paper or leather of the size of the flanch are placed between it and the plates. The flanch C is fastened to the other by screws, 2, 2, 2, &c. Before they are screwed tight the plates are drawn in concentrically, by means of wedges 3, 3, 3, &c., in order to close the joints. 10 It must be observed that the joints made in the form B, Figure 4, must be turned according to the direction of the revolving motion of the saw. Saws composed of three or more pieces are intended to be fitted and fastened in the manner before described. If composed of two pieces—see Figure 3—the joining edge must be hollow at the part A of the plate and sharp at B : see 15 ⊕ A and ⊕ B.

SPECIFICATION OF THE MACHINERY.

The improvements in the machinery for sawing timber in an easy and expeditious manner consist in the modes of laying and holding the piece of wood in the carriage or drag, in the facility of shifting the saw from one cut to another, 20 and in the practicability of sawing both ways, either towards or from the saw or saws. The Figures 5, 6, 7, and 8 represent the machinery fitted up with one circular saw only ; the circular saw A is adjusted upon a spindle of a cylindrical form, which turns within rodings C, C. The saw is intended to be turned with either a strap or a band moved by any power (water, wind, steam, horses, or men). 25 The drum D, Figure 5 and 6, is to receive a strap. The log or piece of timber (see E, Figure 5, 6, 7, 8, is placed upon a drag or carrige F, and held fast by means of clamps G. The carriage or drag is moved to and from the saw by the handle or crank H, Figures 5, 6, and 8, communicating by the assistance of cog wheels to a pinion J, Figure 5 and 7, which engages in a rack K, Figure 7. 30 The drag or carriage is furnished with rollers in order to ease its longitudinal motion, and is intended to be moved by hand in order to accelerate or stop it at pleasure. The length of the carriage or drag is according to the size of the timber intended to be converted by such machinery. When the saw has performed one cut it is shifted to the next in the following manner :—Supposing, 35 for example, that the saw has been through the first cut 1—see the log E, Figure 5 ; the saw is then moved collaterally to the next cut 2, and so on, and kept to its place by means of a screw N, Figure 5 and 6 ; this screw N is complete at one end to the extremity of the cylindrical shafts, and passes in

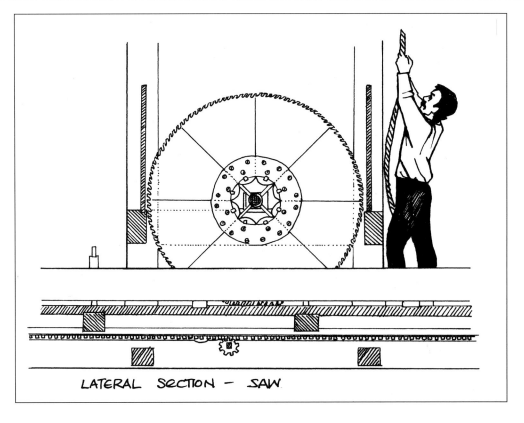

LATERAL SECTION — SAW

130

a direction parallel to it through the rodings C +, Figures 5, 6, and 8. This manner of sawing timber requires no fastening to the log when it is intended to be slabed only, excepting, however, when the log is crooked, in which case it may be forced to become straighter by the assistance of the clamps G, G, Figures 5, 6, 7, 8. M, M, Figure 8, represent circular wedges intended to follow the cut opened by the saw, and by that means to ease the friction and to steady the piece of wood. The circular wedges more collaterally when shifted, in order to meet with the next cut of the saw. The drag or carriage might be moved (if found more advantageous) by the machinery which gives motion to the saw. The Figures 9 and 10 represent the manner of adjusting several saws on one spindle, by which means a piece of wood may be partly or entirely converted at one operation; in that case the flanches of the saws are fixed upon an iron drum A, Figures 9 and 10, and kept firm and close with each other by four bolts 1, 2, 3, 4, Figure 10. In order to lower saw or saws when they wear away, the side rails P, P, Figures 9 and 10, may be depressed by means of the wedges Q, Q, Figures 9 and 10. It must be observed that the log or piece of wood does not lay close upon the drag or carriage; it is raised by some pieces placed crosswise : see O, O, O, Figures 5, 6, 7, and 8. The circular wedges described at M, M, Figure 8, are also represented at M, Figure 11 ; they revolve by the motion of the log, and keep each piece, and consequently the whole log, steady. Figure 12 represents an instrument composed with one or several plates of metal, to be used in lieu of the circular wedge or wedges; the distance between these plates is regulated by that between the saws. Figure 13 represents the breadth of the plates described above.

In witness whereof, I, the said Marc Isambard Brunel, have hereunto set my hand and seal, the Seventh day of June, in the year of our Lord One thousand eight hundred and five.

M° I. BRUNEL. (L.S.)

AND BE IT REMEMBERED, that on the same Seventh day of June, in the year above mentioned, the aforesaid Marc Isambard Brunel came before our Lord the King in His Chancery, and acknowledged the Specification aforesaid, and all and every thing therein contained, in form above written. And also the Specification aforesaid was stamped according to the tenor of the statute in that case made and provided.

Inrolled the same Seventh day of June, in the year above written.

D. HOWARD,
a Master Extra. in Chancery.

LONDON:
Printed by GEORGE EDWARD EYRE and WILLIAM SPOTTISWOODE,
Printers to the Queen's most Excellent Majesty.

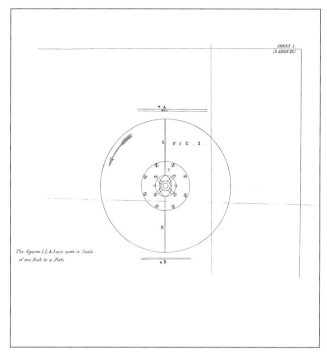

SHEET I
(2 SHEETS)

F I G . 3 .

The figures 1, 2, & 3 are upon a Scale of one Inch to a Foot.

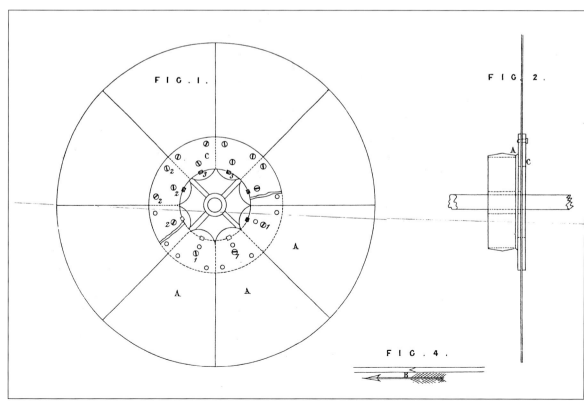

F I G . I .

F I G . 2 .

F I G . 4 .

A.D. 1806 . . . . . . N° 2968.

Cutting Veneers.

BRUNEL'S SPECIFICATION.

TO ALL TO WHOM THESE PRESENTS SHALL COME, I, MARC ISAMBARD BRUNEL, of Portsea, in the County of Southampton, Gentleman, send greeting.

WHEREAS His most Excellent Majesty King George the Third did, by His
5 Letters Patent under the Great Seal of the United Kingdom of Great Britain and Ireland, bearing date at Westminster, the Twenty-third day of September, in the forty-sixth year of His reign, give and grant unto me, the said Marc Isambard Brunel, my exors, admors, and assigns, His especial licence, full power, sole priviledge and authority, that I, the said Marc Isambard Brunel,
10 my exors, admors, and assigns, should and lawfully might, during the term of years therein mentioned, make, use, exercise, and vend, within England, Wales, and the Town of Berwick-upon-Tweed, my Invention of "**A NEW MODE OF CUTTING VENEERS OR THIN BOARDS BY MACHINERY;**" in which said Letters Patent there is contained a proviso obliging me, the said Marc
15 Isambard Brunel, by an instrument in writing under my hand and seal, to give a particular description of the nature of my said Invention, and in what manner the same is to be performed, to be inrolled in His Majesty's High Court of Chancery within six calendar months after the date of the said recited

---

2   A.D. 1806.—N° 2968.

Brunel's Mode of Cutting Veneers, &c.

Letters Patent, as in and by the same, relation being thereunto had, may more fully and at large appear.

NOW KNOW YE, that in compliance with the said proviso, I, the said Marc Isambard Brunel, do hereby declare that my said Invention of cutting veneers or thin boards consists in separating the wood by means of a sharp 5 instrument, forming part of an engine, which I specify as follows:—

The Figure 1 represents the plan of the engine; Figure 2, the longitudinal elevation; Figure 3, a transversal section; and Figure 4, a transversal elevation. A, Fig. 1, 2, 3, and 4, represents the cutter of the engine. This cutter, which might be made of one single piece, is, on account of its length, 10 represented as being composed of several pieces or plates of steel 1, 2, 3, 4, &c., Fig. 1, held together and fastened by means of screws 6, 7, 8, 9, &c., Fig. 1 and 2, and of a clamp C, Fig. 3, to a frame B, Fig. 1, 2, 3. It must be observed, that these pieces or plates of steel can be pushed out in proportion as they wear out. This frame B forms a slider, which is to move along the two 15 rails D, D, Fig. 1 and 3. As the two rails D, D, require a considerable degree of strength, they extend, by means of brackets, along each side of a pipe E, Fig. 1, 2, 3, together with which they form one solid body. This pipe, having a flanch at each extremity, is fastened to two standards F, F, Fig. 1, 2, 3, 4, which are strongly bolted through their base to a platform G, 20 Fig. 1, 2, and 4. The frame or slider B is connected, by means of a rod f, Fig. 1 and 2, to the machinery by which it is to be put and kept in motion. The part of the engine which carries the wood is composed as follows:—H, Fig. 1, 2, 4, and 5, represents a cast-iron bed bolted to the platform G, Fig. 1 and 4. The upper edge of each side of this bed terminates into a pro- 25 jecting angular rail J, J, Fig. 2 and 5; K, Fig. 1, 2, 3, 4, 5, 6, & 7, repre- sents a cast-iron frame, which may be denominated a carriage, intended to slide on the bed H, by the assistance of a screw L, and rack M, Fig. 3 and 5. This sliding motion of the carriage, guided by two clamps N, N, Fig. 2 and 5, is to propel the wood towards the cutter. When the veneer or thin board has 30 been separated from the piece of wood by the operation of the cutter, the car- riage is removed back in order to clear the wood from under the cutter. This piece of wood is then to be elevated proportionally to the thickness of the veneer which is next to be cut by means of a parellell motion disposed as follows:—The table O, Fig. 1, 2, 3, and 4, upon which the piece of wood is 35 placed, is supported by four screws P, P, two of which are seen in the Figures 2, 3, 4, and 6. These screws P, P, pass through four nuts Q, Q, Q, Q, Fig. 7, let in sockets, which make part of the carriage K. It is evident that by turning these nuts Q, Q, the screws P, P, and consequently the table O,

---

A.D. 1806.—N° 2968.   3

Brunel's Mode of Cutting Veneers, &c.

will be elevated or lowed at pleasure. The four nuts are turned all together and at the same time, by the assistance of endless screws R, R, Fig. 7, which, being fixed on the same spindle T, Fig. 2 and 7, act upon the wheels S, S, Fig. 1, 2, 3, 4, 6, and 7. These wheels S, S, are connected with the nuts
5 Q, Q. The two screws, placed at a and b, Fig. 7, are right-hand, and the two others, at c and d, are left-hand screws (see also R, R, Fig. 6). In lieu of a long cutter, similar to that which is herewith described, another might be sub- stituted (see Fig. 8). A, Fig. 8, represents the cutter, B the frame or slider; both answer to the section of the same parts represented under the cor-
10 responding letters in Fig. 3. This slider B is in this case intended to be moved from one end to the other of the rails D, D. It is obvious that the cutter, whether long or short, requires to be kept perfectly flat and true with respect to propelling motion of the carriage K and the parrellel motion of the slider B, and also very sharp. In order to obtain these three points, I have
15 added to the engine a lap U, Fig. 2, 3, 4, and 5, upon which the cutter is to be ground when requisite. V, V, Fig. 5, represent two uprights fixed on the front of the carriage K. The frame X, X, Fig. 5, of this lap, is supported by means of two steady pins let into the uprights. It is elevated or lowered at pleasure by the assistance of the screws Y, Y, Fig. 5. The lap U is brought
20 under the cutter by sliding back the carriage as much as is necessary.

The engine before specified is used and managed in the following maner:— The piece or pieces of wood to be cut into veneers or thin boards are placed and fastened on the table O by means of cement or glue (in the latter case the top of the table O ought to be lined with wood). The slider B being supposed
25 in motion, the workman attending the engine adjusts at first the table to a proper degree of elevation, propels the carriage by the assistance of the wheel Z, Figures 1, 2, and 4. The workman, guided by the apparent effect of the cutter, continues to force the carriage until the veneer is entirely separated. He then moves back the carriage, with the assistance of the same
30 wheel Z, and prepares for another cut, by elevating the table as much as is necessary. This is accomplished by turning the spindle T, Fig. 2 and 7, with the assistance of the handle e, Fig. 1, 2, and 4, which being done he can proceed as before.

In witness whereof, I, the said Marc Isambard Brunel, have hereunto set
35 my hand and seal, the Twenty-third day of March, in the year of our Lord One thousand eight hundred and seven.

         M^c I. BRUNEL.   (L.S.)

---

4   A.D. 1806.—N° 2968.

Brunel's Mode of Cutting Veneers, &c.

AND BE IT REMEMBERED, that on the Twenty-third day of March, in the year of our Lord 1807, the aforesaid Marc Isambard Brunel came before our said Lord the King in His Chancery, and acknowledged the Specification aforesaid, and all and every thing therein contained and spe- cified, in form above written. And also the Specification aforesaid was 5 stampt according to the tenor of the Statute made for that purpose.

Inrolled the Twenty-third day of March, in the year of our Lord One thousand eight hundred and seven.

LONDON:
Printed by GEORGE EDWARD EYRE and WILLIAM SPOTTISWOODE,
Printers to the Queen's most Excellent Majesty.

A.D. 1808 . . . . . . N° 3116.

Circular Saws.

BRUNEL'S SPECIFICATION.

TO ALL TO WHOM THESE PRESENTS SHALL COME, I, MARC ISAMBARD BRUNEL, of Chelsea, in the County of Middlesex, Gentleman, send greeting.

WHEREAS His most Excellent Majesty King George the Third did, by
5 His Letters Patent under the Great Seal of the United Kingdom of Great Britain and Ireland, bearing date at Westminster, the Fourteenth day of March, in the forty-eighth year of His reign, give and grant unto me, the said Marc Isambard Brunel, my exörs, admörs, and assigns, His especial licence, full power, sole priviledge and authority, that I, the said Marc
10 Isambard Brunell, my exörs, admörs, and assigns, during the term of years therein mentioned, should and lawfully might make, use, exercise, and vend, within England, Wales, and the Town of Berwick-upon-Tweed, my Invention of "CERTAIN IMPROVEMENTS ON CIRCULAR SAWS FOR SAWING WOOD IN AN EASY AND EXPEDITIOUS MANNER;" in which said Letters Patent there is contained
15 a proviso obliging me, the said Marc Isambard Brunel, by an instrument in writing under my hand and seal, to cause a particular description of the nature of my said Invention, and in what manner the same is to be performed, to be inrolled in His Majesty's High Court of Chancery within six calendar months after the date of the said recited Letters Patent, as in and
20 by the same, relation being thereunto had, may more fully and at large appear.

NOW KNOW YE, that in compliance with the said proviso, I, the said Marc Isambard Brunel, do hereby declare that my said Invention is described in manner following, that is to say :—

---

2 A.D. 1808.—N° 3116.

Brunel's Improvements on Circular Saws.

The Certain Improvements on Circular Saws for Sawing Wood in an Easy and Expeditious Manner are intended to cut out thin boards or slips with as little waste as appears practicable. In order to obtain that end a saw requires to possess two essential qualities, viz¹, a very thin edge, and at the same time a great degree of steadiness. The certain improvements above mentioned 5 consist in the construction of a circular saw or circular saws which are intended to unite both qualities, and thereby render the sawing easy and expeditious.

The annexed Drawings shew the plan, elevation, and sections of a circular saw or circular saws composed of one or more pieces. The saw represented 10 in the plan and elevations is ten feet diameter, and is composed of several plates of steel C, C, C, C, C, C, Fig. 2, fixed to a circular frame or wheel A, A, Fig. 2, 3, and 4. The extremity of this circular frame terminates into a tapered rim, as is represented at full size at A, Fig. 1. The plates of steel C, C, Fig. 2, form segments fixed by means of a circular clamp or clamps B, 15 Fig. 1, screwed or rivetted to the rim A. The surface of the segments, laid as they are represented at C, Fig. 1 and 2, is of a conical form, which terminates into a thin edge D, Fig. 1. This edge is cut, as a common circular saw, into teeth, which may be set at pleasure in the usual manner. The conical side of the plates or segments is hollowed out towards the extreme 20 edge, in order to render the passage of the saw easy and free from friction. In the act of sawing, the board or slip separated by the saw is kept off to prevent its rubbing against the plates, see I, Fig. 1, 3, and 4, either by hand or by the assistance of some wedge fixed to the machinery. The circular frame on the edge of which the segments C, C, C, C, C, C, Fig. 2, are 25 screwed is fixed to a shaft E, Fig. 3 and 4, which is to be set in motion by such means as may appear the most eligible. In the annexed Drawings it is intended to be put in motion by a steam engine, and connected to it by means of a strap revolving on a drum F, Fig. 3 and 4. The manner of using this saw is as follows :—The log or piece of wood to be cut is laid against and 30 fastened to a table H, H, Fig. 2, 3, and 4. The table is fixed on a table carriage G, G, Fig. 2, 3, and 4, which has a progressive and retrograde motion. By sliding in a longitudinal direction on a frame J, J, Fig. 2, 3, and 4, after each cut the log is, by the retrograde motion of the carriage, cleared from the saw; and in order to prepare the next cut, the table is advanced by 35 a parallel and lateral motion, regulated by two dividing wheels K, K, Fig. 2, 3, and 4, which determines the thickness of the board to be cut. The progressive and retrograde motions of the carriage is produced by means of a pinion M, Fig. 2 and 4, fixed on a spindle N, Fig. 3 and 4. This pinion, by

---

A.D. 1808.—N° 3116. 3

Brunel's Improvements on Circular Saws.

the assistance of an intermediate cog wheel O, Fig. 2 and 4, which engages into a rack P, P, Fig. 2, 3, and 4, is to give motion to the carriage. The spindle N, Fig. 3 and 4, may be moved either by hand or any other manner. In this Drawing it is intended to be moved by the steam engine by the assis-
5 tance of straps revolving on two drums Q, Q, Fig. 3 and 4. One of these drums is to produce the progressive, and the other the retrograde, motion.

The Fig. 5 represents a section of a circular saw on the same principle as that before described. This saw is of one single plate, and fixed to a metal chuck A, to which it is fastened by a metal ring or clamp B, screwed to the
10 chuck A. The extremity of this saw may be shaped into a thin edge of the form of that represented at D, Fig. 1. C represents at full size a section of a part of the saw terminating into an edge on which teeth are cut and set in the usual manner.

Fig. 6 represents a section of a circular saw of one single plate, which is
15 entirely flat on one side and convex on the other. This plate is fixed to a metal chuck in the manner described in Fig. 5. A represents at full size a section of a part of the plate of the saw terminating into a thin edge.

In witness whereof, I, the said Marc Isambard Brunel, have hereunto set my hand and seal, the Fourteenth day of September, in the year
20 of our Lord One thousand eight hundred and eight.

Mᶜ I. BRUNEL. (L.S.)

AND BE IT REMEMBERED, that on the Fourteenth day of September, in the year of our Lord 1808, the aforesaid Marc Isambard Brunel came before our Lord the King in His Chancery, and acknowledged the Specifi-
25 cation aforesaid, and all and every thing therein contained and specified, in form above written. And also the Specification aforesaid was stampt according to the tenor of the Statute made for that purpose.

Inrolled the Fourteenth day of September, in the year of our Lord One thousand eight hundred and eight.

LONDON:
Printed by GEORGE EDWARD EYRE and WILLIAM SPOTTISWOODE,
Printers to the Queen's most Excellent Majesty.

A.D. 1810 . . . . . . Nº 3369.

Shoes and Boots.

BRUNEL'S SPECIFICATION.

TO ALL TO WHOM THESE PRESENTS SHALL COME, I, MARC ISAMBARD BRUNEL, of Chelsea, in the County of Middlesex, Gentleman, send greeting.

WHEREAS His most Excellent Majesty King George the Third did, by 5 His Letters Patent under the Great Seal of the United Kingdom of Great Britain and Ireland, bearing date at Westminster, the Second day of August, in the fiftieth year of His reign, give and grant unto me, the said Marc Isambard Brunel, my exŏrs, adñiŏrs, and assigns, His especial licence, full power, sole priviledges and authority, that I, the said Mark Isambard Brunel, 10 my exŏrs, adñiŏrs, and assigns, should and lawfully might, during the term of years therein expressed, make, use, exercise, and vend, within England, Wales, and the Town of Berwick-upon-Tweed, my Invention of "CERTAIN MACHINERY FOR THE PURPOSE OF MAKING OR MANUFACTURING SHOES AND BOOTS;" in which said Letters Patent there is contained a proviso, obliging me, the 15 said Mark Isambard Brunel, by an instrument in writing under my hand and seal, to cause a particular description of the nature of my said Invention, and in what manner the same is to be performed, to be enrolled in His Majesty's High Court of Chancery within six calendar months after the date of the said recited Letters Patent, as in and by the same, relation being thereunto had, 20 may more fully and at large appear.

NOW KNOW YE, that in compliance with the said proviso, I, the said Marc Isambard Brunel, do hereby declare that the nature of my said Invention, and in what manner the same is to be performed, are described and

---

ascertained by the Drawings hereunto annexed, and the following description thereof, that is to say:—

I do declare that my Invention for making shoes and boots is intended chiefly for manufacturing a certain description of shoes and boots, the soles of which are fastened to the upper leather or closing by means of metallic pins or 5 nails. *Fig.* 1, 2, 3, represent an apparatus intended for pressing, and also for cutting the soles to their proper form, although the cutting out of the soles and heels to their proper form, and likewise the boring of the holes for the nails or pins, might have been done at one or two operations by the assistance of a press with stamping instruments, yet, from the accidents to which such 10 complicated instruments are liable, and, at the same time, from the difficulty attending the keeping of them in proper working repair, I have given the preference to this apparatus, which by its simplicity will obviate all inconveniences.

The piece or pieces of leather intended to be used for the sole of shoes or 15 boots is laid, as seen at A (Fig. 1 and 2), between the plates B and C; the upper one B opens at pleasure; the leather thus laid is pressed as strongly as may be required, by the assistance of a screw and nut D, Fig. 1 and 2, and when in that state it is to be cut all round by means of a common knife to the form of the plate B, Fig. 3. The leather thus shaped is ready to be 20 applied to and united with the upper leather or closing.

The machine represented by the Figs. 4, 5, and 6, is intended for uniting the sole with the upper leather or closing; the upper leather or closing having been previously prepared, as far as is usually done, is laid and well stretched in the usual way upon a last of cast iron, see A, Fig. 5, and in that state held fast 25 close to the last by means of metal clamps 1, 2, 3, 4, 5, 6, 7, 8, Fig. 6, or 1, 2, 3, 4, and 5, Fig. 5. These clamps are kept steady to their respective places by means of pauls G, G, Fig. 4 and 5, acting upon joints and butting against clamps. The piece of leather denominated the inner sole (the section of which is seen at full size at D, Fig. 9) is then laid upon the last, and the 30 projecting edges F, F, of the closing G, G, are flattened down, and when in that state they are ready to receive the sole. The sole is to be placed carefully upon the last, then pressed down and held fast by the assistance of a clamp B, Fig. 4, 5, 6, and kept steady in its place with a link H, Fig. 5 and 6.

The next operations relate to the letting in and driving the metallic nails or 35 pins, which I propose to have done as follows:—The rod D, Fig. 4 and 5, is intended to drive in the nails or pins, and to open, at one and the same time, the holes through the leather. This rod is to act in such a manner that for every nail or pin forced down by the assistance of the rod a hole is bored by

---

means of the point K, Fig. 5, for the reception of the contiguous nail, see Fig. 11 (at full size). A represents the lower end of the rod; B represents a point acting as an awl to open the holes; C represents a pin or nail ready to be forced down by the rod A; this rod, represented at D, Fig. 4, is suspended 5 to the extremity of a short lever L, by which it can be moved up and down; but as some degree of force is requisite to drive the nails or pins, the lever M is intended to strike on the top of the rod; this lever is put in action by the assistance of the treadle N. The clamp B, Fig. 4, 5, and 6, is by about one half of an inch smaller (all round) than the sole, and its sides from a guide against which the 10 lower end of the rod D, Fig. 4 and 5, is kept during the operation of driving in the nails or pins, except that both extremities of the sole are left to be finished without the assistance of the clamp. The bed D, Fig. 4, 5, and 6, is fixed upon an horizontal slide F, Fig. 4 and 6, which is intended to carry or move at pleasure, by means of the lever E, Fig. 6, the last from right to 15 left, in order that the rod may work as near as possible under its perpendicular line. The bed D, Fig. 4, 5, and 6, has a circular horizontal motion upon the slide F, Fig. 4 and 6, for the same purpose. Figs. 7 and 8 represents a clamp of a different construction, intended to press down and hold the sole in its place before the nails or pins are driven in; a section of this clamp is seen at 20 full size at B, Fig. 9; above it there is a plate C fixed upon the clamp; this plate C at Fig. 8 is cut all round its edge with as many notches as there are pins to be driven in the sole, and forms a more complete guide than the one before described, as it determines the place and distance to be observed in putting the nails or pins. The projecting angle A, Fig. 10, which corresponds 25 with the awl A, Fig. 9, is intended to be dropt in the notches of the plate C. When all the nails or pins have been let and driven in their places the shoe is so far finished as to require nothing but the usual triming and polishing before it is taken off.

The process to be observed for making boots is exactly the same as described 30 before, but the machine differs in some degree from that represented in the Fig. 4, 5, and 6. The shank of the last must be of a convenient height to allow sufficient room for the leg of the boot, and the clamps 1, 2, 3, 4, 5, 6, 7, and 8 must be of a corresponding size.

As my Invention for making shoes and boots consists likewise in using, 35 in lieu of metallic nails or pins, pegs or pins of either leather, bone, whalebone, catgut, or any other animal or vegetable substance fit for it, it is obvious that the inner ends of pegs or pins of such substances, when driven in, cannot (like metal) be turned in and secured by the usual method of clinching. In order, therefore, to give to these pegs or pins a

---

proper and sufficient holding, both inside and outside, I propose to have them driven in quite through the inner sole, so far as to exceed it by at least $\frac{1}{16}$ of an inch. The last is to have, for that purpose, a groove all round, and near its edge, to admit of the point or awl B, Fig. 11, penetrating freely and without any obstruction. The pegs or pins which I intend to 5 substitute in lieu of metallic nails or pins are to be let in in the manner before specified, with the exception that they are not to be driven with so great a force. When the shoe, together with the heel, have been completely fastened with pegs or pins, I propose to have the outer ends of these pegs or pins seared smooth by means of a hot iron, and afterwards hammered in very close 10 and tight. The shoe or boot, when in that state, is fit to be taken from the last.

In order to facilitate the taking off of the shoe or boot made in the manner before described, with all the inner ends of the pegs or pins projecting out, I propose to have a last made, as represented in Fig. 12 and 13; the part A 15 moves in and out at pleasure, and when drawn in, as in Fig. 13, the size of the last is sufficiently reduced to admit of the shoe being easily removed; such last is to be used in lieu of the solid one represented at A, Fig. 5. The ends of the pegs or pins projecting inside of the shoe or boot are to be flattened down or seared by means of heated iron; great care is to be taken, when 20 performing that operation, not to damage the closing. The shoe or boot so far made is ready to be polished and trimmed in the usual manner.

In witness whereof, I, the said Marc Isambard Brunel, have hereunto set my hand and seal, the Second day of February, in the year of our Lord One thousand eight hundred and eleven.

Mᶜ Iˢ BRUNEL. (L.S.) 25

AND BE IT REMEMBERED, that on the Second day of February, in the year of our Lord 1811, the aforesaid Marc Isambard Brunel came before our said Lord the King in His Chancery, and acknowledged the Specification aforesaid, and all and every thing therein contained and specified, in form above written. And also the Specification aforesaid was stampt according to 30 the tenor of the Statute made for that purpose.

Inrolled the Second day of February, in the year of our Lord One thousand eight hundred and eleven.

LONDON:
Printed by GEORGE EDWARD EYRE and WILLIAM SPOTTISWOODE,
Printers to the Queen's most Excellent Majesty.

A.D. 1813 . . . . . . N° 3643.

Saw Mills.

BRUNEL'S SPECIFICATION.

TO ALL TO WHOM THESE PRESENTS SHALL COME, I, Marc Isambard Brunel, of Chelsea, in the County of Middlesex, Civil Engineer, send greeting.

WHEREAS His most Excellent Majesty King George the Third did, by His Letters Patent under the Great Seal of the United Kingdom of Great Britain and Ireland, bearing date at Westminster, the Twenty-sixth day of January, in the fifty-third year of His reign, give and grant unto me, the said Marc Isambard Brunel, my exŏrs, admŏrs, and assigns, His especial licence, full power, sole privilege and authority, that I, the said Marc Isambard Brunel, my exŏrs, admŏrs, and assigns, should and lawfully might, during the term of years therein mention, make, use, exercise, and vend, within England, Wales, and the Town of Berwick-upon-Tweed, my Invention of "CERTAIN IMPROVEMENTS IN SAW MILLS;" in which said Letters Patent there is contained a proviso that if I, the said Marc Isambard Brunel, shall not particularly describe and ascertain the nature of my said Invention, and in what manner the same is to be performed, by an instrument in writing under my hand and seal, and cause the same to be inrolled in His Majesty's High Court of Chancery within two calendar month next and immediately after the date of the said Letters Patent, that then the said Letters Patent, and all liberties and advantages whatsoever thereby granted, shall utterly cease, determine, and become void, as in and by the same, relation being thereunto had, may more fully and at large appear.

NOW KNOW YE, that in compliance with the said proviso, I, the said Mark Isambard Brunel, do hereby declare that the nature of my said Invention, and the manner in which the same is to be performed, are particularly described and ascertained in and by the Drawings hereunto annexed, and the following description thereof (that is to say) :— 5

Fig. 1 represents the front elevation of a saw frame, sliding on three points A, A, B, as the stretching of a number of saws in a frame has a tendency to bend in the heads of the frame, and consequently of affecting the sides in the same manner. The two bolts or ties C, C, are intended to form a counteraction, which will keep the uprights D parallel to each other. The 10 lower slide or rod E is intended to guide the reciprocating motion of the frame. This rod slides within a socket B, and in the direction it is represented in Fig. 2. It is not parallel with the uprights D. This deviation, however small, gives to the frame, when in action, a proportionate degree of oscillation. 15

Fig. 2 represents a lateral section of the frame seen in Fig. 1, and shews at the same time its inclination with regard to the angle it forms with the carriage on which the timber to be sawed is laid. This frame is to alter its inclination at every other stroke. The alteration is given to it by the aid of the eccentric triangular piece F, which is put in motion by the interposition of 20 two cog wheels G, H, Figs. 1, 2, and 4, when the eccentric F is brought to the opposite point to that in which it is now represented. It does in the action of passing from one point to the other rise by the aid of the swinging frame within which in revolves by the lever K, and inclines forward the moveable carriage L, into which the socket B is fixed. This alternative motion alters 25 the inclination of the frame, as it is desirable that the saws should not press against the wood when they ascend. I propose, in order to accomplish this, to have the socket A, A, Figs. 1 and 2, fixed to a double hinge or knuckle M, Figs. 2 and 6. When the frame begins to ascend it carries along with it the sockets A, which being confined by the hinge M, are drawn back as much as 30 is found requisite to relieve the pression of the saws against the wood. When the frame descends, the sockets are returned to their former position. This receding motion of the frame, which is intended chiefly to relieve the friction of the face of the saws, may be effected by the aid of a slide similar to that represented at A, Fig. 7. In lieu of a socket a slide acting upon a double 35 joint will recede as the side D of the saw frame ascends, and returns to its former position when it descends. In this case a roller or spring slide is to be applied in front in order to confine the side D close to the face of the

---

moveable slide. The back motion of the frame, the chief object of which is to relieve the friction of the saws when ascending, is to be affected in the following manner, which in some cases is preferable to the two former before described, see Fig. 8 :—D represents one of the uprights of the frame sliding through the socket A. This socket (or both if they could be seen, is fixed 5 on the frame L, which frame rests on a center C, on which it has a motion, an eccentric wheel B revolving in a manner corresponding with the alternative motion of the saw frame; that is, when the frame ascends the narrower part of the eccentric presents itself to the roller E, and its wider part where it descends; by this alternative motion the frame is moved forward and back- 10 ward at pleasure. A saw frame, fixed in the manner represented in Fig. 8, has the advantage that it may be laid horizontally for the purpose of whetting and setting the screws with ease by the aid of an instrument which will be here-after specified. Two or more saws might be carried in one single hook, repre-sented in Figures 9, and stretched into the saw frame, with either screws or 15 wedges in the manner specified by me in a former occasion. In order to equalize the tightness of each of the several saws held by a single hook, small wedges may be used, as represented in the Figure 9. As the saws I make use of in the frame here described are stretched to their proper degree of tightness by means of a lever, it is desirable that when the same saws require to be rectified 20 and whetted by the assistance of the apparatus represented in Figs. 10, 11, and 12, which apparatus I have specified in a former occasion, they should have a degree of tightness nearly equal; in order to produce this effect, I propose to add to this apparatus a lever A, Fig. 10, resembling the beam of a steelyard. The lever, by the interposition of the parts D, E, act on the 25 hook B, which stretches the saws at pleasure by turning the handle H. To this lever a weight F is applied, and moved at pleasure to any distance from the fulcrum C. The instrument represented in Figs. 13, 14, and 15 is intended to assist in whetting the saws; it is added to the moveable car-riage G, laid on the apparatus represented in Figs. 10 and 11. The Fig. 13 30 represents to the lateral view of the instrument, Fig. 15 its plan, and Fig. 14 its section. It is adjusted by means of a screw to any height, according to the edge of the saw, when they are fixed in the apparatus before mentioned. A, Fig. 14, represents one of the moveable side pieces seen in the Fig. 15 ; each of these side pieces carry a small rest B, of very hard steel, against which 35 every tooth of the saws lay during the operation of whetting. These moveable slide pieces A may be used either singly or together. I prefer only one at a time, as it enables to file with more ease the point of the teeth to any

---

angle ; the side piece A on the side of the teeth to be filed, and the other turned up. These pieces are adjusted by means of small screws C, C, Fig. 15. It is evident that with the aid of this instrument the teeth of the saws will be whetted with great precision and uniformity. The carriage G is stopt to its proper place by means of a small paul D. 5

It being desirable that the teeth of circular saws should be whetted with an equal degree of precision, I propose to make use of an instrument constructed on the same principle, adapted to the size of the saw. Circular files might be used instead of strait ones, but I give the preference to strait ones. When a great number of strait saws are fixed in frames, I propose, instead 10 of taking them off to be set and whetted, to perform the setting and whetting in place either on the frame itself or on an apparatus capable of receiving the same. This apparatus is represented in the Figs. 16, 17, 18, and 19. The Figure 16 represents the plan of the instrument ; 17, 15 a longitudinal section ; 18, a longitudinal elevation ; and 19, a transversal section.

The saw frame, as seen in part as represented at a, being laid between the slides A, A, Figs. 16 and 19, the carriage B, B, Figs. 16, 18, and 19, is brought over, and is moved at pleasure from end to end. The carriage is composed of slides connected with each other, and framed together by bolts C 20 and E. The small socket, with a projecting arm D, is intended to slide in a transverse direction on the bolt C. At the lower point of this arm a small file is fixed, and in order to regulate the depth of the fileing, the upper extre-mity of the arm rests on the second bolt E, on which it is adjusted at pleasure with a screw F. It is obvious that by moving the arm D transversely the 25 file will act upon the teeth.

In order to guard against the jarring of the saws during the operation of filing, I propose to introduce some intermediate pieces of the form repre-sented at G, Figs. 16 and 17. These pieces are intended to move with the carriage. 30

The various improvements herein specified refer only to those parts of the Drawings which are either coloured or shaded.

In witness whereof, I, the said Marc Isambard Brunel, have hereunto set my hand and seal, this Twenty-sixth day of March, in the year of our Lord One thousand eight hundred and thirteen. 35

Mᶜ Jᴰ BRUNEL. (L.S.)

Brunel's Improvements in Saw Mills.

AND BE IT REMEMBERED, that on the Twenty-sixth day of March, in the year of our Lord 1813, the aforesaid Marc Isambard Brunel came before our said Lord the King in His Chancery, and acknowledged the Specification aforesaid, and all and every thing therein contained and specified, in form above written.   And also the Specification aforesaid was stampt according to the tenor of the Statute made for that purpose.

Inrolled the Twenty-sixth day of March, in the year of our Lord One thousand eight hundred and thirteen.

LONDON:
Printed by GEORGE EDWARD EYRE and WILLIAM SPOTTISWOODE,
Printers to the Queen's most Excellent Majesty.

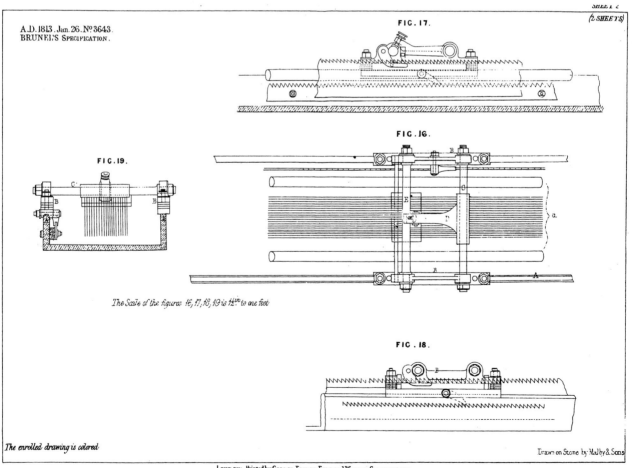

FIG. 17.

FIG. 16.

FIG. 19.

A.D. 1813. Jan. 26. Nº 3643.
BRUNEL'S SPECIFICATION.

The Scale of the figures 16, 17, 18, 19 is 1½ᵗʰ to one foot

FIG. 18.

The enrolled drawing is colored

Drawn on Stone by Malby & Sons

LONDON: Printed by GEORGE EDWARD EYRE and WILLIAM SPOTTISWOODE,
Printers to the Queen's most Excellent Majesty. 1856.

A.D. 1814 . . . . N° 3791.

Rendering Leather Durable.

BRUNEL'S SPECIFICATION.

TO ALL TO WHOM THESE PRESENTS SHALL COME, I, MARC ISAMBARD BRUNEL, of Chelsea, in the County of Middlesex, Civil Engineer, send greeting.

WHEREAS His most Excellent Majesty King George the Third did, by
5 His Letters Patent under the Great Seal of Great Britain and Ireland, bearing date at Westminster, the Twelfth day of March, in the fifty-fourth year of His reign, give and grant unto me, the said Marc Isambard Brunel, my exōrs, admōrs, and assigns, His special licence, full power, sole privilege and authority, that I, the said Marc Isambard Brunel, my exōrs, admōrs, and
10 assigns, during the term of years therein expressed, should and lawfully might make, use, exercise, and vend, within England, Wales, and the Town of Berwick-upon-Tweed, my Invention of "A NEW METHOD OF GIVING ADDITIONAL DURABILITY TO CERTAIN DESCRIPTIONS OF LEATHER;" in which said Letters Patent there is contained a proviso that if I, the said Marc Isambard Brunel,
15 shall not particularly describe and ascertain the nature of my said Invention, and in what manner the same is to be performed, by an instrument in writing under my hand and seal, and cause the same to be inrolled in His Majesty's High Court of Chancery within six calendar months next and immediately after the date of the said Letters Patent, that then the said Letters Patent
20 should be void, as in and by the same, relation being thereunto had, may more fully and at large appear.

NOW KNOW YE, that in compliance with the said proviso, I, the said Marc Isambard Brunel, do hereby declare that the nature of my said Inven-

---

tion, and the manner in which the same is to be performed, are particularly described and ascertained as follows (that is to say):—

The method I propose for giving additional durability to certain description of leather, is chiefly applicable to the leather used for making the soles of shoes and boots. The method consists of studding the leather with small nails or pins, 5 after it has been previously saturated or impregnated with a composition of neats' feet oil mixed with tar in the proportion of nine or ten to one. The studding is independent, however, of the application of the composition above mentioned, but it is material that the leather should be studded with nails or pins before it is used, as the points may then be clinched with ease; therefore the operation 10 of studding may be performed either by hand or machinery. If the nails or pins are to be let in by the aid of machinery, the instrument intended to drive them into the leather consists of a small drift similar to a punching press, by means of which the nails or pins are forced into the leather. The nail or pin is to be dropt into a socket, wherein it is kept in a proper direction 15 to receive the blow or pressure from the press or hammer. As nails or pins might be made and let into the leather at one operation and more expeditiously than by a previous one, I give the preference to such mode. These nails or pins are to be cut by the aid of a machine, on the principle of those already known for cutting brads or nails; but instead of allowing the nail to fall 20 indiscriminately into a receiver, I propose to make it drop into a small funnel or proper pipe, and to have it conveyed into a socket, wherein, by means of a drift or hammer or any other fit instrument, it is to be forced into the leather held fast under the socket. It is desirable that the nails or pins should be let in rather obliquely, in order that they may not be affected by 25 the perpendicular pressure. Heels and soles thus studded, when intended for the particular purpose of repairing shoes or boots of any description, I propose to have one row of longer nails or pins along and near the edges, and exceeding the inner surface to a sufficient degree, for the purpose of being applied and fastened at once and with ease to the parts already existing. A 30 sort of studding may be formed solid like a curved comb, which may be let in at one single operation, imitating thereby the horseshoe-like fence used for heels and toes of shoes, or, by breaking the connection between every point, one or more rows might be let in at one single blow.

In witness whereof, I, the said Marc Isambard Brunel, have hereunto 35 set my hand and seal, the Twelfth day of September, in the year of our Lord One thousand eight hundred and fourteen.

Mᶜ Jᴰ BRUNEL. (L.S.)

AND BE IT REMEMBERED, that on the Twelfth day of September, in the year of our Lord One thousand eight hundred and fourteen, the aforesaid Marc Isambard Brunel came before our said Lord the King in His Chancery, and acknowleged the Specification aforesaid, and all and every thing therein
5 contained and specified, in form above written. And also the Specification aforesaid was stampt according to the tenor of the Statute made for that purpose.

Inrolled the Twelfth day of September, in the year of our Lord One thousand eight hundred and fourteen.

CAMPBELL

LONDON:
Printed by GEORGE EDWARD EYRE and WILLIAM SPOTTISWOODE,
Printers to the Queen's most Excellent Majesty.

A.D. 1818 . . . . . . N° 4204.

Forming Tunnels or Drifts under Ground.

BRUNEL'S SPECIFICATION.

TO ALL TO WHOM THESE PRESENTS SHALL COME, I, Marc Isambard Brunel, of Lindsay Row, Chelsea, in the County of Middlesex, Civil Engineer, send greeting.

WHEREAS His most Excellent Majesty King George the Third did, by His Letters Patent under the Great Seal of the United Kingdom of Great Britain and Ireland, bearing date at Westminster, the Twentieth day of January, in the fifty-eighth year of His reign, give and grant unto me, the said Marc Isambard Brunel, my exors, admors, and assigns, His special licence, full power, sole privilege and authority, that I, the said Marc Isambard Brunel, my exors, admors, and assigns, should and lawfully might, during the term of years therein mentioned, make, use, exercise, and vend, within England, Wales, and the Town of Berwick-upon-Tweed, my Invention of " THE METHOD OR METHODS OF FORMING TUNNELS OR DRIFTS UNDER GROUND ;" in which said Letters Patent there is contained a proviso that if I, the said Marc Isambard Brunel, shall not particularly describe and ascertain the nature of my said Invention, and in what manner the same is to be performed, by an instrument in writing under my hand and seal, and cause the same to be inrolled in His Majesty's High Court of Chancery within six calendar months next and immediately after the date of the said Letters Patent, that then the said Letters Patent, and all liberties and advantages whatsoever thereby granted, shall utterly cease, determine, and become void, as in and by the same, relation being thereunto had, may more fully and at large appear.

A.D. 1818.—N° 4204.

Brunel's Method of Forming Tunnels or Drifts under Ground.

NOW KNOW YE, that in compliance with the said proviso, I, the said Mark Isambard Brunel, do hereby declare that the nature of my said Invention, and the manner in which the same is to be performed, are particularly described and ascertained in and by the Drawings hereunto annexed, and the following description thereof (that is to say) :—

I shall premise by observing, that the chief difficulties to be overcome in the execution of tunnels under the bed of great rivers lies in the insufficiency of the means of forming the excavation. The great desideratum, therefore, consists in finding efficacious means of opening the ground in such a manner that no more earth shall be displaced than is to be filled by the shell or body of the tunnel, and that the work shall be effected with certainty.

The first method I shall describe for obtaining this desirable result, is applicable to a tunnel of large dimensions as well as to a simple drift or a driftway. In the formation of a drift under the bed of a river, too much attention cannot be paid to the mode of securing the excavation against the breaking down of the earth. It is on that account that I propose to resort to the use of a casing or a cell, intended to be forced forward before the timbering which is generally applied to secure the work. This cell may be similar to one of those represented in Fig. 1, see letter C. The workman thus inclosed and sheltered may work with ease and in perfect security. It is obvious that the smaller the opening of a drift, the easier and the more secure the operation of making the excavation must be. A drift on dimensions not exceeding three feet in breadth by 6 feet in height, forms an opening of 18 feet area; whereas the body of a tunnel on dimensions sufficiently capacious to admit of a free passage for two carriages abreast cannot be less than 22 feet diameter, consequently about 20 times as large as the opening of a small drift. One of the modes which I propose to follow for the purpose of forming excavations suitable to tunnels of large dimensions consists in rendering the operation nearly similar to that of forming a small drift, as being the most easy and the most secure way of proceeding. The apparatus represented in the Fig. 1, 2, 3, & 4, is one of the nature above described, and whereby a large excavation suitable to the dimensions of the proposed tunnel may be made. This apparatus is intended to precede the body or shell of the tunnel, and it is represented as if the work had already been commenced with a part of the tunnel *a, a, a, a,* Fig. 2 & 4, formed behind it. Fig. 1 represents a transversal view of an apparatus composed of small cells A, B, C, D, E, F, G, H, J, K, lying alongside of and parallel with each other. Each cell may be forced forward independently of the contiguous one, so that each workman is supposed to operate in a small drift indepen-

---

A.D. 1818.—N° 4204.

Brunel's Method of Forming Tunnels or Drifts under Ground.

dently of the adjacent one. The front of the work is protected by small boards L, L, L, L, which the workman applies as he finds most convenient. Several men may work at the same time with perfect security and without being liable to any obstruction from each other.

Fig. 2 represents a longitudinal section of the apparatus, wherein a work man is seen at work. Each cell is to be moved or forced forward by any mechanical aid suitable to the purpose, but I give the preference to hydraulic presses M, M, Fig. 2, which are made to abut against a strong framing N, N, N, fixed within the body or shell of the tunnel. When the ground has been removed and the several cells have been forced forward to a sufficient distance, then a space P, P, corresponding with that distance, is left between the ends of the cell and the shell of the tunnel *a, a, a, a.* But in order to prevent the breaking down of the earth, or the eruption of a large body of water to each of the cells, and on that side of each which is exposed to the pressure of the earth, I apply strong iron plates extending beyond the cell, so as to overlap the shell of the tunnel previously made, thus protecting the space already cleared. The body or shell of the tunnel may be made of brick or masonry, but I prefer to make it of cast iron, which I propose to line afterwards with brickwork or masonry.

Fig. 3 represents a transvers sectional view of the tunnel at the line *b, b,* Fig. 4, shewing the framing which is to form the abatement for the hydraulic presses.

Fig. 4 is a plan exhibiting the internal arrangements of the cells, the hydraulic presses M, M, M, M, and the framework N, N, N, forming the abatement of those presses. Each cell in the longitudinal direction thereof is formed into a prismatic figure such as adapts itself to the situation which it respectively occupies in the area which is intended to be excavated. In order to facilitate the progressive movement of the cells I introduce friction rollers between the opposite sides of all the cells, one row of which, for the sake of exemplification, is represented at P, Fig. 2. As the construction of the hydraulic presses is well understood by mechanics in general, and as the application and use of the said presses and framework forming the abatement to the same must be sufficiently apparent from the preceding description thereof and the various Figures in the annexed Drawings, they require no further explanation; I have only to add, that after so much of the earth, as before described, has been removed by the workman, and after the cells have been forced forward by the aid of the presses into the position as represented in Fig. 2 and 4, and also after another portion of the shells of the tunnel has been added so as to occupy or fill up the space *p, p,* it then becomes necessary

---

A.D. 1818.—N° 4204.

Brunel's Method of Forming Tunnels or Drifts under Ground.

that the framework or abatement N, N, N, should be moved forward through a space equal to that through which the cells had previously been moved. The said framework having been so moved, it must again be firmly fixed in its new situation, so as to resist the reaction of the presses as the repetition of their aid becomes necessary to press or force forward the cells on the removal of fresh portions of earth. As no competent workman would be at a loss to find adequate means to secure the abutment in its place, so as that it should not recede or give way by the opposed action of the presses in pushing forward the cells, I need not here point them out; I would, however, explain the method which I employ for advancing or moving the abatement forward whenever that becomes necessary. It is obvious that the piston rods of the presses must extend to and be in immediate contact with the abatement, but I also make them fast to the said abatement, and by making the said presses double-acting, or, in other words, in such a manner as that the force of the water may be employed as occasion requires on either side of the pistons of the said presses, I obtain, from the agency of the said presses when so constructed, not only the means of pushing forward the cells, but also of drawing forward the said abatement.

Before I proceed to describe my other method of forming tunnels or drifts, I would premise that the leading features of this other method consists in rendering the operation nearly similar to that of forming one or more small drifts or driftways; the combination of mechanical expedients by means of which I perform the same, I denominate a teredo or auger, from its great analogy to that instrument and also the vermes, known under the name of Teredo Navalis. This insect is capable of perforating the toughest timber by the power and organization of its augur-like head worked by the motion of the body inclosed within its tubular cell, which cell may be supposed to represent a tunnel. The head, which forms part of the mechanical combination alluded to before, has not, however, received the confirmation of an auger for the purpose of cutting into and through the ground, but the protruding lip forms the orifice through which the earth is to be removed by manual aid. This head forms likewise a shield that is to protect the work against the breaking down of the earth. In this instance the shell of the tunnel is formed by a systematic combination of hollow pieces of cast iron which, when united together by bolts or screws, presents externally somewhat the appearance of a riband wound edge to edge round a cylinder throughout its whole length, as seen in Fig. 9 from A to B, when it is evident the riband would take a direction similar to that of the thread of a screw.

Fig. 9 is a representation of the exterior of a part of the tunnel already formed, as from A to B and B to C; it represents the exterior of the apparatus which I employ in forming the excavation. Fig. 10 is an end view of the head of the apparatus, in which two orifices are brought to sight, one at D and the other at E. These orifices may be considered as the mouths of two teredos through each of which the earth is to be removed; Fig. 8 is an internal view of the same. In both views 8 & 10 the end of the larger teredo is represented as being formed of a number of radiant pieces extending from the circumference of the tunnel to the cylinder forming the body of the lesser teredo. The cylindrical body of the larger teredo is composed of a number of pieces resembling externally staves. To the end of each of these staves is affixed one of the radiant pieces $a, a, a, a,$ &c. These staves with the radiant pieces attached to them are capable of being moved forward in the longitudinal direction of the tunnel one at a time. In order to overcome the friction of the staves sliding against the earth and against each other, I employ of hydraulic presses one press to each stave, attached in the manner represented at 6, b, Fig. 6. Fig. 5 & 11 represent sectional views of the body of the greater teredo, and illustrate the form, disposition, and union of the staves. Fig. 6 represents a longitudinal section of Fig. 9, explaining the internal construction of the apparatus. This apparatus may be divided into three separate heads, namely, the body of the larger teredo already described, the body of the lesser teredo with its shaft and the internal drum K, Fig. 5, 6, and 11. This drum first serves to give and preserve the cylindrical form of the body of the teredo, next it supports the staves forming that body, and to the said drum the hydraulic presses marked $b, b,$ are fixed. In Fig. 11, the internal edges of the staves are seen in contact with and bearing upon the friction wheels c, c.

Having already said that the staves and radiant pieces attached to them are moveable, and the means by which they are moved, I now proceed to explain the object for which they are moveable, and the order in which they are to be moved. It is evident from the nature of the thing that the orifice or mouth of the teredo should always be presented to the earth that is to be removed; but as the whole apparatus does not move round so as to produce this effect, it is necessary that the staves with the radiant pieces should be made subservient to that end, this is rendered practicable by the order in which they can move so as to preserve at all times the character of the teredo, not by turning the teredo itself, but by giving a progressive motion to the protruding lip of the teredo, which remains to be described. This lip is composed of an iron plate sufficiently strengthened and of the shape seen at $d, e,$ D, & $f,$ Fig. 10, and also of the portion of a cylinder extending beyond and over

lapping the end of the opposite staves. This lip is capable of being moved round the body of the lesser teredo, and may be forced forwards as the work proceeds. When this lip has been moved through a space equal to the width of a stave, the stave next in succession must be forced forward in the order already described and contained in the operation to the end of the work. Other expedients might be pointed out which the nature of the ground might render necessary so as to reduce the aperture of the orifice or closing it entirely. The lesser teredo, so much resembling the common auger or a boring bit, requires no further description, as its application is too obvious. When the lesser teredo is in action, the workman stands in the body of the teredo, and applies through the orifice the tools or instruments necessary to displace the earth.

Fig. 7 is a representation of a tunnel when completed, having a small drain O, as seen in Fig. 6 & 7, underneath it.

Lastly, I declare that I do not claim any of the mechanical implements I make use of in their detached character as forming any part of my Invention, but my Invention consists in the combination and application of the expedients already described.

In witness whereof, I, the said Marc Isambard Brunel, have hereunto set my hand and seal, this Twentieth day of July, in the year of our Lord One thousand eight hundred and eighteen.

Mᶜ J. BRUNEL. (L.S.)

AND BE IT REMEMBERED, that on the Twentieth day of July, in the year of our Lord One thousand eight hundred and eighteen, the aforesaid Marc Isambard Brunell came before our said Lord the King in His Chancery, and acknowledged the Specification aforesaid, and all and every thing therein contained and specified, in form above written. And also the Specification aforesaid was stamped according to the tenor of the Statutes made for that purpose.

Inrolled the Twentieth day of July, in the year of our Lord One thousand eight hundred and eighteen.

Redhill: Printed for Her Majesty's Stationery Office, by Malcomson & Co., Ltd.

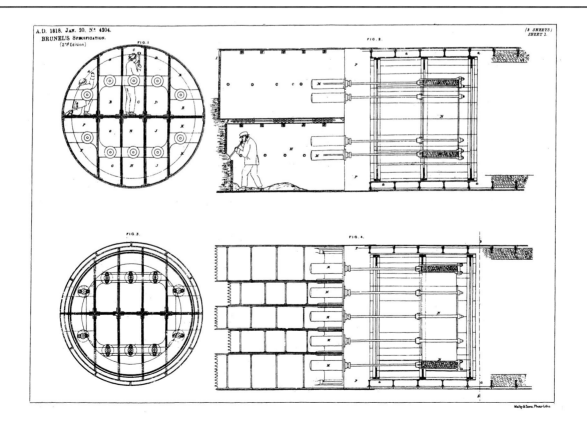

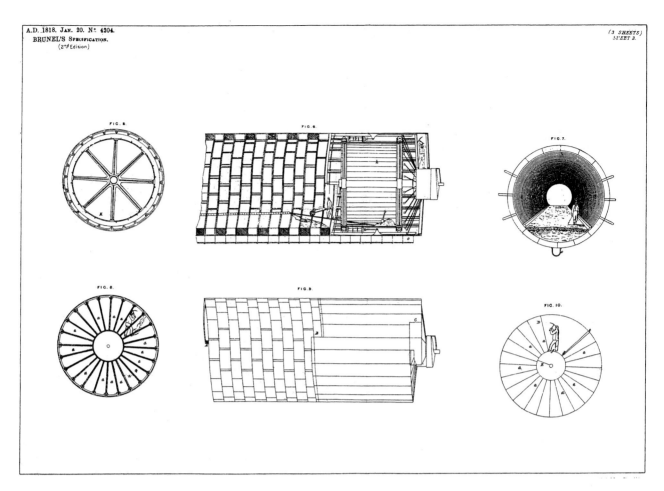

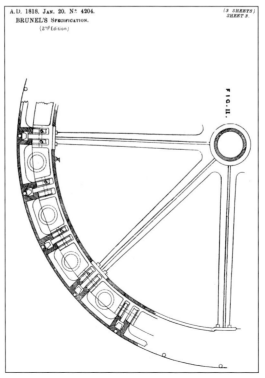

A.D. 1818 . . . . . . N° 4301.

Manufacture of Tin Foil.

BRUNEL'S SPECIFICATION.

TO ALL TO WHOM THESE PRESENTS SHALL COME, I, Marc Isambard Brunel, of Chelsea, in the County of Middlesex, Civil Engineer, send greeting.

WHEREAS His most Excellent Majesty King George the Third did, by His
5 Letters Patent under the Great Seal of the United Kingdom of Great Britain and Ireland, bearing date at Westminster, the Fifth day of December, in the fifty-ninth year of His reign, give and grant unto me, the said Marc Isambard Brunel, my exōrs, adōrs, and assigns, His especial licence, full power, sole privilege and authority, that I, the said Marc Isambard Brunel, my exōrs,
10 adōrs, and assigns, during the term of years therein mentioned, should and lawfully might make, use, exercise, and vend, within England, Wales, and the Town of Berwick-upon-Tweed, my Invention of "**A New Species of Tin Foil, capable of being Crystallized in Large, Varied, and Beautiful Crystalliza-tion**;" in which said Letters Patent there is contained a proviso, that if I, the
15 said Marc Isambard Brunel, shall not particularly describe and ascertain the nature of my said Invention, and in what manner the same is to be performed, by an instrument in writing under my hand and seal, and cause the same to be inrolled in His Majesty's High Court of Chancery within six calendar months next and immediately after the date of the said Letters
20 Patent, that then the said Letters Patent, and all liberties and advantages whatsoever thereby granted, shall utterly cease, determine, and become void, as in and by the same, relation being thereunto had, will more fully and at large appear.

NOW KNOW YE, that in compliance with the said proviso, I, the said
25 Marc Isambard Brunel, do hereby declare that the nature of my said Invention of a new species of tin foil, capable of being crystallized in large,

---

varied, and beautiful crystallization, and the manner in which the same is to be performed, are particularly described and ascertained by the following description thereof, that is to say:—

To convey a correct idea of my Invention, it must be observed, first, that tin in a state of fusion assumes in cooling a crystalline conformation, but if the 5 surface is scraped or rubbed, or if the piece obtained from fusion is rolled or beat, the crystalline conformation is broken, sandy, and confused. The tin foil commonly known and used in the looking-glass trade in particular, being pre-pared by rolling and beating, is incapable of producing a large crystallization. In order to restore it to that state in which it can assume a crystalline con- 10 formation, and become capable of being crystallized in large, varied, and beautiful crystallization, it must previously be melted. The Invention consists, therefore, in the process and in the means which I employ for the purpose of melting sheets of tin, and particularly the tin foil, which, in general, is little more than $\frac{1}{100}$ part of an inch in thickness, and which may be several feet 15 in surface. The sheet of tin or of tin foil intended to be operated upon is laid and spread upon a table or plate of iron, cast or wrought, heated to a tempera-ture a few degrees under that at which tin melts. This plate or table is either of metal, glass, stone, or any other suitable substance. The surface must be very smooth and very flat; great care is to be taken that no air is left under 20 the sheet of tin, when it has been laid down perfectly smooth, and as close as it appears on the back of a looking glass. I then apply an additional degree of heat over the surface of the tin by means of a gas flame, which I use in preference to any other means of heating, because the gas conveyed by means of flexible pipes renders the flame as manageable as a pencil. I am able, by 25 that means, to produce with certainty of effect a large, varied, and beautiful crystallization. The flame must be applied with great precaution, but practice will shew how to moderate that heat and regulate its application. When the operation is over, the sheet assumes a new character, it is then a new species of tin foil, which it was my object to obtain; the separation from the plate 30 or table takes place spontaneously by merely blowing on the surface of the foil; the tin being thus restored to its natural state, possesses the crystalline conformation which is necessary to answer my purpose. This crystalline con-formation is not, however, suitable until it is elicited by the action of acids, which are applied according to the process which is practised on tin plates, that is to 35 say, take one part, by measure, of sulphuric acid, and dilute it with five parts of water. Take also one part of nitrous acid of commerce, and dilute it with an equal bulk of water, and keep each of the mixtures separate. Then take ten parts of the sulphuric acid, diluted in the manner before stated, and

---

mix it with one part of the diluted nitric acid, and then apply this mixed acid to the surface of the tin foil with a soft brush or a spunge, and repeat the application of the said mixed acid until the result you expect proves satis-factory. When this has been done, the foil must be well washed with clean
5 water and immediately dried. It may then be covered with a varnish or japan, more or less transparent, colourless or coloured at pleasure. The table or plate might be heated in an oven or on a stove, or in any other way, but as the heat must indispensably be to a given degree all over its surface, I have given the preference to the apparatus described hereafter. This apparatus
10 consists in a large pan on a trough of cast iron, of such dimensions as to admit of the plate or table being immersed in it, or being laid over its content. Under this pan or trough I have one or more fires, with their respective flues and registers. The trough is filled with a metallic composition, say, $\frac{2}{3}$ tin and $\frac{1}{3}$ lead, which will fuse and remain fluid at a temperature many degrees
15 below that at which tin will melt. The plate or table is laid on the fluid metal, over which it naturally floats and keeps itself in an horizontal position. Oil, mercury, sand, and other substances might have been used as a medium for heating the plate or table, but I give the preference to a metallic com-position, and the greater the body of it in fusion the more regularly the heat
20 will be diffused and sustained. One or two thermometers applied to this apparatus will prove very useful to indicate the degrees of heat and to regulate the same by the aid of the registers.

In witness whereof, I, the said Marc Isambard Brunel, party hereto, have hereunto set my hand and seal, this Fifth day of June, in the
23 year of our Lord One thousand eight hundred and nineteen.

Mᶜ I. (L.S.) BRUNEL.

AND BE IT REMEMBERED, that on the Fifth day of June, in the year of our Lord 1819, the aforesaid Marc Isambard Brunel came before our said Lord the King in His Chancery, and acknowledged the Specification aforesaid,
30 and all and every thing therein contained and specified, in form above written. And also the Specification aforesaid was stamped according to the tenor of the Statute made for that purpose.

Inrolled the Fifth day of June, in the year of our Lord One thousand eight hundred and nineteen.

LONDON:
Printed by George Edward Eyre and William Spottiswoode,
Printers to the Queen's most Excellent Majesty.

A.D. 1820 . . . . . . N° 4434.

Stereotype Printing Plates.

BRUNEL'S SPECIFICATION.

TO ALL TO WHOM THESE PRESENTS SHALL COME, I, MARC ISAMBARD BRUNEL, of Chelsea, in the County of Middlesex, Engineer, send greeting.

WHEREAS His late most Excellent Majesty King George the Third did, by His Letters Patent under the Great Seal of the United Kingdom of Great Britain and Ireland, bearing date at Westminster, the Twenty-fifth day of January, in the sixtieth year of His reign, give and grant unto me, the said Marc Isambard Brunel, my exors, admors, and assigns, His especial licence, full power, sole privilege, and authority, that I, the said Marc Isambard Brunel my exors, admors, and assigns, should and lawfully might make, use, exercise, and vend, within England, Wales, and the Town of Berwick-upon-Tweed, during the term of years therein mentioned, my Invention of "CERTAIN IMPROVEMENTS IN MAKING STEREOTYPE PLATES;" in which said Letters Patent there is contained a proviso that if I, the said Marc Isambard Brunel, shall not particularly describe and ascertain the nature of my said Invention, and in what manner the same is to be performed, by an instrument in writing under my hand and seal, and cause the same to be inrolled in His Majesty's High Court of Chancery within six calendar months next and immediately

---

2   A.D. 1820.—N° 4434.

Brunel's Improvements in Making Stereotype Plates.

after the date of the said Letters Patent, that then the said Letters Patent, and all liberties and advantages whatsoever thereby granted, shall utterly cease, determine, and become void, as in and by the same (relation being thereunto had) will more fully and at large appear.

NOW KNOW YE, that in compliance with the said proviso, I, the said Marc Isambard Brunel, do hereby declare that the nature of my said Invention, and the manner in which the same is to be performed, are described and specified in manner following, that is to say:—

It consists in the means and methods which I adopt,—

First, for taking moulds from the original plate of types or woodcuts; and, secondly, for obtaining the cast, or the printing plate, from the said mould.

As the various processes which are commonly practised in manufacturing stereotype plates could not be rendered sufficiently expeditious to answer the object I had in view, namely, that of multiplying printing plates for the purpose of accelerating the printing of daily papers, I have, in order to attain that desideratum, adopted the following means and methods in which my improvements consist, and which are equally applicable to the making of stereotype plates in general in an easy and expeditious manner. By the method in question the mould is made by compression somewhat in the way that an impression might be taken from types or woodcuts by means of wax, and for that purpose I make use of a suitable composition, but that which I prefer consists of pipe clay 7 parts, chalk or burnt clay pounded very fine 12 parts, and starch one part in bulk, not by weight. The whole being mixed with water is made into a paste of the consistency of thick putty. In that state it is fit to take an impression from types, and I accordingly use it as follows; that is, I spread it over a metal plate, the size of which ought to exceed the plate of types by half an inch at least. I adopt in preference a steel plate as thin as a common tenon saw, which plate is perforated with small holes, say $\frac{1}{16}$ of an inch diameter, and about half an inch apart from each other. The composition and plate being so prepared the next process is the spreading of the composition over the steel plate in such a manner as to afford a proportionate supply to the different sizes of types, as well as to the distances and vacant spaces. This is done by the following means, that is, the original plate of types or woodcuts, or both together, from which a mould is intended to be taken, must previously be well secured in a chace or galley. The instrument I use being in form and character like a galley, I shall denominate it a galley. The bottom part of the galley in question should be very flat, and the whole very strong. The upper edge of the sides should be nearly on a

---

A.D. 1820.—N° 4434.   3

Brunel's Improvements in Making Stereotype Plates.

level with the surface of the types. The steel plate should be connected with and adjusted to the galley by means of two centers acting like a hinge. The plate may in that way act as the lid of a box, and may be brought down over the surface of the types. These implements being so disposed, I proceed by laying a sufficient quantity of the composition over the inner surface of the steel plate, which I denominate the mould plate. I then spread a piece of skin or parchment in a wet state over the surface of the composition. I further lay over the surface of the plate of types a covering consisting of several folds of thin calico or other suitable stuff stretched in a very light frame. The mould plate thus covered with the composition, and with the skin or parchment, is then laid down upon the surface covered by the calico, and pressed or rubbed with some degree of force, so as to effect the distribution of the composition over the mould plate; a roller passed several times over the back of the mould plate answers very well for the purpose. The use of the skin and other covering is to prevent immediate contact with the type plate while the composition is forced to diffuse itself over the surface of the mould plate. The excess of the composition oozes out partly at the sides, and partly through the small holes perforated all over the mould plate. After this first operation the composition on the mould plate exhibits all the prominent features of the impression. The skin and calico covering are then taken off, and in lieu of them I substitute two leaves of thin paper, both over the type plate. The pressure is then repeated until the impression appears more distinct. One of the two leaves is then taken off from the surface of the mould to which they have adhered and the pressure repeated. The last leaf is stripped off in its turn, and before laying the mould down the face of the types should be rubbed over with a brush which has been slightly touched with oil. The mould plate is then turned down again, and the composition now being in immediate contact with the types receive by a gentle action that finish which is requisite to obtain a perfect impression. The flexibility of the mould plate contributes very materially to facilitate the operations. The thinner the mould is reduced by the last action of rubbing or pressing the better, provided the thickness of the composition which remains at the bottom of the letters is even and uniform. It cannot, indeed, be too thin; it would even be desirable that the steel mould plate should come in contact with the force of the types. The use of the roller in producing the mould by compression might be done away with, and the operations might be effected by means of a press similar to a printing press; on the other hand the mould plate being flexible instead of its being connected with and adapted to the galley by means of two centers acting like an hinge, as has been specified before, might be applied to and fastened round

---

4   A.D. 1820.—N° 4434.

Brunel's Improvements in Making Stereotype Plates.

a roller, say 12ⁱⁿ. diameter, and four or five inches broad (these dimensions may be very arbitrary), in which case the impression would be obtained by either sliding the galley under the roller, or by carrying the roller over the types contained in the galley; in the first of which cases the roller should revolve on an axis bearing on two standards, fixed on a bed proportional in its dimensions to those of the galley. These two standards should be at such a distance from each other, and the center of the axis should have the means of being adjusted sufficiently high to admit of the galley being moved to and fro under the said roller. The parrallelism of this reciprocating action should be maintained by lateral guides fixed to the bed in question. The parts being so disposed are ready to operate. The steel mould plate having been previously coated with the composition, and covered with the skin or parchment, and the calico or stuff covering having been laid over the types, the galley may then be brought into action under the roller until the surface of the composition, which now assumes the character of a cylindrical mould, exhibits the prominent features of the impression; the excess of the composition oozes out at the sides, or is pressed out by the revolving action of the roller. The skin or parchment and the other coverings are next to be removed, and replaced by the two leaves of thin paper, as has been stated before, and when after repeating the action of the galley, the impression appears as deep and as distinct as is required, the paper leaves are stripped off in succession. The composition is now brought in contact with the types, which are to be previously brushed over, as has been described before, and in that state the reciprocating movement of the galley is repeated, and the roller adjusted at the same time, until the impression is as deep and as perfect as is required. It is necessary to observe, that in order to insure the whole effect, which should result from this mechanical arrangement, the roller and galley are to be connected together by means of a wheel with teeth and a rack. The wheel is to be fixed on one extremity of the roller, and the rack fitted and adjusted to one of the edges of the galley, so as to combine their respective action. It is easy matter to understand how the action may be reversed, so as to effect that which is adverted to in the second case, namely, that of carrying the roller over the types contained in the galley; the axis of the roller should extend at both sides and terminate into two handles. A tooth wheel of a fine pitch is to be adapted to each flat end of the roller, and meet with a rack of a corresponding division, fastened on each edge of the galley. The lateral action of the roller is to be confined by means of two flanches extending beyond the diameter of the toothed wheels. The roller so disposed may be rolled to and fro over the face of the types with ease and precision. The mould plates being fastened round it, as has been said before, every

other successive operation connected with the various coverings and with the types being attended to, an impression may be obtained by this method in as perfect a manner as may be required. And after having effected it as directed in this as well as in the first case, the mould plate is taken off from the roller, and
5 as it returns to a plain surface, it produces a flat mould, which was the object anticipated from either of these two modes. A mould taken by either of the methods herein specified is easily dried by exposure to the air, or by a moderate heat in a stove, which heat should in no instance exceed 500°. The mould is by that means ready prepared in a very expeditious way to receive
10 the metallic composition of which the stereotype plate is intended to be made. And in order to obtain the cast, I proceed as follows: that is to say, the mould plate is placed with the face upwards in an instrument resembling a tray, which I accordingly denominate a tray, of dimensions fit to hold the steel mould plate. The sides of the tray are about one inch high, and the bottom
15 is open in holes of about one inch in diameter as close as possible. These apertures are intended to admit water or other fluid, which is injected for the purpose of cooling the mould plate. The metallic compound of which the stereotype is to be made may be same as used in the ordinary way; but when expedition is an essential point, I take in preference the well known alloy,
20 consisting of bismuth 10 pounds, lead 6 pounds, and tin 4 pounds, which alloy is heated to about 400°. When the cast is to be obtained I proceed as follows; that is, I place the mould plate in the tray with the face of the mould upwards. I then apply above the surface of the mould plate a thick plate of cast iron, say ¾ of an inch in thickness. This plate, which I intend to regulate
25 the thickness of the stereotype plate, is disposed in manner following:—First, this regulating plate must be of dimensions corresponding exactly with the size which the stereotype is to be. Its four sides must have prominent edges projecting from the surface as much as the thickness which is to be given to the stereotype plate. At each corner of the regulating plate there is an
30 adjusting screw intended to support it in its place. I then place this plate with the prominent edges downwards over the mould plate, taking care that it corresponds with its outlines both in breadth and length. This regulating plate rests then upon the four adjusting screws. The tray being so disposed requires to be heated to about 200° before the metal can be poured in. I
35 now proceed by pouring a sufficient quantity of fluid metal into the tray. The metal makes its way round and under the regulating plate, which may be raised by hand if required. I immediately inject, by means of a flexible pipe connected with a small reservoir of water or other fluid, a sufficient quantity of such water or other fluid through the apertures at the bottom of

the tray, which strikes the bottom of the mould plate, and cools the metal in it. This is productive of a good effect although the thickness of the metal in fusion is but very inconsiderable. The cast is soon ready to be removed, and while it is still hot the extra metal round the regulating plate breaks off almost of itself, leaving the stereotype plate of its proper dimensions. 5 The composition is then washed off, and the plate may be considered to be fit for the press, and it is blocked up as may be required. This I do by having plates of proper sizes and thickness, over which the stereotype plates are fixed by means of wax hardened with some rosin. As it is very desirable to have a great pressure of metal when the finest casts are to be obtained, 10 in these cases I proceed in the following manner; that is, I place the tray with the mould plate and regulating plate in a metal chamber which can be hermetically closed. The chamber I make use of is a cast-iron cylinder placed horizontally, closed at one end, and having a moveable door plate so constructed as to be capable of instant application, and of great 15 resistance. On the upper part of the chamber there is a small orifice destined to admit the metal into the chamber. The chamber is connected by means of a pipe furnished with a stopcock to a vessel in which air has been condensed to a sufficient degree. I also adapt to this chamber a small forcing pump intended to throw in through the under part water 20 or any other cooling fluid. The chamber being so disposed and fit to receive the tray and mould plate, is prepared for ordinary use. I proceed as follows; that is, I enclose the tray with its contents, which have been previously heated, in the chamber, which is immediately shut up, and I pour the metal through the orifice, which should be instantaneously stopped herme- 25 tically. The metal soon makes its way under the regulating plate and spreads itself over the surface of the mould. I immediately open the communication with the condensed air, which produces the effect of a head of metal, and I inject water, or other cooling fluid, for the purpose of setting the metal in the cavities of the mould. Having the advantage of a 30 chamber, which admits of being hermetically closed, I can make use of it to obtain first a vacuum, a method which has been, I understand, practised by others, but not in connection with nor immediately followed by the expansive power of condensed air, which is productive of the best effect, and which, when coupled with the means of cooling the metal, must accelerate very materially 35 the manufacture of stereotype plates in general. As the mould plate retains, even after it is completely dry, a sufficient degree of flexibility to admit of the mould being shaped into segments or portions of a cylindrical form, and would produce in this manner stereotype plates corresponding with that form, I

propose effecting this if it should be required by having a box of cast iron, of a depth exceeding by at least one inch the breadth of the mould plate, the sides of which would be formed to a given radius, and lay parallel with each other, and no farther apart than is necessary for the stereotype plate. The
5 mould plate being placed in it edgewise, and the back being forced against the concave side of the box, the metal may then be poured in, and the stereotype plate will be found, when cooled and set, shaped to the given radius, or nearly so. It is necessary to heat the box before the metal is run in it.
It is an easy matter to conceive that the means and methods I have
10 herein described are susceptible of being modified, and may even be accomplished by a variety of ways; but as the nature and originality of my improvements consist in, first, taking from the original plate of types moulds which are calculated to answer my purpose, and, secondly, for obtaining a cast from the said moulds, I have confined my Specification to a description of those
15 means and methods which I have found to answer best in practice, considering that expedition is a most essential point to be attended to. I shall, however, add, that a plate, possessing all the properties of a stereotype plate, may be made in a very perfect manner by a kind of metal plating laid over a substance of the nature of japan or of thick varnish. And for this purpose I take
20 some shell-lac, and having melted it, I spread it over an iron plate ⅛ of an inch, or thereabouts, in thickness. This coating is hardened by exposure to heat, until it acquires a consistency sufficient to oppose some resistance in the mould. I immerse the plate so coated in type metal in fusion, just as is practised in the operation of tinning. The plate comes out tinned as it were
25 all over the coated surface, and as this substance retains its heat for a sufficient time, the metal coating remains also in a fluid state so long as the heat continues above the melting point. I then apply the metal plate to the mould plate, and press both together. The mould plate should be previously heated to a degree somewhat below the melting point of the type metal employed.
30 The plate and mould are left to cool under the press. The impression thus obtained is remarkably perfect. The same plate may be made to take another or several successive impressions, by repeating at each time the immersion into type metal, and proceeding as stated before. This mode of obtaining stereotype plates is very expeditious, and may prove of great benefit.
35 In witness whereof, I, the said Marc Isambard Brunel, have hereunto set my hand and seal, this Twenty-fifth day of July, in the year of our Lord One thousand eight hundred and twenty.

Mᶜ. Iᴰ. (L.S.) BRUNEL.

AND BE IT REMEMBERED, that on the Twenty-fifth day of July, in the year of our Lord 1820, the aforesaid Marc Isambard Brunel came before our said Lord the King in His Chancery, and acknowledged the Specification aforesaid, and all and every thing therein contained and specified, in form above written. And also the Specification aforesaid was stamped according 5 to the tenor of the Statute made for that purpose.

Inrolled the Twenty-fifth day of July, in the year of our Lord One thousand eight hundred and twenty.

Redhill: Printed for His Majesty's Stationery Office, by Love & Malcomson, Ltd.

A.D. 1820 . . . . N° 4522.

Copying Presses.

BRUNEL'S SPECIFICATION.

TO ALL TO WHOM THESE PRESENTS SHALL COME, I, Marc Isambard Brunel, of Chelsea, in the County of Middlesex, Civil Engineer, send greeting.

WHEREAS His most Excellent Majesty King George the Fourth did, by
5 His Letters Patent under the Great Seal of that part of the United Kingdom of Great Britain and Ireland called England, bearing date at Westminster, the Twenty-second day of December, in the first year of His reign, give and grant unto me, the said Marc Isambard Brunel, my exors, adnors, and assigns, His especial licence, full power, sole privilege and authority, that I,
10 the said Marc Isambard Brunel, my exors, adnors, and assigns, during the term of years therein mentioned, should and lawfully might make, use, exercise, and vend, within England, Wales, and the Town of Berwick-upon-Tweed, my Invention of "A Pocket Copying Press, and also certain Improvements on Copying Presses;" in which said Letters Patent there is contained a
15 proviso that if I, the said Marc Isambard Brunel, shall not particularly describe and ascertain the nature of my said Invention, and in what manner the same is to be performed, by an instrument in writing under my hand and seal, and cause the same to be inrolled in His Majesty's High Court of Chancery within six calendar months next and immediately after the date
20 of the said Letters Patent, that then the said Letters Patent, and all liberties and advantages whatsoever thereby granted, shall utterly cease, determine, and become void, as in and by the same, relation being thereunto had, will more fully and at large appear.

NOW KNOW YE, that in compliance with the said proviso, I, the said Marc Isambard Brunell, do hereby declare and ascertain the nature of my said Invention, and the manner in which the same is performed, by a description and Drawing thereof (that is to say):—

That my Invention doth consist in a light pocket press for transferring 5 writing by means of a damped medium, without the necessity of using wetting or drying books, and for other purposes.

DESCRIPTION OF THE DRAWING.

Fig. 1 represents a longitudinal section of the press when closed or in the act of pressing. A is the bottom of the press, made of the best gun metal, at 10 each end of which is a small standard or upright of the same metal; C, D, the upper end of the standard; C forms a joint for the brass handle or lever; E, G, is a second lever made of steel, moving from its fulcrum at X, resting at or near its center on the screw head H, and acted upon at its extremity L by that part of the handle or lever E at F. B is the convex bottom of a wooden casing, 15 the further side of which is in this section coloured red. This convex bottom moves up and down between the standards C, D, and has a thin steel plate spring J curved so as to correspond with the convexity of the bottom, and attached to it by the screw H. The two ends of this spring rest in apertures 20 cut in the standards for their reception. By this means when the two levers E and G act upon the screw head at H and press down the wooden casing, the spring being kept up at each end, is forced into a straight line, and then if the pressure on the screw head be removed by the elevation of the handle or lever E, as shewn at Fig. 4, the spring resumes its curved form, raising the 25 wooden casing up with it. M is merely a hollow case in which to keep the damper. Figs. 2 and 3 represent end views of the press. I shews the damper, which will be hereafter described in the case M; the sides of the wooden casing are represented at B, B. Fig. 4 represents the press when open or when the pressure on the screw head H is removed by the handle or 30 lever E having been raised. In this Figure it will be seen that the spring j has resumed its curved form, and that the wooden casing B is considerably raised from the bottom of the press. Fig. 5 is a section of the press, shewing the transferring book passing under it for pressure. O represents the transferring book which consists of blank leaves of very thin paper in the ordinary 35 way. Fig. 6 is a plan of the press shewing the form and thickness of the lever G. Fig. 7 is a longitudinal section of the press shewing the wooden casing which conceals the levers, &c. Fig. 8 is a transverse section of the handle or lever E. Fig. 9 is a longitudinal section of the damper, which

---

consists of a small metallic cylinder around which a number of sheets of calico, fine linen, or other ligh material of the size of the paper used is rolled. This cylinder of the calico rolled round it is damped and kept in a tube d, d, closed at one end, and with a cap or cork at the other end with a sponge Q attached to it,
5 to be used for moistening the calico. The transfer of writing by means of the press is performed in the following manner:—Put the original writing into the transferring box, then turn over it as many blank leaves of the book as there are pages in the original writing; next unroll from the cylinder of the damper as many sheets of the damp calico, and lay each of them over each of
10 the blank leaves, and on each of these put a sheet of oiled paper, shut up the book, and it is ready to go under the press; having been drawn through the press, at the same time undergoing a succession of pressures, the writing will be found transferred. The materials of which this press is made may be varied at pleasure, and a screw or wedge may be introduced to communicate
15 motion to the lever G. The dimensions also of the press may be adapted to particular situations. But a light pocket press for transferring writing by means of a damp medium in the manner here described being entirely new, and never before used in this Kingdom, to the best of my knowledge and belief, I am desirous to maintain this my exclusive right and privilege to the same.

20 In witness whereof, I, the said Marc Isambard Brunel, have hereunto set my hand and seal, this Twenty-second day of June, in the year of our Lord One thousand eight hundred and twenty-one.

M° I° (L.S.) BRUNEL.

AND BE IT REMEMBERED, that on the Twenty-second day of June,
25 in the year of our Lord 1821, the aforesaid Marc Isambard Brunel came before our said Lord the King in His Chancery, and acknowledged the Specification aforesaid, and all and every thing therein contained and specified, in form above written. And also the Specification aforesaid was stamped according to the tenor of the Statute made for that purpose.

30 Inrolled the Twenty-second day of June, in the year of our Lord One thousand eight hundred and twenty-one.

LONDON:
Printed by George Edward Eyre and William Spottiswoode,
Printers to the Queen's most Excellent Majesty.

A.D. 1822 . . . . N° 4683

Marine Steam Engines.

BRUNEL'S SPECIFICATION.

TO ALL TO WHOM THESE PRESENTS SHALL COME, I, Marc Isambard Brunel, of Chelsea, in the County of Middlesex, Engineer, send greeting.

WHEREAS His most Excellent Majesty King George the Fourth did, by His Letters Patent under the Great Seal of that part of the United Kingdom of Great Britain and Ireland called England, bearing date at Westminster, the Twenty-sixth day of June, 1822, in the third year of His reign, give and grant unto me, the said Marc Isambard Brunel, my exōrs, admōrs, and assigns, His especial licence, full power, sole privilege and authority, that I, the said Marc Isambard Brunel, my exōrs, admōrs, and assigns, during the term of years therein mentioned, should and lawfully might make, use, exercise, and vend, within that part of the United Kingdom of Great Britain and Ireland called England, the Dominion of Wales, and Town of Berwick-upon-Tweed, my Invention of "Certain Improvements on Steam Engines ;" in which said Letters Patent there is contained a proviso that if I, the said Marc Isambard Brunel, shall not particularly describe and ascertain the nature of my said Invention, and in what manner the same is to be performed, by an instrument in writing under my hand and seal, and cause the same to be inrolled in His Majesty's High Court of Chancery within six calendar months next and immediately after the date of the said Letters Patent, that then the said Letters Patent, and all liberties and advantages whatsoever thereby granted, shall utterly cease, determine, and become void, as in and by the same, (relation being thereunto had), will more fully and at large appear :

NOW KNOW YE, that in compliance with the said proviso, I, the said Marc Isambard Brunel, do hereby declare that the nature of my said Invention, and the manner in which the same is to be performed, are particularly described and ascertained in and by the Drawings hereunto annexed, and the following description thereof (that is to say) :—

I do declare that my improvements on steam engines are limited to those used for the purpose of navigation ; first, in the manner of disposing and combining two engines so as to apply their powers in the most direct way for producing a rotary action ; secondly, in the method of moderating and regulating the movements of these engines, and producing an equable action in the movement of these engines ; thirdly, in the mode of condensing the steam ; fourthly, in certain adaptations to the boilers.

In the first place, the disposition I adopt to produce the rotary action is by placing the line of the powers in the direction to be at right angles, or nearly so, with each other. Though the right angle is the most favorable position for the respective powers, I do not confine myself to that angle only, as I may contract or increase to a certain extent, as necessity may require. In the Fig. 1, the engines being nearly at right angles with each other, the crank is a solid double crank, as represented at A, Figs. 2 and 5, on which both engines act. The marine engine represented with its parts by the Figures from Number 1 to Number 23 inclusively as the plan of condensation, I adopt for sea service every part of the said engines, therefore, disposed accordingly. Fig. 1 represents the lateral elevation of the said engine ; Fig. 2 a transverse section of the vessel, shewing the position of the engine and the boilers ; Fig. 3 a section of the steam cylinder and of the moderator ; Fig. 4 a ground plot of the engine with the boilers in their respective places ; Fig. 5 a view of the steam cylinder, shewing the connexion with crank A, with the moderator C, and with the air pump B. These Figs. exhibit the principle characters of the engine, combining the four points adverted to. The direction in which the powers act on the crank is too obvious to require further explanation. With respect to the regulating and moderating of the engine, it is effected by an hydraulic apparatus C, Fig. 3, and 5, consisting of a pump, the piston of which forces the water or any other fluid which might be employed through an orifice, which can be contracted or enlarged by the action of a governor, as in the engine now in question. This pump is connected with the condenser. It has a suction pipe B, Fig. 1 and 3, and also an eduction orifice F, Fig. 6. In the Fig. 6 is represented the regulating plug A, destined to contract the arrear of the orifice B. It must be observed that the orifice B is not to be closed entirely. It is proposed to allow a waste equal to about the

---

$\frac{1}{30}$ part of the orifice, in order to avoid shocks by the sudden closing of the plug. In order to facilitate the action of this plug under all degrees of pressure, it is connected by a stem with a small piston C, fitted into a close chamber ; the stem of the plug being hollow allows of a certain quantity of the forced liquid to escape through it, and to act on the back of this piston ; and the plug being thus balanced between two equal pressures, is easily acted upon by the governor. The governor is connected with the rotary action of the engine by means of small cog wheels B, B, Fig. 2, which I prefer to any other method of connexion. As the common governor usually applied to land engines cannot act regularly at sea, I modify the same in a way that will fit it for that particular service. The Fig. 6 represents the marine or locomotive governor as at rest, that is, when the engine works at the given rate. To fit this governor for that peculiar service, it is indispensable that all effect of gravitating or any other power, except centrifugal force, be neutralised. This is done in the manner represented in Fig. 6 ; and as a substitute for gravitation, which in the land engines is used to close the balls, I make use of a spiral spring D, Fig. 6. Thus, as seen in Fig. 6, when the balls are extended as at E, E, the plug will close the orifice B to the utmost degree required, which is such as to counteract the whole of the power of the engine, even if both wheels were out of the water, and thus the action of the engine will be regulated. I need not point out the precise mode of connecting the governor with the moderator, because such connexion must depend upon other arrangements, and is too well understood on land engines to require further details in this.

The next point is the condensation. As sea water is found to be so destructive to the boilers and machinery, I propose to use fresh water only for the purpose of generating steam, and by condensing the whole of the steam that is generated, and returning the water thus obtained without mixing the water of condensation with it, the quantity of fresh water in the boilers will always remain nearly the same. The Figs. 7, 8, 9, 10, represent the condenser, which consists of a combination of pipes surrounded with cold water. Any other contrivance that would present a very extensive surface to the action of the cold water and the steam might answer equally well, but I give the preference to this combination, which consists, first, of an assemblage of large pipes A, Fig. 7 and 8, which collectively form a spacious chamber, wherein the steam is let through the pipes B¹, B³, Fig. 7 and 4. B¹ is the steam pipe, and B³ the waste pipe. That is, when there is an excess of steam, instead of allowing it to escape, as is now the case, I propose to have a safety valve connected with that pipe, by which that excess will ender the condensors. As there is a perfect command of the power of the engine by means of the moderater, I propose to have the paddle

---

wheels shafts connected or unconnected, at pleasure, by the means usually employed, and to let the steam engine work out the steam as it happens to be generated. The smaller pipes B, B, B, Fig. 9, which, with the large one A, A, shew their disposition in the condensers, are so arranged so as to form independant chambers adapted to the main one, A, and fixed to it by means of a screw C, passing through the middle pipe. The Fig. 10 represents the manner these small pipes are combined together, and how they are let in their respective places. B represents the plan of a set of pipes ; C, the cups into which they are let ; and D, the seat or socket of the adjustment. All these pipes being enclosed in an iron reservoir supplied with cold water, which is continually changed, the steam entering the pipes is condensed, and the water thus obtained is returned into the boiler by a small pump B, Fig. 8. As it is necessary that a constant supply of cold water should be applied round the pipes, the pump of the moderator draws off the reservoir at every stroke a certain quantity of the heated water, and thus allows of an equal proportion of cold water to enter into it ; and to remove the air that might enter the condenser, there is an air pump B, Fig. 5, which is connected by means of a pipe with one of the vertical chambers of the condenser, as at D, Fig. 4. In this manner the condensation is carried on and the vacuum preserved. With respect to the boilers, it being a desideratum that the weight of metal should be as little as possible without reducing the necessary capacities for the water and steam, with these objects in view, I propose the adaptation for steam rooms of one or more room or rooms of sufficient capacity to boilers constructed in that form which possess the greatest strength, namely, the cylindrical form, with spherical ends, as represented in the Figs. 11, 12, 13, 14, and 15. In the middle of the boiler, and without cutting out the metal at that place, but merely perforating it with small holes, I propose to raise a cylindric or other formed steam room, the height of which is not limited, and if two or more rooms are found necessary they may be added in the same manner. The cylindrical body of the boiler is intended to be kept entirely full of water, and if there is but one steam room there will be very little agitation of the water when at sea. A steam room may be formed by means of cylindrical or any other form chamber laid the parellel direction with the body of the boiler, but I give the preference to the first form. It is to be observed, that with the mode of condensation now proposed there will be no excess of water, and if there should be any waste the boiler cannot be affected by the reduction of one third of the water, provided the steam pipes are made of a yielding metal, as copper. The connection of the steam pipes as intended to be made at the end of the vertical pipe A, Fig. 11. The Fig. 13 represents a longitudinal section of the boiler, and Fig. 14 a section of the plan, both shewing the disposition of the fire places and of the flues. The Fig. 15 represent a transverse section of the boilers and flues. I further propose to adapt to the boilers two coal feeders, with hoppers, as represented at A, A, Fig. 12. These are intended to supply

5 coals through several apertures B, Fig. 13. The same boilers and feeders are seen in Fig. 2. Fig. 16 and 17 represent the coal feeder, consisting chiefly of a cast iron cylinder A, having several cavities C C, C, corresponding with the apertures B, B, B, &c., Fig. 13. In the revolving action of this cylinder the cavities carry a certain quantity of ground coals, and distributed over the fire.

10 The movement is intended to be communicated from the steam engine to the spindle C, C, Fig. 2, which acts by means of endless screws on the axis of the feeding cylinder. Fig. 18 represents a transverse section of the said cylinder on a larger scale. The Figure 5² represents a section on a double scale of the parts D, D, D, in the Figure 5. The Figs. 19¹, 19², 20, 21, 22, and 23,

15 represent certain details relating to the steam piston and steam cylinder. Fig. 19¹ represents part of the head beam and one of the rollers that are destined to direct the parallel movement of the piston rod between two parallel strait edges. Fig. 19² represents the convex part of the piston. Fig. 20 represents a sectional view of the steam cylinders, with the piston and rod. A,

20 steam pipe ; B, convex side of the piston ; D, concave side of the same ; E, E, the packing, explained in Fig. 22 ; F, hollow piston rod ; G, head beam ; H, H, connecting rods ; J, air pump rod ; K, moderator pump rod ; L, L, guide rollers ; M, spring supporters applied at the extremities of the head beam. These supporters are destined to relieve the weight of the piston

25 in its inclined position. The Fig. 21 represents a side view of the spring supporters. Fig. 22 represents the packing of the piston, consisting of a combination of cotton or any suitable substance, A, and of metal, B. The metal spring rings B operate as a metallic packing. 23 represents that part of the piston which corresponds with Fig. 22. It is to be observed that

30 I do not claim those parts that are specified in the Figs. 19¹, 19², 20, 21, 22, and 23, as parts of my improvements, but as necessary agents for the purpose of rendering the action of the steam piston as free from friction and wear as appears practicable. I have not given the dimensions of the steam passages, or explained the nature or the action of the valves, slides, or return pumps,

35 because those parts do not materially differ from the same parts in other steam engines. The blowing valves, stop cocks, &c., are not noticed. As in the navigation of rivers the common mode of condensing may be preferred to the new one which I have explained, it is necessary to shew the adaptation of a common condenser and air pump to the marine steam engine constructed on the

principle I propose. In this case the pump of the moderator has no other function to perform but that of a regulator. The Figure 24 represents an engine drawn on a larger scale than in the preceding Figure 1. The action of this engine differs but little from that of the other. The

5 cranks, instead of being in a single piece, are made in the usual way, because the angle exceeds a right angle so much as to require this adaptation, and in this case the two crank pins are connected with a link in the ordinary manner. A, A, represents the moderator, the section of which is seen in Fig. 26. As the movement of the water is alternate, the

10 plug must be acted upon at both ends in the same manner, as has been described before, and made on the principle which is represented in Fig. 31. It will be necessary to add to the moderator a small pipe, about a quarter of an inch bore, destined to draw in and let out a small quantity of water to compensate for the piston rod, which is alternately in and out of the body of the

15 pump. B, B, represent the condensor (see also B, Fig. 26). C represents one of the air pumps, both which are seen at C, C, 26. These air pumps are intended to act as forcing pumps, and both ways, in order to render the resistance uniform. D, pipe leading from the steam cylinder to the condensor. E, eduction pipe. The Figure 27 represents a plan of the disposition of the

20 moderator (A), of the air pumps (C, C), and of the apparatus (D, D,) destined to counteracting the weight of the piston and rod, in order to prevent the wearing of the lower side of the piston and cylinder. Fig. 28 is a transverse section of the moderator. Fig. 29 represents a natural view of the apparatus D, D, described in Fig. 27. As it is important to point out how

25 the moderator may be applied to steam engines already constructed, particularly those wherein the steam cylinder is vertical, I should propose in that case to adopt a moderator like the one represented in Figures 24 and 25, over the piston rod, in which case it would extend above the deck. Any competent engineer would easily find the means of adapting an apparatus of this descrip-

30 tion to an engine already constructed, and of adapting the governor already described to the shafts. It is desirous that the rotary movement of the governor should be as two to one at least to that of the shaft. As it may be extremely inconvenient to have the moderators above the deck, to prevent this I should propose one on the plan represented in Figs. 30 and 31, in

35 which the piston rod is made hollow, and constitutes the moderator itself. The section area of the piston rod, as represented Fig. 30, is to that of the cylinder in the proportion of 30 to one, consequently the diminution of the power is very inconsiderable. The rod A of this moderator is firmly fixed at B, and the piston C of the moderator is therefore stationary. The Fig. 31 represents a section of this

piston upon a larger scale. The water or any other suitable fluid must pass through the several apertures D, D, D, D, and the plug E is destined to contract the orifice F. This plug is connected with a small rod G, Figs. 30 and 31. This small rod is acted upon by the governor, to which it is con-

5 nected at the head H. In order to facilitate the action of this plug under the alternate pressures to which it is exposed, it has, like that represented in Fig. 25, two small pistons J, J, fitted into corresponding chambers. The stern is hollow. When the water is forced in the direction of the black arrows, a small orifice through the stem allows the water to pass in the back of the lower

10 piston J, and thus keeps it between two equal pressures, and when the water is forced in the direction of the red arrows produces a similar effect on the back of the piston J, and, vice versa, the water is forced alternately from one side to the other, opposing such resistance as the plug E may occasion by the power of the governor. The lower rod K is a mere compensator ; its diameter

15 must be equal to the upper one A. It is to be observed that I lay no claim of improvement whatever to the condensor or to the air pumps, which constitute a part of the marine engine, Fig. 24, as my improvements in this engine are limited to the application of the powers in the most direct way for producing a rotary action by placing the lines of the powers nearly at right angles

20 with each other, to the moderator with its governor, and to boilers of the construction and with the adaptations which I have proposed and specified. The contracted space to which steam engines applied to navigation, particularly in rivers, are often limited, might not very readily admit of the manner of disposing and combining the two engines and their parts, as I have before

25 represented. Then, under such circumstances, I would propose their being placed as they are represented in the Figures 32, 33, 34, 35, and 36. The same letters correspond to the same things. The Fig. 32 represents a lateral elevation of the engine. Fig. 33 represents its component parts open to view. A, A, the steam cylinders ; B, B, B, B, steam passages and slide boxes ; C,

30 air pump ; D, D, condensors ; E, piston rod with a moderator, represented in Figs. 30 and 31 ; F, F, steam pipes ; G, crank for the air pump ; H, H, pipes leading from the condensor into the air pump, with foot valves as usual ; J, J, steam slides rods ; K, K, Figs. 34 and 35, the eccentrics ; L, L, the main cranks. The governor is to be connected with the paddles shafts, and made

35 to act upon the extremities of the small rods at N, N, Fig. 34. When the navigation is confined to rivers, the moderator, the governor, and the mode of condensation which I have proposed are quite unnecessary, consequently a common piston rod is sufficient. O, O, Fig. 32, represent the outriggers, destined to support the weight of the steam piston. These outriggers corre-

spond with the one represented in Fig. 21, and are the same purpose as explained in the description of that Figure. My improvement in the mode of condensing the steam may be applied to the engine in common use, but as the adaptation of it must depend upon the construction and position of the engine,

5 it is impossible for me to point out the mode of adapting it under all circumstances. It is only necessary to state that here must be two separate pumps, one for the water of condensation and the other for the air, as stated in describing my method of condensation. My improvements on boilers are likewise applicable to the engines now in common use.

In witness whereof, I, the said Marc Isambard Brunel, have hereunto 10 set my hand and seal, this Twenty-sixth day of December, in the year of our Lord One thousand eight hundred and twenty-two.

M^c I^d　(l.s.)　BRUNEL.

AND BE IT REMEMBERED, that on the Twenty-sixth day of December, in the year of our Lord 1822, the aforesaid Marc Isambard Brunel came before 15 our said Lord the King in His Chancery, and acknowledged the Specification aforesaid, and all and every thing therein contained and specified, in form above written. And also the Specification aforesaid was stamped according to the tenor of the Statute made for that purpose.

Inrolled the Twenty-sixth day of December, in the year of our Lord One thousand eight hundred and twenty-two.

LONDON :
Printed by George Edward Eyre and William Spottiswoode.
Printers to the Queen's most Excellent Majesty.

A.D. 1825 . . . . . . Nº 5212.

Gas Engines.

BRUNEL'S SPECIFICATION.

TO ALL TO WHOM THESE PRESENTS SHALL COME, I, Marc Isambard Brunel, of Bridge Street, Blackfriars, in the City of London, Esquire, send greeting.

WHEREAS His present most Excellent Majesty King George the Fourth,
5 by His Letters Patent under the Great Seal of Great Britain, bearing date at Westminster, the Sixteenth day of July, One thousand eight hundred and twenty-five, in the sixth year of His reign, did, for Himself, His heirs and successors, give and grant unto me, the said Marc Isambard Brunel, His especial licence, that I, the said Mark Isambard Brunel, my exōrs, adñiors,
10 and assigns, or such others as I, the said Marc Isambard Brunel, my exōrs, adñiors, or assigns, should at any time agree with, and no others, from time to time and at all times during the term of years therein expressed, should and lawfully might make, use, exercise and vend, within England, Wales, and the Town of Berwick upon Tweed, and also in all His Majesty's Colonies and
15 Plantations abroad, my Invention of "CERTAIN MECHANICAL ARRANGEMENTS FOR OBTAINING POWERS FROM CERTAIN FLUIDS, AND FOR APPLYING THE SAME TO VARIOUS USEFUL PURPOSES;" in which said Letters Patent is contained a proviso, obliging me, the said Marc Isambard Brunel, by an instrument in writing under my hand and seal, particularly to describe and ascertain the nature of my said
20 Invention, and in what manner the same is to be performed, and to cause the same to be inrolled in His Majesty's High Court of Chancery within six calendar months next and immediately after the date of the said in part recited Letters Patent, as in and by the same, reference being thereunto had, will more fully and at large appear.

NOW KNOW YE, that in compliance with the said proviso, I, the said Marc Isambard Brunel, do hereby declare that the nature of my said Invention, and the manner in which the same is to be performed, are particularly described and ascertained in and by the Drawings hereunto annexed, and the following description thereof (that is to say) :—
5

I do declare that my Invention of Certain Mechanical Arrangements for obtaining Powers from certain Fluids, and for applying the same to various Mechanical Purposes, is described in the following Specification. The fluids or liquids from which the powers in question are to be obtained are those that result from the condensation of certain gases, which, at the ordinary 10 pressure and temperature of the atmosphere, always remain in the gaseous or aeriform state. There are several gases which come under this description, some of which are treated upon by Mr. Faraday in the papers read before the Royal Society of London in One thousand eight hundred and twenty-three. I give the preference to carbonic acid gas. This gas at the tem- 15 perature of freezing water requires a pressure of about thirty atmospheres to condense and retain it in a liquid state. It may be obtained by decomposing any carbonate by the action of any of the common acids. The mode of obtaining the liquid from the gas is by forming the gas under a gasometer, and condensing it afterwards by means of a condensing pump into a vessel, and 20 continuing the operation until it passes to the liquid state.

Having explained the way of obtaining the gas, and the liquid resulting from it, I shall proceed now to describe the various parts of mechanical arrangements intended for obtaining the power, and applying it to various mechanical purposes.
25

These mechanical arrangements consist chiefly of two cylindric vessels, denominated receivers, into which the liquid has been previously formed by the action of the pump; one receiver is shewn at A, A, and the other A¹, A¹, Fig. 5 and 6. The Fig. 1, A, A, is an enlarged section, and Fig. 2 a plan of the same. The same letters designate the same things in the different Figures. 30 The communication from the pump to the receiver is through the orifice O, Fig. 1, which can be stopped at pleasure by the plug or stop-cock P, Fig. 1 and 2. When the receiver has been charged with the liquid and closed, a pipe D is applied to it at K, Fig. 1 and 2, and connected at X with a vessel denominated the expansion vessel, B, B, Fig. 3 and 4. L, L, Fig. 1 and 3, is a 35 lining of wood or other non-conductor of heat, to prevent the absorption which would otherwise be occasioned by the thick substance of the metal. This expansion vessel is connected through a pipe G, Figs. 3, 5, and 6, to a working cylinder represented at H, Figs. 5 and 6. Figs. 5 and 6 represent two receivers,

---

A, A, and A¹, A¹, connected with the working cylinder H through the intermediate expansion vessels B, B, B¹, B¹. The Figs. 3 and 4, B, B, represent an expansion vessel on an enlarged scale. These vessels contain oil or any other suitable fluid, shewn at b, b, Fig. 3, as a medium between the gas and the
5 piston Q, Fig. 5. Before explaining the action of these parts I shall describe how the power is obtained in the receiver. The receiver consists, as represented in Fig. 1, of a strong metal cylinder, in the interior of which there are one or more thin metal vessels; tubes are most convenient, as represented at T, T, T, Figs. 1 and 2. The joints of these tubes through the top and bottom
10 of the condenser are made perfectly tight; this I do by packings rather than by soldering. The use of these tubes is to apply alternately heat and cold to the liquid contained in the receiver without altering very sensibly the temperature of the cylinder. The operation of heating and cooling through the thin tubes T, T, T, Fig. 1 and 2, may be effected with warm water, steam, or any
15 other heating medium, and cold water or any other cooling medium. For this purpose the tubes T, T, T, Fig. 5, are united at the top and bottom by a chamber and cock C, C, C, C. By the opening of these cocks with the pipes E, E, hot and cold water, which I propose here to use, may be alternately let in and forced through by means of pumps. The cocks may be worked as
20 steam cocks are worked in some steam engines. Now, if hot water, say 120°, is let in through the tubes of the condenser A, A, Figs. 5 and 6, and cold water at the same time through the condenser A¹, A¹, the liquid in the first receiver will operate with a force of about 90 atmospheres, while the liquid in the receiver A¹, A¹, will only exert a force of 40 or 50 atmospheres. The
25 difference between these two pressures will therefore be the acting power, which, through the medium of the oil, will operate upon the piston Q in the cylinder H, Fig. 5. It is easy to comprehend that by letting hot water through the receiver A¹, A¹, and cold water at the same time through the opposite one A, A, a re-action will take place, which will produce in the work-
30 ing cylinder H an alternate movement of the piston Q, applicable by the rod I to various mechanical purposes, as required. It is to be observed that the use of the gasometer and of the forcing pumps is simply for obtaining the gas and for charging the receiver with the liquid. When the receiver is once charged and has been closed with the stop-cock P, Fig. 1 and 2, the gasometer and
35 forcing pumps are to be disconnected from the receiver by unscrewing the pipe D at the screw joint K, Fig. 1 and 2. The same pipe may, however, be used as the means of connecting the receiver with the expansion vessel, or another suitable pipe may be substituted to it at the orifice O, Fig. 1, and the other extremity adapted to the expansion vessel at X, Figs. 3, 4, 5, and 6. I have

---

intentionally avoided adapting two distinct pipes to the receiver, viz., one for the pumps, and another for connecting the receiver with the expansion vessel, because if there were two orifices two stop-cocks would be necessary, which should be avoided. It is obvious that there are no difficulties for connecting the forcing pump with both receivers, as the small pipes used for that purpose 5 may be made to reach either in the mechanical arrangements for obtaining powers from the liquids when formed.

I claim as my Invention, the application of the receiver as it is before described, the essential part of which is the internal application of the vessel or tubes destined to convey the heat and cold. Their form and the manner they 10 are applied render them capable of resisting the intense external pressure of the liquid, and yet admit of their being made so thin as to allow of the rapid and complete transmission of heat and cold through them to and from the liquid.

And I further claim the arrangements by which the receivers, acting in 15 opposition to each other, produce, without any aid of intermediate valves, an alternative action, which may be applied to various mechanical purposes.

In witness whereof, I, the said Marc Isambard Brunel, have hereunto set my hand and seal, this 14th day of January, in the year of our Lord One thousand eight hundred and twenty-six. 20

Mᶜ Iᴅ. (L.S.) BRUNEL.

AND BE IT REMEMBERED, that on the Fourteenth day of January, in the year of our Lord 1826, the aforesaid Marc Isambard Brunel came before our said Lord the King in His Chancery, and acknowledged the Specification aforesaid, and all and every thing therein contained and specified, in form 25 above written. And also the Specification aforesaid was stamped according to the tenor of the Statute made for that purpose.

Inrolled the Sixteenth day of January, in the year of our Lord One thousand eight hundred and twenty-six.

LONDON :
Printed by GEORGE EDWARD EYRE and WILLIAM SPOTTISWOODE,
Printers to the Queen's most Excellent Majesty

The Works at the Head of the Tunnel now in progress under the River Thames

This watercolour, painted in 1826, provides an important and unique view of the Rotherhithe shaft and its surroundings at a very early stage in the construction of the Thames Tunnel. The artist is standing near to what is now Grice's Granary, facing eastwards, with the river to his left.

The Rotherhithe shaft can be seen at the centre of the painting. Most contemporary paintings and sketches show the shaft open to the skies, surmounted by Maudslay's 'A'-frame steam engine. It is likely however that the octagonal wooden superstructure had been erected, possibly at Isambard Kingdom Brunel's instigation, shortly after the beginning of tunnelling in November 1825, with the intention of protecting both the shaft and the steam engine from the harsh winter elements.

In the foreground is a smaller structure — a brick well with a square wooden roof. Two workmen can be seen lowering one of their colleagues into the well. The well, 7ft 9in in diameter, had been dug early in 1826 in order to reach the top of the tunnelling shield, which had run into a layer of gravel and sand after only 12ft of tunnelling had been completed, causing an inundation of water. The construction of the well is documented in Henry Law's *Memoir of the Thames Tunnel*, published in 1857.

On the left of the picture can be seen a full-size section of the tunnel built in brick, in bas-relief, onto the side of a house which faced onto Rotherhithe Street on the eastern edge of the site. Painted on the wall of the house is a representation of the shield. Marc Brunel would use this mock-up to demonstrate to visitors the massive scale of the enterprise and to explain the workings of the shield.

George Yates was a topographical artist who worked in London between 1825 and 1837. He is best known for his extensive series of views of both the old and new London Bridges. He also painted a number of views illustrating the rapid development of the Thames riverside including the new Surrey Docks.

Watercolour by George Yates, signed and dated 1826. Southwark Culture & Heritage Services

The Banquet in the Thames Tunnel, 1827

After successfully tunnelling for over 550ft under the River Thames, Marc Brunel suffered his first major setback on the evening of 18 May 1827 when the river suddenly irrupted into the tunnel. This disaster caused digging to be suspended for several months while repairs were carried out.

On 10 November, to celebrate the full resumption of work under the Thames and the fact that the halfway point had been reached, Isambard Kingdom Brunel, now installed as the Resident Engineer, presided over a banquet in the tunnel itself — the first underwater banquet in the world. Marc, who organised the event, was prevented from attending by illness.

The western archway was draped in crimson. A long table covered with white damask and lit by four decorative candelabra supplied by the Portable Gas Company was elaborately set with silver and crystal. Among the 50 specially invited guests were Captain Stevens (Equerry to HRH Duke of Gloucester), Benjamin Hawes (Marc's son-in-law), Captain Codrington and Mr Bandinel of the Foreign Office. In the background, the uniformed band of the Coldstream Guards — Richard Beamish's old regiment — played an air from *Der Freischütz*, the National Anthem, *Rule Britannia* and *See the Conquering Hero Comes*. The guests enthusiastically toasted among others the King, the Duke of Clarence and the Duke of Wellington. The loudest cheers were reserved for a toast to Admiral Sir Edward Codrington, a long-time supporter of Marc Brunel, whose victory over the Turkish fleet at Navarino had just been announced in that night's *Gazette Extraordinary*.

When the applause had subsided, the 120 miners and bricklayers, who were enjoying a more modest feast in the eastern archway, asked for a toast to be offered to their tools. Their chairman then presented Isambard with a pickaxe and a shovel as a token of their esteem.

The atmosphere and drama of this extraordinary event are well captured in Jones' painting, though he adds one curious detail. In the foreground, Isambard can be seen conferring with his father, who did not in fact attend the banquet. Marc was more than happy, however, for Isambard to enjoy the limelight and accept the plaudits on his behalf.

The artist George Jones (1786-1869) was a successful painter of military and historical subjects, including several popular paintings of Nelson's sea battles and Wellington's campaigns during the Peninsular War and Waterloo. Several copies of this painting exist, probably commissioned as souvenirs by guests at the banquet.

Oil painting by George Jones. Ironbridge Gorge Museum Trust

The Brunel Watercolour of 1835

Marc Brunel painted this watercolour in 1835 at a time when his fortunes appeared to be taking an upward turn. After years of indecision and delay, the improved shield had been installed and work had resumed on the tunnel following a new government loan.

The jacketed figure in the rowing boat bears closer examination. The hat is very tall and the profile and posture are unmistakable. Marc has painted his son Isambard. The rower is almost certainly William Hawes, whose brother, Benjamin Hawes, married Isambard's sister Sophia. Isambard and William were great friends who spent much of their spare time rowing and walking together. Their favourite jaunt was to catch the 4 o'clock morning stagecoach to Kingston, then row with the tide down to William's house in Lambeth.

The figure walking away from the viewer in the left-hand arch of the tunnel is thought to be Marc himself. Sixty-six years old, he is no longer a young man. Another tunnel beckons Marc and beckons the man, not the engineer.

On the reverse of the watercolour Marc has illustrated the geological strata through which the miners had to force their way, together with a sketch comparing the massive proportions of the Thames Tunnel with that of Trevithick's tiny driftway of 1808.

Whilst a number of drawings by Marc Brunel from the 1820s are known, those of the 1830s are rare, and this example of a cross-sectional view of the Thames Tunnel, by his hand and of this date, is probably unique.

 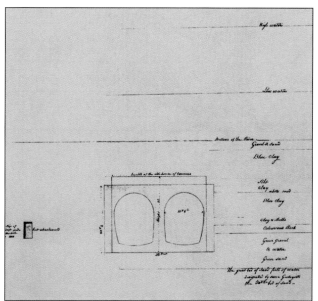

Lithographs of this watercolour are available from Brunel Museum, Rotherhithe, London.

Marc Brunel (1769-1849)

Comparative Chronology

1769 **Marc Brunel born at Hacqueville, Normandy on 25 April.**

James Watt patents his Improved Steam Engine.

Napoleon Bonaparte born.

Richard Arkwright erects first spinning mill.

Arthur Wesley (Wellesley after 1798), 1st Duke of Wellington, born in Dublin.

1770 The composer, Ludwig van Beethoven born in Bonn, Germany according to official records.

William Wordsworth, poet, born in Cockermouth, England.

Navy's entire stock of blocks at Portsmouth destroyed by fire, Walter Taylor II awarded contract to replace.

Nelson joins Navy as midshipman.

Elizabeth Fry, English Quaker prison reformer, born in Norwich.

1771 James Cook discovers New South Wales.

Spinning Jenny patented by Hargreaves.

Gainsborough: *Blue Boy*.

Walter Scott born in Scotland.

Henry Maudslay born in Woolwich, London.

1773 Boston Tea Party.

Battersea Bridge completed and opened to the public.

1774 Joseph Priestley discovers oxygen.

Karl Scheele discovers chlorine.

1775 Beginning of American War of Independence.

Soho Engineering Works founded in Birmingham by Watt and Boulton in partnership.

The author Jane Austen born.

Walter Taylor II's widow patents his blockmaking machines and methods.

J. M. W. Turner, English landscape painter, born in London.

1776 Jenner discovers the principle of vaccination.

Marc Brunel's mother dies.

American Declaration of Independence.

Gibbon: *Decline & Fall of the Roman Empire*.

1778 France joins American war.

François Voltaire, French author, dies in Paris.

La Scala built in Milan.

Marc Brunel joins the College of Gisors.

Jean-Jacques Rousseau, Swiss-French philosopher dies.

1779 Abraham Darby constructs the world's first cast-iron bridge over the River Severn near Coalbrookdale in Shropshire.

Samuel Crompton invents spinning mule.

1780 Gordon Riots in London.

Marc Brunel transferred to the Seminary of St Niçaise in Rouen.

First Derby held at Epsom.

1781 George Stephenson, railway engineer, born in Wylam near Newcastle.

British surrender at Yorktown.

1783 American Independence signed.

First manned flight in a hot-air balloon.

'Pitt the Younger' Prime Minister of England.

Lancelot (Capability) Brown, English landscape designer, dies.

1784 Joseph Bramah exhibits his 'Unpickable Lock' in his Holborn shop window.

Death of Dr Samuel Johnson.

1785 Edmund Cartwright invents the power loom.

1786 **Marc Brunel joins the French Navy as an officer cadet and sails to the French West Indies.**

Davy Crockett born, Tennessee, 17 August.

Mozart: *Marriage of Figaro*.

1787 Voyage of HMS *Bounty* begins on 1 December.

1788 First British convicts sent to Botany Bay.

'Improved' horse-drawn threshing machine patented.

Lord Byron born in London.

1789 French Revolution: Bastille stormed on 14 July.

George Washington becomes first President of USA.

28 April, mutiny on the *Bounty*.

1790 Benjamin Franklin dies.

Marc Brunel designs a coffee-husking machine in Guadaloupe.

John Howard, prison reformer, dies of typhus in the Crimea.

1791 Louis XVI and Marie-Antoinette arrested at Varennes.

The composer Wolfgang Amadeus Mozart dies in poverty.

Charles Babbage, pioneer of modern computers, born in Teignmouth, Devon.

Dr Joseph Ignace Guillotin's 'humane machine' for decapitation first used in France.

Thomas Paine: *The Rights of Man*.

Death of Abraham Darby of 'Ironbridge' fame.

1792　Republic declared in France.

　　　Denmark becomes first country to prohibit slave trade.

　　　Death of Sir Richard Arkwright, inventor of mechanical spinning.

　　　Sophie Kingdom leaves Portsmouth for France.

　　　Robespierre at height of his powers, calls for 'eternal vigilance' against opposition.

　　　Gioacchino Antonio Rossini, Italian composer, born in Pesaro.

1793　Louis XVI executed on guillotine.

　　　Marc Brunel meets Sophie Kingdom for the first time.

　　　John Clare, English poet, born in Helpston, Northamptonshire.

　　　Nelson's first meeting with Emma Hamilton in Naples.

　　　Marc Brunel flees to USA on the *Liberty*, 7 July, and arrives in New York in September.

1794　Robespierre executed.

　　　Thomas Mudge, clockmaker, dies in Plymouth.

　　　Napoleon enters Brussels.

1795　First hydraulic press invented by Joseph Bramah.

　　　Sir Charles Barry, English architect, born in London.

　　　Thomas Carlyle born in Ecclefechan, Scotland.

　　　Death of Josiah Wedgwood, English potter.

1796　Jenner produces smallpox vaccine.

　　　Marc Brunel takes American citizenship and becomes Chief Engineer of New York.

1797　Franz Schubert born in Vienna.

　　　Henry Maudslay sets up business on his own account in Wells Street.

　　　Nelson loses right arm in attack on Santa Cruz.

1798　Lithography invented.

　　　Coal-gas lighting patented in England.

　　　Opening of the New York 'Park' or 'Bowery' Theater.

　　　Nelson's victory at Aboukir Bay; French fleet destroyed.

1799　Income Tax introduced in Britain.

　　　Marc Brunel files first patent No 2305 for his 'Polygraph', 11 April.

　　　Birth of John Stringfellow, inventor of a steam-powered aeroplane, 50 years before the Wright Brothers.

　　　Marc Brunel and Sophie Kingdom married on 1 November at St Andrew's, Holborn.

　　　Napoleon becomes First Consul of France.

1800　Volta makes first battery.

1801　First census in Britain.

　　　Fox taken into partnership with Walter Taylor III of Southampton.

　　　William Pitt 'the Younger' resigns.

　　　Murder of Czar Paul of Russia, accession of Alexander I.

　　　Marc Brunel files patent No 2478 for 'Ships' Blocks' on 10 February.

　　　The Act of Union creates the United Kingdom of Great Britain and Ireland.

1802　First steamship launched: *Charlotte Dundas*.

　　　Victor Marie Hugo born in France.

　　　First railway engine with high-pressure boiler built at Coalbrookdale for Richard Trevithick.

　　　Peace of Amiens (France and Britain).

　　　Marc Brunel files patent No 2663 for 'Trimmings and Borders for Muslins.'

1803　Beethoven dedicates *Eroica* symphony to Napoleon.

　　　Walter Taylor III dies in Southampton.

　　　Robert Stephenson, structural engineer, born in Willington Quay.

1804　Pope crowns Napoleon Emperor.

　　　Beethoven deletes *Eroica* dedication to Napoleon in disgust when he becomes Emperor.

　　　Trevithick's first steam rail locomotive runs in Wales.

　　　Bryan Donkin invents first automatic paper-making machine.

　　　Benjamin Disraeli born in London.

1805　**Marc Brunel files patent No 2844 for 'Saws and Machinery'.**

　　　Nelson wins battle of Trafalgar and dies 21 October.

　　　Napoleon crowned king of Italy.

　　　Hans Christian Andersen, Danish author, born in Odense.

1806　Isambard Kingdom Brunel born 9th April.

　　　Napoleon occupies Berlin and Warsaw.

　　　Marc Brunel files patent No 2968 for 'Cutting Veneers'.

　　　William Pitt dies in penury.

1807　Slave trade abolished in British Empire.

　　　George Stubbs, English painter and etcher, dies.

1808　Abdication of Charles IV of Spain.

　　　Peninsular War begins.

　　　Bounty mutineers discovered on Pitcairn Island.

　　　Marc Brunel files patent No 3116 for 'Circular Saws'.

1809　British defeat at Corunna, death of Sir John Moore.

　　　Charles Darwin born in Shrewsbury, 12 February.

　　　Alfred Lord Tennyson born in England.

　　　Napoleon annexes Papal States and enters Vienna.

　　　Louis Braille born in France.

1810　**Marc Brunel files patent No 3369 for 'Shoes and Boots'.**

　　　Napoleon marries Archduchess Marie Louise.

　　　Wellington in command of British forces in Peninsular War.

1811　**Marc Brunel employed by the government to design a sawmill at Chatham.**

　　　George III mental breakdown, Prince Regent installed.

　　　Luddite riots against machines.

　　　First steam-powered cylinder printing press demonstrated in London by Friedrich Koenig.

1812 Retreat from Moscow. Destruction of Napoleon's 'Grand Army'.

Marc Brunel's first experiments with steam navigation.

British Prime Minister Spencer Perceval assassinated.

Augustus Welby Northmore Pugin, English architect, born in London.

Charles Dickens, novelist, born Portsmouth, 7 February.

1813 Jane Austen: *Pride and Prejudice.*

French defeated by Wellington at Vitoria.

Beethoven: *Fidelio.*

Marc Brunel files patent No 3643 for 'Sawmills'.

1814 Abdication of Napoleon, sent to Elba.

Marc Brunel files patent No 3791 for 'Rendering Leather Durable'.

Louis XVIII restored as king of France.

Stephenson's first effective steam locomotive.

Marc Brunel elected Fellow of the Royal Society.

Joseph Bramah dies in London, aged 66.

1815 Napoleon escapes from Elba and forms new army.

Battle of Waterloo.

Davy lamp invented.

Corn Law introduced in Britain.

Treaty of Vienna; France pays 150 million francs compensation to Austria.

Anthony Trollope, English novelist, born in London.

1816 Stethoscope invented in France.

Rossini: *The Barber of Seville.*

Charlotte Brontë born in Yorkshire.

Death of Charles Stanhope, 3rd Earl Stanhope. English scientist & inventor of the 'Stanhope Press'.

1817 Jane Austen dies.

William Cobbett flees to USA

1818 **Marc Brunel files patent No 4204 for 'Tunnelling Shield'.**

Marc Brunel files patent No 4301 for 'Manufacture of Tinfoil'.

1819 Peterloo massacre in Manchester.

Sir Joseph William Bazalgette, English engineer, born in Enfield, Middlesex.

George Eliot (Mary Ann Evans) born in Arbury, Warwickshire.

Prince Albert, later Prince Consort to Queen Victoria, born near Coburg.

1820 **Marc Brunel files patent No 4522 for 'Copying Presses'.**

Marc Brunel files patent No 4434 for 'Stereotype Printing Plates'.

Marc Brunel's eldest daughter Sophia, married to Benjamin Hawes.

USA purchases Florida.

Schubert: *Trout* quintet.

Death of George III.

Sir Walter Scott: *Ivanhoe.* Florence Nightingale, English nurse, born.

1821 First Macadamised roads appear.

Marc Brunel imprisoned for debt in the King's Bench prison.

First street lighting in Britain (Pall Mall).

Palace Theater, the Bowery, New York destroyed by fire.

Faraday discovers electro-magnetic rotation.

Death of Napoleon at St Helena, 5 May.

John Constable: *The Hay Wain.*

1822 Famine in Ireland.

Sir Charles Napier navigates the first iron steamer from London to Le Havre.

Marc Brunel files patent No 4683 for 'Marine Steam Engines'.

1823 Beethoven: *Ninth Symphony.*

Telford's Caledonian canal completed.

Faraday liquefies chlorine.

Marc Brunel proposes tunnel under the Thames at a public meeting.

1824 Charles Dickens's father imprisoned for debt in the Marshalsea prison.

Lord Byron dies of marsh fever at Missolonghi.

1825 16th February Marc Brunel begins work on the first tunnel under the Thames.

First railway opened: Stockton to Darlington.

Accession of Czar Nicholas I of Russia.

Marc Brunel files patent No 5212 for 'Gas Engines'.

Faraday discovers benzene.

Trade Unions legalised in England.

1826 First crossing of Atlantic under steam by Dutch ship *Curacao.*

Menai suspension bridge opened.

1827 Death of Beethoven in Vienna.

Wellington becomes Prime Minister.

First irruption of the River Thames during the building of the Thames Tunnel.

1828 Webster: *American Dictionary of the English Language.*

Schubert dies in Vienna, 19 November.

Count Leo Nikolayevich Tolstoy born in Russia.

Tunnel sealed and work abandoned after second irruption.

1829 Metropolitan Police Force established.

'Braille' invented in France by Louis Braille.

Sewing machine invented in France.

Stephenson's *Rocket.*

1830 George IV dies 26 June.

William IV crowned king of England.

Joseph Smith founds Mormon Church.

Berlioz: *Symphonie Fantastique.*

Isambard Kingdom Brunel submits plans for the Clifton Bridge over the Avon Gorge, Bristol.

1831 Faraday discovers electromagnetic induction.

Major cholera epidemic in England.

Darwin begins voyage on *Beagle*.

Henry Maudslay dies 15 February.

New London Bridge opened.

1832 Reform Act.

Sir Walter Scott dies.

Marc Brunel elected Vice-President of the Royal Society.

1833 First British Factory Act: child labour under age nine banned.

William Wilberforce, anti-slave-trade activist, dies.

Alfred Bernhard Nobel, Swedish chemist and inventor of dynamite, born in Stockholm.

Kensal Green cemetery opened.

Richard Trevithick, civil engineer, dies in poverty.

1834 Tolpuddle Martyrs transported to Australia.

Fox Talbot: First photographic negative.

Chopin: *Preludes*.

William Morris, English craftsman, poet and leading Socialist, born in London.

Death of Thomas Telford, civil engineer.

Houses of Parliament burned down.

1835 The word 'socialism' introduced into the language.

Work resumed on Thames Tunnel.

Death of William Cobbett, author of *Rural Rides* and champion of the poor.

Mark Twain born in Florida, Missouri.

1836 Isambard Kingdom Brunel marries Mary Horsley.

Administration of all lighthouses in England and Wales transferred to Trinity House by Act of Parliament.

Chartist movement founded.

Battle of the Alamo.

1837 Queen Victoria crowned.

Third and fourth irruptions in the Thames Tunnel.

Emma, Marc's younger daughter, marries Reverend Frank Harrison.

Charles Dickens: *Oliver Twist*

First message sent by telegraph.

Death of John Constable, English landscape painter.

1838 I. K. Brunel: *Great Western*.

Sunday Times founded.

Georges Bizet, French composer, born near Paris.

Death of Charles Maurice de Talleyrand-Périgord, close associate of Napoleon.

Morse Code invented.

1839 First Opium War in China.

Chartist riots in Birmingham and Newport.

First Daguerreotype photograph.

Faraday: Theory of electromagnetism.

1840 Penny post instituted.

Queen Victoria marries Prince Albert.

First bicycle produced in Scotland.

Composer Peter Ilyich Tchaikovsky born in Russia.

Thomas Hardy born in Dorset.

1841 Hong Kong acquired by Britain.

Marc Brunel knighted by Queen Victoria.

Thames Tunnel opened to pedestrian traffic.

Punch founded.

1842 Great potato famine in Ireland begins.

Sir Marc Brunel suffers first stroke, 7 November.

Marshalsea prison closed down.

1843 Thames Tunnel officially opened, 25 March.

1844 First Co-operative Society formed in Rochdale

Alexander Dumas: *The Count of Monte Cristo*.

Alexandra, later Queen Consort of Edward VII, born in Copenhagen.

1845 I. K. Brunel builds the *Great Britain*, the first screw-driven iron ship.

Sir Marc Brunel suffers second stroke.

Elizabeth Fry, English Quaker prison reformer, dies.

Thomas John Barnardo, founder of Dr Barnardo's Homes, born in Dublin.

1846 Planet Neptune discovered.

Corn Laws repealed.

Great Britain runs aground in Ireland (Dundrum Bay).

1847 Factory Act: 10-hour day instituted.

Charlotte Brontë: *Jane Eyre*.

Emily Bronte: *Wuthering Heights*.

Thackeray's *Vanity Fair*.

British Museum opened.

1848 Gold discovered in California.

Karl Marx and Friedrich Engels publish Communist Manifesto.

End of the Chartist Movement.

French Republic proclaimed.

1849 **Death of Sir Marc Brunel, 12 December; interred in Kensal Green cemetery.**

Bibliography

Primary sources

Manuscript sources

Bristol University Archives

Sir Marc Isambard Brunel Diaries, 1822, 1823.
Thames Tunnel Minutes of Occurences, 1825-1827
Letters from Marc Isambard Brunel to Isambard Kingdom Brunel, 1834-1839
Isambard Kingdom Brunel, Personal Diary 1827-1829
Isambard Kingdom Brunel, Private Diaries 1824-1826

Institution of Civil Engineers Archives

Sir Marc Isambard Brunel, Journals 1824-1843 (with the exception of 1842 which is missing).

Printed sources

Newspapers and periodicals

The Illustrated London News, 5 August 1843
The Times, 2 August 1842; 27 March 1843; 13 December 1849
Gentleman's Magazine, (84) (2) (1814), pp285-6
Gentleman's Magazine, (19) (1) (1843), p420
Punch, (5) (1843), p61
Punch, (18) (1850), p110

Secondary sources

Books

Andress, David, *The Terror: Civil war in the French Revolution* (Little, Brown, 2005)

Atterbury, P. and Wainwright, C. (eds), *Pugin: A Gothic Passion* (V & A Museum, 1994)

Bagust, Harold, *London Through the Ages* (Thornhill Press, 1982)

Beamish, Richard, *Memoir of the Life of Sir Marc Brunel* (Longmans, 1862)

Besant, Walter, *The History of London* (Longmans, 1894)

Brunel, Isambard, *The Life of Isambard Kingdom Brunel* (Longmans, 1870)

Buchanan, Angus, *Brunel : The Life and Times of Isambard Kingdom Brunel* (Hambledon & London, 2002)

Burrows, E. G. and Wallace, M., *Gotham: A History of New York City to 1898* (Oxford University Press, 1998)

Clapham, J. H., *The Early Railway Age 1820-1850* (Cambridge, 1930)

Clements, Paul, *Marc Isambard Brunel* (Longmans, 1970)

Cléry, Jean-Baptiste, *A Journal of The Terror* (Folio Society, 1955)

Clowes, W. B., *Family Business 1803-1953* (William Clowes & Sons, 1969)

Clunn, Harold P., *The Face of London* (Phoenix House, 1951)

Coad, Jonathan, *The Portsmouth Blockmills* (English Heritage, 2005)

Connelly, O. (ed), *Historical Dictionary of Napoleonic France, 1799-1815* (London, 1985)

Curl, James Stevens (ed), *Kensal Green Cemetery* (Phillimore, 2001)

Dixon, W. H., *The London Prisons* (London, 1850)

Dugan, Sally, *Men of Iron: Brunel, Stephenson and the Inventions that Shaped the Modern World* (Macmillan, 2003)

Ellmers C. and Werner A., *Dockland Life: A Pictorial History of London's Docks 1860-1970* (Mainstream Publishing, 1991)

Evans, John, *An Excursion to Windsor in July 1810* (Sherwood, 1817)

Flint, Stamford Raffles (ed), *Mudge Memoirs* (Netherton & Worth, 1883)

Ford, Boris (ed), *Cambridge Cultural History of England* (CUP, 1995)

Gaunt, William, *Chelsea* (Batsford, 1954)

Godfrey, J. L., *Revolutionary Justice; a study of the organisation, personnel and procedures of the Paris Revolutionary Tribunal* (Chapel Hill, 1951)

Greer, D., *The Incidence of The Terror in the French Revolution, a statistical interpretation* (Harvard University Press, 1935)

Grieg, T. Watson, *Ladies' Old-Fashioned Shoes* (1885) and *Supplement* (1889)

Hadfield, C., *World Canals: Inland Navigation Past and Present* (David & Charles, 1986)

Halliday, Stephen, *Making the Metropolis: Creators of Victoria's London* (Breedon Books, 2003)

Hay, Peter, *Brunel: His Achievements in the Transport Revolution* (Osprey, 1973)

Hodge, James, *Richard Trevithick, 1771-1833* (Shire Publications, 1973)

Hogg, O. F. G., *The Royal Arsenal* (London University Press, 1963)

Hood, Thomas, *The Complete Works of Thomas Hood* (Ward, Lock & Co, undated)

Horne, R. S., *The Blockmills in HM Dockyard Portsmouth* (Portsmouth Royal Dockyard Historical Society, 1968)

Inwood, Stephen, *A History of London* (Macmillan, 1998)

Jones, Colin, *The Longman Companion to the French Revolution* (Longmans, 1988)

Jones, S. K., *Brunel in South Wales* (Tempus, 2005)

Knight, Charles, *London — Pictorially Illustrated* (Charles Knight, 1842)

Lampe, David, *The Tunnel* (Harrap, 1963)

MacDougall, Philip, PhD., *The Saw Mills of Chatham Dockyard* (Chatham Dockyard Historical Society, 2002)

Mathewson, A. and Laval, D., *Brunel's Tunnel and Where it Led* (Brunel Exhibition, Rotherhithe, 1992)

Mayhew, Henry, *London Labour & the London Poor* (Griffin, Bohn & Co, 1860)

Moran, James, *Printing Presses: History and Development from the Fifteenth Century to Modern Times* (Faber & Faber, 1973)

Morison, Stanley, *Printing 'The Times' 1785-1935,* (Printing House Square, 1953)

Noble, Celia Brunel, *The Brunels — Father and Son* (Cobden-Sanderson, 1938)

Palmer, P. S., *History of Lake Champlain* (Albany, 1866)

Palmer, R. R., *Twelve Who Ruled: The Year of the Terror in the French Revolution* (Princeton University Press, 1941)

Pannell, J. P. M., *The Taylors of Southampton* (Institution of Mechanical Engineers, 1955)

Pannell, J. P. M., *Old Southampton Shores* (David & Charles, 1967)

Pennant, Thomas, *Some Account of London* (Faulder, 1813)

Phillips, Sir Richard, *A Morning's Walk from London to Kew* (J. & C. Adlard, 1820)

Pudney, John, *Brunel & His World* (Thames & Hudson, 1974)

Ramsey, Sherwood, *Historic Battersea* (1913)

Rolt, L. T. C., *Isambard Kingdom Brunel* (Longmans, 1957)

Simmonds, H. S., *All About Battersea* (1882)

Sims, George R., (ed), *Living London,* (Cassell & Co, 1903)

Skelton, *Southampton Guide* (1818)

Smiles, Samuel, *Industrial Biography* (John Murray, 1882)

Steel, David, *Elements of Rigging and Seamanship* (1794)

Stokes, I. N. P., *Iconography of Manhattan Island,* (Robert H. Dodd, 1915-28)

Swann, J., *Shoes* (Batsford, 1982)

Sylvester, N. B., *History of Saratoga County, New York,* (1878)

Thomas, John Meurig, *Michael Faraday and the Royal Institution: The Genius of Man and Place* (Adam Hilger, 1991)

Tomlinson, H. M., *London River* (Cassell, 1921)

Tour du Pin, Madame de la, *Escape from the Terror* (written in 1820 but not published until 1906 in France)

Trease, Geoffrey, *Samuel Pepys and his World* (Thames & Hudson, 1972)

Trevelyan, G. M., *English Social History* (Longmans, 1944)

Vaughan, Adrian, *Isambard Kingdom Brunel* (John Murray, 1991)

Vincent, W. T., *The Records of Woolwich District* (1890)

Wade, R. C., *American Urban History* (Oxford University Press, 1981)

Walford, Edward, *Old & New London* (Cassell, Petter Galpin, *c*1890)

Walford, Edward, *Village London* (Cassell, Petter Galpin, *c*1883).

Warner, Jessica, *John The Painter* (2004)

Withey, H., and Withey, E., *Biographical Dictionary of American Architects* (New Age Publishing, 1956)

Wolmar, Christian, *The Subterranean Railway* (Atlantic Books, 2004)

Chapters in edited volume

J. Saunders, 'The Thames Tunnel', in Charles Knight (ed), *London Pictorially Illustrated,* (Charles Knight, 1842)

Journal articles

H. W. Dickinson, 'The Taylors of Southampton: Their Ships' Blocks, Circular Saw, and Ships' Pumps', *Transactions of the Newcomen Society* (29) (1953-5), pp169-178

Joshua Field, Report on the Thames Tunnel, *Institution of Civil Engineers Minutes of Proceedings* (1) (1837-41), p41

Cynthia Gladwyn, 'The Isambard Brunels', *Proceedings of the Institution of Civil Engineers* (50) (1971), pp1-14

Useful Addresses

Bristol University Library
Tyndall Avenue
Bristol
BS8 1TL
Tel: 0117 928 9000
www.bristol.ac.uk

Brunel Museum
Railway Avenue
Rotherhithe
London
SE16 4LF
Tel: 020 7231 3840
www.brunelenginehouse.org.uk

Brunel University
Uxbridge
Middx
UB8 3PH
www.brunel.ac.uk

Chatham Dockyard Historical Society
The Museum
Chatham Historical Dockyard
Chatham
Kent
ME4 4TE
www.cdhs.org.uk

The Guildhall Library
Aldermanbury
London
EC2P 2FJ
Tel: 020 7332 1869
www.cityoflondon.gov.uk

Institution of Civil Engineers
1 Great George Street
London
SW1P 3AA
Tel: 020 7665 2043

The Maudslay Society
c/o Dr J. M. Maciejowski
Pembroke College
Cambridge
CB2 1RF

The National Maritime Museum
Greenwich
London
SE10 9NF
Tel: 020 8858 4422
www.nmm.ac.uk

The National Portrait Gallery
St Martin's Place
London
WC2H 0HE
Tel: 020 7306 0055
www.npg.org.uk

The Naval Dockyards Society
c/o Dr Ann Coates
44 Lindley Avenue
Southsea
PO4 9NU
Tel: 023 9286 3799
www.hants.gov.uk/navaldockyard

The Navy Records Society
c/o R.H.A. Broadhurst
Pangbourne College
Pangbourne
Berks
RG8 8LA

Portsmouth Naval Base Property Trust
19 College Road,
HM Naval Base
Portsmouth
PO1 3LJ
Tel: 023 9282 0921

Portsmouth Royal Dockyard Historical
Trust (Support Group)
c/o R. W. Russell OBE
48 Highfield Avenue
Waterlooville
Hants
PO7 7PX

The Pugin Society
Hon Sec Pam Cole
33 Montcalm House
Westferry Road
London
E14 3SD
Tel: 020 7515 9474
e-mail: opamakapam@aol.com

Royal Institute of British Architects
66 Portland Place
London
W1B 1AD
Tel: 020 7580 5533

The Royal Institution of Great Britain
21 Albemarle Street
London
W1S 4BS
Tel: 020 7409 2992

The Science Museum
Exhibition Road
London
SW7 2DD
Tel: 020 7942 4401
www.nmsi.ac.uk

Index

A

Aberdeen, Earl of, 80
Above Bar Church, Southampton, 108
Abutments, 52
l'Académie de l'Industrie de Paris, 81
l'Académie Royale des Sciences, Belles Lettres, et Arts de Rouen, 81
Administered, Administration, 121
Admiralty, 23, 26, 28, 32, 35, 37, 40, 47, 51, 57, 110
Admiralty, Lords Commissioners of the, 25, 32, 37, 102
Admission fee, 96, 111
Aitken, James (John the Painter), 32
Albanian chief, 92
Albany, 17
Alexander I, Czar of Russia, 52, 56
Althorp, 59, 86
American citizenship, Certificate of, 19
American(s), 91, 100, 113
Amiens, 59
Ancien Regime, 98
Approach roads, 65, 99
Aqueducts, 61
Arbuthnot, Mr, 57
Arch(es), 80, 87
Armstrong, William, 67, 68, 71
Atlantic, 62
Avery Architectural & Fine Arts Society, 10
Avon Gorge, Bristol, 84
Aylesbury, 59

B

Babbage, Charles, 30, 106
Babington, 89
Bacon, Mr, 50
Bacquancourt, de, 23
Bailing machine, 110
Baldwin, Mrs, 75
Ball, 79, 80
Bandinel, Mr, 57
Bank of England, 95
Bankrupt, 38, 55, 117
Bargate, Southampton, 108
Barge House, Lambeth, 54
Bargees, Barges, 39, 55, 104
Barlow, Professor Peter, 83
Barlow, Sir Robert, 49
Baron de Waggenheim, 95
Barrel-making, 36
Barry, Sir Charles, 61
Bastille, 21, 107
Bath, 105
Battersea, 9, 36, 37, 38, 46, 51, 53, 55
Battersea bridge, 36, 39
Battersea sawmill, 37, 43, 44, 51, 58, 60, 61
Beamish, Richard, 10, 15, 22, 31, 35, 39, 51, 52, 56, 67, 70, 71, 72, 73, 75, 77, 78, 79, 80, 83, 85, 87, 88
Becket, Thomas, 62
Belt-centring, 109
Benedictine, 21
Benefit (of the Rules), 116
Bentham, Jeremy, 10
Bentham, Sir Samuel, 24, 28, 30, 31, 32, 34, 35, 47, 48, 109, 110, 112
Bentinck, Captain, 26
Bermondsey, 59
Bertram, 76
Birmingham, 59
Blake, William, 108
Blenheim, 59
Blockmaking machinery, 20, 24, 28, 31, 34, 35, 36, 37, 38, 44, 50, 66, 99, 108, 110
Blocks, hand-made, 37
Blue plaque, 100
Board of Ordnance, 25
Bodmin, 63
Boilers, 43, 72
Bolton, Lancashire, 113
Boot factory, 44, 52, 55, 67
Boots (and Shoes), 44, 45, 51
Bordeaux, 14,
Bore, The Great, 80
Boring machine(s), 109, 110
Borings, 65, 66, 82
Boron, 59
Borthwick, James, 37, 38, 39
Borthwick, Patrick, 38
Boulton & Watt, 25, 32
Boulton, Matthew, 12
Bourbon, 84
Bowery Theater, 10, 18
Box tunnel, 97
Boyd, Mr W., 38

Bramah, Joseph, 23, 38, 65, 66, 113
Bramah Lock, 113
Bramah, Timothy, 65, 67
Breach (of the Rules), 121
Breguet, Abraham Louis, 58
Bribery, 110, 112, 116
Bricklayers, 69, 70, 73, 76, 82
Brickwork, 49, 69, 71, 72, 84, 89, 90, 104
Bridge Street, Blackfriars, 66
Bridges, 38, 52, 55, 56, 58, 60, 61, 62, 81, 84, 85, 97
Brighton, 68, 86
Bristol, 98, 105
British Embassy, 34
British Guyana, 59
British Museum, 86
Brooshooft, Mr, 115, 116, 117, 118, 119, 120, 121
Brunel Archive Collection, 76
Brunel University, 86
Brunel, Charles Ange, 15
Brunel, Emma, 36, 68, 74, 88
Brunel, Harriet, 52
Brunel, Isambard Kingdom, 9, 10, 36, 38, 50, 52, 54, 56, 58, 60, 66, 67, 68, 71, 72, 73, 75, 76, 77, 78, 79, 80, 83, 84, 85, 86, 87, 88, 97, 98, 100, 113
Brunel, Jean, 11
Brunel, Jean Charles, 13, 14
Brunel, Sophia (daughter), 10, 28, 54, 56, 74, 75
Brunel, Sophie (wife), 10, 24, 30, 32, 36, 52, 53, 56, 58, 68, 74, 75, 88, 97, 98, 99
Buchanan, Professor Angus, 9
Buckingham Palace, 95
Buckingham, Duchess of, 21
Burnet, Mr., 9, 53, 115, 118
Burr, James, 31, 34, 111

C

Caen, 54
Calais, 21, 52, 65
Calff. Claes Cornelisz Jonge, 109
Cambridge, Duke and Duchess of, 67, 80
Canadian company, 83
Canal(s), 48, 49, 50, 52, 59, 60, 61, 82
Candles, 59, 60, 67
Captain Zaitsch, 95
Carlyle, Jane and Thomas, 38
Carpentier, François, 14, 15, 20, 21
Carriage(s), 48, 61
Cast iron, 78, 113
Castlereagh, Lord Robert, 44, 45
Catherine the Great, 28
Cavity, 87, 89, 104
Celebration Dinner, 78
Cement, 65, 67, 76, 84
Centering, 81
Chalk, 48, 49, 52, 62, 99
Channel tunnel, 65, 99
Chapman, William, 62
Chateau Gaillard, 12
Chatham, 35, 40, 46, 48, 50, 51, 52, 65, 67, 97
Chatham Dockyard, 47, 48, 50, 65, 97
Chelsea, 9, 36, 37, 38, 66
Chester, 60
Chief Engineer, 19, 71, 84
Chimney, 49, 97, 103
Chipping(s), 42
Chronometer, 15
Chummage, 114, 115, 118
Chum-master, 114, 115, 119, 120
Circular saw(s), 37, 38, 39, 67, 109
City (of London), 39, 46
City of London Tavern, 65
Civil War, 19
Clacton, 51
Clark, Tierney, 83
Clay, 44, 58, 62, 65, 67, 68, 70, 72, 74, 75, 77, 78, 79, 88, 89, 104
Clement, Joseph, 24, 113
Clements, Paul, 9
Clerks, 114, 115, 117
Clifton, 84, 97
Cloak, 22
Coach(es), 105, 106, 111
Coaks, 26, 33, 111
Coal(s), 67, 117
Codrington, Admiral Sir Edward, 56, 57
Coffer dams, 65
Coldstream Guards, 76, 78
Colechurch, Peter of, 62
Collins, 79, 80
Colthurst, Joseph, 87

Committee(s), 107, 114, 115, 116, 117, 118, 119, 120
Commons, House of, 88
Commune, 14, 21, 107
Compiegne, 21
Conservatoire des Arts et des Métiers, 51
Constantinople, 92
Contraband, 55
Contract labour, 71
Contract work, 82, 83
Convicts, 40, 43, 50
Copper, 110, 111
Cork Jail Bridge, 81, 85
Cork, Ireland, 81, 85, 105
Cornwall, Cornish, 52, 60, 62, 63, 64
Corruption, 28, 34, 44, 56, 111, 112, 121
Corunna, 44
Cotton balls, 87
Council's Rate Books, 53
Countershaft, 109
Courbevoie, 15
Cow Court, Rotherhithe, 66, 67
Coxson, Mr & Mrs, 105
Crane, 48, 50
Crawford, Andrew, 87
Creditors, 53, 57, 116, 119, 121
Crew, R.H., 42
Crichton, Sir Alexander, 92
Crimea, 54
Criminal class(es), 55, 116
Crossing-sweeper, 111
Crown & Anchor inn, 85
Cugnot, Nicolas, 12
Curl, Professor James Stevens, 9
Cylinder Press, 54
Czar, 52, 57
Czar Alexander I, 35

D

Dampier, Mr Justice, 120
Day-rules, 116
Deal, 52
de Lesseps, 59
Dee, 60
Defoe, Daniel, 62
Delabigarre, Monsieur, 55
Deptford, 25, 35, 36, 37, 40, 56, 108, 110
Diana, Princess of Wales, 20
Dickinson, H.W., 24, 110
Dictionary of National Biography, 9
Discharge(s), 114, 121
Diving Bell, 75, 77, 104
Diving Bell crane, 60
Dock Committee, 59
Docks, 34, 42, 55, 111
Dockyard Apprentices Museum, 32
Document(s), 106
Dodd, Ralph, 62
Donkin, Bryan, 34, 65, 66, 67, 71, 72
Dover, 52
Dowling, 78
Draft Riots, 10, 19
Dragoons, 111
Drain(s), 71, 75, 81, 83, 87, 110
Drifts, Driftway, 62, 64, 81, 82
Drunkenness, 40, 114
Dry dock(s), 111
du Mât river, 58
Dublin, 85
Dugan, Sally, 9
Duke Street, 98
Dulague, Professor Vincent François Jean Noel, 14
Dundas, Robert, 80
Dundonald, Lord Thomas, 44
Dunkirk, 21
Dunsterville, William, 31
Dupont, Gainsborough, 109
Durham, Lord, 83
Dutch Fleet, 110
Dutch Government, 58
Dyce, William, 38

E

Eagle Steamers, 51
East India Company, 37, 108
East India Wharf, 58
East London Railway Company, 99
Edgworth, Mr, 46
Edinburgh, 37, 38, 86, 87
Edinburgh Encyclopaedia, 65
Edinburgh Exhibition of Arts, 87
Egypt, 92, 106

Electric trains, 99
Elements of Rigging and Seamanship, 108
Elezabeth Jonas, 40
Ellacombe, Mr, 50
Embezzlement, 42
Emoluments, 114, 116, 121
Emperor Claudius, 62
Engine House, 72, 75, 76, 100
Essex, 40
Evans, Rev.William, 114
Excise laws, 117
Explosions, 89
Explosive gas, 89
Extortion, 120

F

Fair(s), 99
Falmouth, 22
Faraday, Michael, 30, 89
Farringdon Vegetable Market, 38
Farthing, J. H., 36, 37, 39, 45, 51, 53, 55, 58
Fenton, Murray & Wood, 39
Ferdinand, Duke, 95
Ferme Brunel, 11, 12
Ferries, Ferrymen, 38, 39
Field, Joshua, 35, 46, 67, 113
Fine Arts Commission, 54
Finland, 38
Fir ties, 84, 85
Fire, 62, 89
Firedamp, 89
Flares, 64
Fleet Prison, 38, 55, 114
Fleet river, 38
Flying buttresses, 58
Fog, 104
Folkestone, 65
Fontanka canal, 28
Footbridge, 59, 103
Footway Bridge, 61, 103
Foundation plate, 43
Frames, 69, 70, 71, 72, 73, 75, 76, 77, 78, 79, 80, 88, 89, 90, 109
France, 50, 52, 54, 55, 56, 60, 86, 92, 107, 111
Francis & White, 67, 84
Fraud, Fraudulent, 116, 119
French, 52, 55, 61
French Revolution, 14, 21, 61, 107
French accent, 97
French prisoners-of-war, 40, 50

G

Gainsborough, Thomas, 109
Gambling, 119
Gantry, 48
Gaols, Gaoler(s), 121
Gas, 80,87, 89, 91, 96
Gay-Lussac, Professor Joseph-Louis, 59
Gentleman's Magazine, 39, 46
Gibbet(s), 111
Giddy, Davies, 62, 64
Gilbert, Davies, 63
Gisors, 12
Gladwyn, Lady, 22
Glasgow, 86
Glass cylinder(s), 110
Gloucester, Duke of, 67, 79
Goodrich, Simon, 31, 35, 110
Goodwin, 75
Gordon Riots, 55
Gordon, Lewis, 87, 88
Government Corps of Engineers, 55
Graham, George, 15
Grand Duke Michael of Russia, 95
Grand Junction Canal Company, 81
Grand Opening of the Tunnel, 91
Grand Surrey Canal Company, 59
Gravatt, Colonel, 72
Gravatt, Mr, 72, 73, 75, 76, 78, 83, 104
Gravel, 67, 70, 71, 73, 75, 76, 77, 81, 88, 89, 104
Gravelines, 20, 21
Gravesend, 58, 62
Gravesend, 58, 62
Great Britain, The (ship), 98
Great Harry, 40
Great Lakes, 17
Great Western Railway (GWR), 50, 86, 97
Griffith, John, 61,
Grosvenor, 60
Guadeloupe, 14
Guildhall Library, 114
Guillotine, 20, 21, 107
Gun-tackle blocks, 25
Guy's Hospital, 89

H

Hacqueville, 9, 85
Hamilton, Alexander, 20
Hammersmith, 59
Hampstead, 59
Hand pumps, 72
Hand-filing, 113
Hanwell, 86
Harris, Mr, 78
Harrison, Rev Frank, 88
Harrison, Thomas, 60, 61
Hawer, William, 69
Hawes, Benjamin, 54, 57, 75, 92
Hawkins, R. F., 65
Hazel rods, 104
Heating, 59
Hegley, John, 100
Hellyar, P., 49
Henri Quatre Lycée, 54
Heritage Trust, National, 100
Highwaymen, 111
Highway Rate Books (Wandsworth), 9
HMS *Victory*, 35
Holl, Edward, 49
Hollingsworth, 58, 61
Holy Roman Empire, 22
Hood, Thomas, 80
Hooke, Robert, 109
Hoops, Hoop iron, 84, 85
Horse mill, 109
Horse-drawn vehicles, 111
Horsemongerlane, 120
Horsley, Mary, 88
Horsley, William, 88
Hotel de Ville, 107
House of Lords, 83
Hoy, J. A., 96
Huddersfield Canal Company, 59
Hudson River, 17
Hudson's Bay, 62
Huggins, 73
Hulk(s), 40, 43
Humour, sense of, 97
Hutton, Dr Charles, 64, 82
Hydraulics, 113

I

Île de Bourbon, 58
Illogan, 63
Illustrated London News, 91
Income(s), 96, 114, 116
Industrial revolution, 31, 113
Infantry, Corps of, 110
Infirmary, 117,118
Inflammation of the eyes, 89
Iniquity, 111
Inspector-General of Naval Works, 110
Institute of France, 59
Institution of Civil Engineers, 89
Institution of Mechanical Engineers, 69, 110
Invalid Department, 45
Iodine, 59
Ireland, 72, 85, 107
Irish, 73, 107
Iron & Steel Institute, London, 110
Iron door, 81
Iron piles, 87
Iron plates, 72
Iron rods, 77, 104
Iron ties, 85, 84
Irruption(s), 77, 78, 81, 87, 88, 89, 90
Irving & Brown coal wharf, 89
Isle of Wight, 111
Itchen (river), 108

J

Jacks, 67, 69
Jaundice, 118
Jerusalem, 108
Jessop, William, 40, 64, 82
Johnson, Dr Samuel, 15
Joliffe & Banks, 65
Jones, George, 100
Jones, William Esq., 114, 116, 119, 120, 121

K

Kemble, 73
Kendal, Henry Edward, 61
Kensal Green Cemetery, 9, 61, 99
Killarney, Ireland, 85, 105
King Edgar, 40
King Edward VII, 40
King George II, 42, 55
King Henry VIII, 40
King's Bench Prison, 9, 55, 56, 57, 58, 84, 114, 115, 116, 117
King's Shropshire Light Infantry, 44
Kingdom, Elizabeth, 15, 58
Kingdom, Sophie, 9, 15, 20, 21, 22
Kingdom, William, 15, 85, 109, 110
Kingston, 38, 59
Knight, Charles, 70, 71, 73, 78, 96
Knighted, 90
Knitting machine, 52
Koenig, 53, 54

L

La Croix, 55
Laboratory, The, 40
Lake Champlain, 17, 18
Lake Oneida, 17
Lake Ontario, 17
Land's End, 60
Lane, 73
Lasts, 44
Lathe(s), 110, 113
Laundy, Margaret, 113
Le Havre, 14, 15, 16
Leather, 44
Leather stock, 113
Leeches, 70
Lefèbvre, Abbé, 11
Legislature, 121
Leith, 37, 38, 39
Levant, 16
Lever escapement, 15
Lignum vitae, 24, 26, 34, 35, 111
Limehouse, 62, 81
Lincoln, 97
Lincoln, Abraham, 19
Lincolnshire, 61
Lindsey House, 38
Lindsey Row, Chelsea, 38
Links, 64
Liverpool, 58, 59, 61, 84, 103
Lock(s), 61, 113
London, 38, 51, 55, 56, 58, 60, 61, 62, 65, 86, 105, 111
London – Pictorially Illustrated, 70, 71, 96
London Directory, 37
London Underground System, 65, 99
Longdon, 88
Longitudinal sleepers, 86
Longuemar, Monsieur and Madame, 15
Lords, House of, 88
Louis XVI, 14, 20, 107
Louis XVIII, 52
Lower boxes, 88
Lunatic asylum, 89

M

Macadam(ise), 106
Macfarlane, Charles, 98
Machine-tool industry, 67, 113
Magical cures, 39
Mahogany, 37, 52
Maidenhead, 85, 86
Malaria, 40
Mangin, Joseph, 18
Manning, Anne, 38
Mansfield, Lord, 117, 120
Maréchal de Castries, 14
Margate, 51
Marine steam engines, 113
Marseilles, 14
Marshal (of prison), 56, 114, 115, 116, 117, 119, 120, 121
Marshalsea Prison, 55, 104
Martin, John, 38, 78
Master of the Crown Office, 120
Masterman, J., 80
Mathews, Thomas, 58
Maudslay, Henry, 23, 24, 28, 31, 32, 34, 35, 36, 37, 39, 44, 51, 66, 67, 68,80, 102, 110, 113
Maudslay, Sons & Field, 35
Maudslay, Thomas Henry, 113
Maudslay, William, 113
Mawdsley Hall, Ormskirk, 113
Mayerne, Sir Theodore, 38
Mayflower, 100
Mayo, 76
Mayor, 110
McDougall, Dr Philip, 48
Meda, 107
Medway canal, 67
Medway channel, 51
Medway, River, 48
Melville, Lord, 102, 118
Memoir of the Life of Sir Marc Isambard Brunel, 77
Merchant Marine, Board of Revision for the, 37
Merchant Navy (Fleet), 36
Merchant vessels, 37, 42
Messer, 25, 108
Metalwork, 113
Metric system, 107
Metropolitan Railway Company, 99
Micrometer, 23, 113
Middle boxes, 88
Mildmay, Sir H., 110
Miles, 73
Miller's Pond, 108
Milton Ironworks, 58
Miners, 70, 71, 73, 74, 76
'Misfortune, the', 56
Mobile crane, 49
Mohawk River, 17
Mole, The, 62
Moncrieff, 37

Monge, Gaspard, 16
Montague, Mr, 65
Montmorency, Duc de, 81
Moore, Sir John, 44
Moreau de St Mery, 17
Morris, Mr, 115, 116, 118, 120, 121
Mortise, 109
Mortising machine, 24
Mudge, Mary Sophie, 15
Mudge, Thomas, 9, 10, 15, 58
Mudge, Thomas Jnr, 10, 15, 58
Mudge, Zachariah, 15
Mudge. Mr, 9, 53, 61
Munday, 73
Murdoch, William, 12, 89
Murray & Wood, 32, 36
Museum, 100

N

Napoleon, 26, 31, 34, 38, 59
Napoleonic wars, 28, 40, 46
Nash, John, 61
Nasmyth, James, 24, 35, 113
National Guard, 14, 15
National interest, 57
Naval Dockyard(s), 32, 42, 111
Navier's Seine Bridge, 56
Navy Board, 24, 25, 31, 32, 33, 34, 35, 35, 37, 39, 42, 43, 44, 48, 50, 51, 108, 111, 112
Nazareth, 21
Nelson, Lord Horatio, 35
Nervous fever, 35
Netherlands, 109
Netley Abbey, 24
Neva, 52, 65
New York, 10, 16, 17, 18, 19, 20, 22
New York City Municipal Archives, 10
New York Historical Society, 10
New York State Historical Association, 10
Newcastle, 86
Newcomen Society, 24
Newcomen, Thomas, 12, 64
Newgate Prison, 38, 55
Newport, Isle of Wight, 24
Newton, John, 108
Newton's Chapel, 108
Nightingale, Florence, 55
Noble, Lady Celia, 10
Nonconformist, 108
Normandy beaches, 35
North British Insurance Company, 38
Northampton Mercury, 44
Northcote, James (Jemmy), 98
Northfleet, 86
Northumberland, Duke of, 67
Norway, 38
Numbering machine, 113

O

Oak shuttering, 69
Oak(s), 34, 37, 38, 47
Oakum, 71, 75
Ode to Mr Brunel, 80
Oil lamps, 64
Ojibbeway Indians, 95
Old Drury, 18
Old Greenshield, 78
Olney Hymns, 108
Opening Ceremony, 91
Open sores, 89
Ordnance, 40, 67
Ordnance, Board of, 39, 42, 44, 59
Outgoings, 114
Overcoats, 42
Overcrowding, 119
Owen, Sir Edward, 80
Oxford Canal Company, 81

P

Paddington, 86
Paddle-steamer, 51
Paddle-tugs, 58
Padstow, 60
Page, Thomas, 87, 88, 89
Painter, John the (James Aitken), 32
Palace of Westminster, 55
Palace Theater, 18
Palmer, Mr, 80
Pamphilon, 77
Panama Canal, 59
Pannell, J.P.M., 24, 25, 110
Pantograph lathe, 44
Pardoe, Miss, 92
Paris, 15, 21, 52, 61, 107
Paris Association, 84
Park Row, 18
Park Street, Westminster, 99
Park Theater, 10, 18
Parkin, George, 49
Parliament, Houses of, 61, 85, 88, 110
Partrich, Nicholas, 40
Passport, 16
Peace Treaty, 18
Pedestrians, 65, 99
Peel, Robert, 67

Peninsular War, 44
Pension(s), 43
Pepys, Samuel, 30
Perfect circle freehand, 97
Perkins, Walter, 108
Petition, 116
Philadelphia, 17
Phillips, Sir Richard, 52
Pick-proof lock, 113
Piecework, 71, 82, 83
Piles, Piling, 89
Pilgrim Fathers, 100
Pin(s), 109
Pinchbeck, 66
Pinckney, 77
Pinhole(s), 109
Piston suction pump, 110
Planing machine, 23, 110
Planks, 48, 52
Plumstead Marshes, 40
Plymouth, 35, 110
Pocket Copying Press, 55
Poisonous gases, 89
Poling boards, 69, 71, 89, 90
Poling screws, 69, 70, 71, 75, 89
Polings, 74, 75, 89
Polygraph, 22
Ponsonby, General, 67
Pontoon(s), 97
Poor Clares, 21
Poor Rate Books (Wandsworth), 9
Portable gas, 78
Porter (beverage), 114, 121
Portsea, 31, 32, 36
Portsmouth, 26, 31, 32, 34, 35, 36, 37, 38, 39, 40, 44, 50, 61, 108, 111, 112
Portsmouth dockyard, 28, 31, 36, 108, 110, 111, 112
Portswood, 108
Portswood Green Volunteers, 110
Portuguese warship, 35
Post Office, 106
Postal services, 61
Poultry, The, Cheapside, 58, 66
Poverty, 114, 117, 118
Price's Candle factory, 60
Prince Albert, 90, 92, 98
Prince Leopold, 67, 95
Prince Lieven, 52, 67
Prince of Orange, 45
Prince of Wales, 40
Princess Dock, 61
Printing Industry, 53
Prison Governor, 56
Prison ships, 40
Prison(s), 43, 55, 58, 114, 115, 117
Prisoner(s), 40, 50, 56, 115, 116, 118
Pritchard & Hoof, 67, 83
Private soldier(s), 110
Privately owned yards, 42
Profit(s), 114, 116, 117, 120, 121
Propelling jacks, 69
Protestantism, 22
Prussian Government, 46
Prussian Regiment, 46
Public house, 119
Public roads, 106
Public tolls, 65
Pugin, Auguste Charles, 61
Pugin, Augustus Welby Northmore, 9, 61
Pulley blocks, ship's, 25, 50, 75
Pumps, 63, 71, 72, 73, 75, 76, 77, 78, 82, 88, 104, 109, 110, 111
Punch (magazine), 92, 99
Puritan, 22
Pyramids, 106

Q

Quadrant, 14
Queen Victoria, 90, 92, 95, 96, 98
Quicksand, 64

R

Rackets, 119
Raffles, Lady, 75
Raft, 77, 104
Railroads, 61
Railway, 48, 85, 105
Railway truck(s), 48
Ramsgate, 105
Rattlesnake Island, 17
Reciprocating saws, 109
Redbridge, Southampton, 110
Redman, 67, 69
Reformation, 107
Regent (vessel), 51, 102
Regulations, 115, 116, 117, 119, 121
Rennie, John, 40, 47, 62, 87
Rent(s), 114, 115, 119
Resident engineer, 67, 71, 80, 87
Réunion, 58
Revolution, 14, 16, 17, 21
Revolution, Council of the, 20
Reynolds, Joshua, 15
Rhine, 58

Richardson, Samuel, 78
Riley, Mr, 72, 73
River Charente, 92
River Lee, 85
River Medway, 48
River Nile, 86
Riverbed, 66, 88, 99
Robbery, Robbing, 111
Roberts, Richard, 24
Robespierre, 15, 17, 20, 21, 107
Rochefoucauld, duc de la, 51
Rochester, 50
Rogers, 75, 76
Rolt, R. T. C., 9
Roman Catholic(ism), 22
Ropemakers, 42
Rosewood, 52
Rotary cylinder press, 53
Rotherhithe, 61, 62, 64, 65, 67, 76, 78, 84, 89,
   90, 92, 99, 100, 104
Rotherhithe Church, 65
Rouen, 14, 15, 20, 21, 25, 55, 59, 81
Rousseau, Jean Jacques, 21, 107
Royal Academy, 61
Royal Academy of Sciences of Stockholm, 81
Royal Albert bridge, 97
Royal Arsenal, 42, 43, 45, 113
Royal Carriage Department, 40, 42
Royal Engineers, 72
Royal Family, 85
Royal Gun Factory, 40
Royal Institution of British Architects, 69
Royal James, The, 40
Royal Society, 35, 69, 85
Royal train, 98
Roysen, Samuel, 118
Rubble, 60
Rules (prison), 56, 114, 115, 116, 117, 119,
   120
Rules of Court, 110, 120
'Run on the keys', 116
Russia, 27, 34, 38, 47, 57, 112
Russian Ambassador, 52
Russian Navy, 28
Russian Treasury, 56

S
Sailing ships, 111
Saint Clare, Order of, 21
Saint David Street, 87
Saint Denys Priory, 108
Saint George's Pier, 61
Saint James's Park, 99
Saint Katharine Docks, 62
Saint Lawrence waterway, 17
Saint Mary Colechurch, 62
Saint Mary's Island, 50
Saint Niçaise Seminary, 13
Saint Ouen, 59
Saint Petersburg, 28, 35, 52, 56, 97
Saint Suzanne bridge, 58
Saint Vincent, Earl, 28
Salary, salaries, 111, 114, 115, 121
Saltash, 97
Sand, 70, 73, 84
Sans culottes, 14, 105
Sansom, 53, 58,
Saunders, J., 70, 73, 78
Saw frames, 43, 48
Sawing floor, 48
Sawing hall, 48
Sawing machines, Machinery, 109
Sawmill(s), 36, 39, 42, 49, 50, 53, 55, 58, 59,
   61
Sawpits, 47, 48
Sawyers, 48, 50
Scantlings, 47, 48
Scarlet runners, 92
Scavengers, 56, 120
Schenectady, 17
Science Museum, London, 32, 110
Scotland, 38, 78
Scotsman, The (newspaper), 78, 87, 88, 95
Scott, Sir Walter, 35
Screw-cutting lathe, 23
Screw-jacks, 67, 69, 72
Screw-propulsion (ships'), 23, 98
Screw-threads, 113
Segment(s), 67
Seppings, Sir Robert, 97
Sergeant-wheelwright, 113
Serpentine, 58
Seven-foot track gauge, 97
Sextant, 22, 104
Sganzin, Monsieur, 58
Shaft(s), 43, 65, 67, 68, 70, 71, 73, 76, 77, 78,
   80, 87, 90, 91, 92, 95, 96, 109
Shaping machine, 110
Shaw, Mr, 55, 58,
Sheave, Sheaf, 109, 111
Sheave-making machine, 24
Sheerness, 35
Sheet Iron, 103
Shield, 65, 66, 67, 68, 69, 70, 71, 73, 74, 76,
   77, 78, 79, 82, 83, 87, 89, 90, 99, 104

Ship & Swan public houses, 89
Ship-breaker's yard, 89
Shippe Inn, 100
Ships' pulley blocks, 23, 109
Ships' pumps, 109
Shivers, 33
Shoemaking, 45, 48
Silt(ing), 40, 47, 70, 71, 72, 75
Silvester, Mr, 59
Simpson, Joseph, 61
Sinclair, Sir John, 80
Sinking tower, 67
Slade, Sir Thomas, 25
Slide-rest lathe, 23
Sliders, 69
Sloane, Hans, 108
Slough, 98
Small Yards, 42
Smart, George, 109
Smith, W. MP, 65, 67, 71, 80, 81, 85
Smithfield Meat Market, 38
Smithsonian Institution, Washington, USA,
   10, 32
Socialism, 107
Société française de Statistique universelle, 81
Société libre d'Emulation de Rouen, 81
Société libre du Commerce et de l'Industrie de
   Rouen, 81
Somerset, Duchess of, 61
Somerset, Lord, 67, 80
Soundings, 65
South America, 35
South Mast Pond, 48
South Stoneham, 108
Southampton, 24, 25, 26, 108, 109, 110, 111
Southampton Water, 108, 111
Southwark, 39, 55, 62
Sovereign of the Seas, The, 40
Spain, Spanish, 60
Spencer, Earl, 20, 28, 30, 35, 39, 56, 57, 59,
   67, 81, 85, 86, 90, 97
Spencer, Lady, 59
Spilsbury, Maria, 108
Spiral staircase, 90
Spithead, 102
Spry, Joan, 15
Stages, 69, 76, 79, 106
Staircase, 76, 80, 91, 92, 95, 119, 120
Stallholders, Stall keepers, 95, 99
Stanhope Press, 53
Stanhope, Lord Charles, 53
Stanley, Hans, 25
Star Chamber, 55
State Education Department of New York
   University at Albany, 10
Staves, 36, 71
Steam, 51, 105, 109
Steam dredger, 47, 63
Steam engine, 43, 46, 48, 51, 52, 61, 63, 67,
   68, 71, 72, 110
Steam locomotives, 61, 99
Steam navigation, 44
Steam propulsion, 61
Steam tugs, 51
Steam vessel(s), 51, 102, 103
Steam wagons, 61
Steening, 75
Stephenson, George, 86
Stephenson, Robert, 86
Stereo plates, 53
Stewart, 67
Stibbert, General Giles, 108
Straights (shoes), 44
Strata, Stratum, 52, 70
Stratford, 97
Straw, 73, 75, 84, 120
Strike, 75
Stroke, 83, 91, 99
Sturge's Bowling Ironworks, 58
Sturgess & Company, 66
Sunderland, 86
Survey, Clerk of the, 43
Suspension bridge(s), 52, 58, 59, 84
Swamps, 40
Swaythling, 108, 110
Swiss Guard, 15
Sydney Harbour, 52
Sykes & Company, 55

T
Taillefer, Louis, 15
Tails, 69, 71
Talleyrand, 17
Tallow, 59, 60
Tamar, 52
Tap, Tap-room, 115
Taps and Dies, 113
Tarpaulin, 77
Tay bridge, 58
Taylor & Martineau, 53, 55
Taylor workshops and factories, 108
Taylor, Captain, 108, 110
Taylor, Elizabeth, 108, 109
Taylor, John, 89, 110
Taylor, Samuel Silver, 28, 30, 31, 108, 110

Taylor, Thomas Ebenezer, 110
Taylor, Walter I, 10, 24, 108
Taylor, Walter II, 10, 24, 25, 36, 108
Taylor, Walter III, 10, 22, 24, 26, 28, 33, 35,
   108, 110
Taylor, William, 65, 110
Taylors of Southampton: Pioneers in
   Mechanical Engineering, 24, 25, 108, 110
Taylors of Southampton: Their Ships' Blocks,
   Circular Saw, and Ships' Pumps, 108, 110
Templer, Mr, 114, 120
Teredo navalis, 65
Terror, the, 20, 21, 107
Thames Archway Company, 62, 64, 82
Thames boatmen, 38
Thames Tunnel, 9, 64, 65, 66, 70, 81, 82, 85,
   86, 87, 88, 89,
Thames Tunnel Company, 61, 65, 66, 67, 72,
   81, 82, 83, 84, 87, 88, 90, 96
Thames Tunnel Minutes of Occurance, 72, 76
Thames, River, 36, 38, 40, 46, 55, 59, 62, 63,
   65, 69, 82, 104, 111
The Times (Newspaper), 24, 53, 54, 76, 80,
   88, 90, 91, 92, 99
Théot, Catherine, 107
Thornton's Act, Mr, 117
Thurman, 17, 18
Tide, 77, 78, 88, 89, 104
Tilbury, 62
Tillett, 76
Timber yards, 43
Tolls, 99
Tompion, Thomas, 15
Tooley Street, 62
Top boxes, 72, 75, 88, 90, 104
Top plate, 71
Topdog, 48
Topsail yard blocks, 25
Totnes, 61, 84
Tower of London, 52, 95
Towing (of ships), 102
Trafalgar (Battle of), 30, 34
Traitor's Gate, 62
Transportation, 40,
Treasurer, 117
Treasury, The, 42, 46, 52, 57, 85, 87, 88
Trevithick, Richard, 12, 61, 63, 64, 81, 82
Trinidad, 58
Tudor House Museum, Southampton, 108,
   109
Tug Captains, 39
Tuileries, 14
Tunnel, 48, 52, 61, 62, 64, 66, 71, 74, 75, 76,
   78, 79, 80, 81, 83, 88, 89, 90, 91, 92, 95,
   96, 97, 98, 99
Tunnel Club, 85
Tunnel pier, 92, 95
Turks, 28
Turner & Montague, 67
Turner, J. M., 38
Turnkeys, 56, 114, 115, 121
Turnpikes, 61, 66, 106
Type (printing), 54

U
Umbrella, 97
Underdog, 48
United States (of America), 91, 101
University of Bristol, 76
University College, London, 110
Unpickable lock (Bramah's), 23
Unskilled labour, 110
Uppage, Colonel, later Major-General, 42, 43
Uuluuich, 40
Uxbridge, 86

V
Vadier, 107
Vanguard, The, 40
Vansittart, Nicholas, 45, 57
Vasie, Roert, 62, 65
Vaughan, Adrian, 9
Vauxhall, 86
Vauxhall Bridge, 86
Veneer(s), 36, 37, 38, 39, 46, 52
Ventilation, 91, 99
Vertical saws, 50
Vertical shaft, 109
Vessel(s), 103, 111
Viaduct, 86
Viceroy of Egypt, 86
Victoria, Queen, 90, 92, 95, 96, 98
Vigo, Bay of, 60
Vinall, Mr, 49
Visitors, 74, 76, 78, 79, 80, 91, 92, 95, 110,
   120, 121
Vistula Bridge at Warsaw, 81

W
Wages, 89, 115
Wales, 61, 106
Walings, 89
Walker, James, 83, 88
Walter, John, 53, 54
Walton-on-Thames, 108, 110

Wandsworth Council, 9
Wapping, 61, 62, 65, 87, 89, 90, 92, 95, 99
Warwick, 59, 97
Washington (city), 18
Washington, George, 20
Watchmen, 114
Water cistern, 43
Water tank, 48
Waterford, Ireland, 85
Watering time, 40
Waterloo, Battle of, 45
Watermen, 62, 66, 92
Waterworks, 52
Watt, James, 12, 64
Wearmouth, 86
Wedge-shaped type, 54
Welby, Catherine, 61
Wellington boots, 44
Wellington, Duke of, 57, 67, 80, 81, 83
Wells Street, London, 113
West Drayton, 86
West India Docks, 62, 75, 104
West Indies, 14, 84, 113
Westgate Street, Southampton, 25, 26, 108
Westminster, 38
Weston Lane, Southampton, 26, 108
Whale oil, 109
Wharf pilings, 89
Wheeled traffic, 99
Whidby, Joseph, 40
Whitworth, Joseph (later Sir), 23, 113
Wide trousers, 42
Williams, Richard, 89
Windlass, 38
Wollaston, Dr, 56, 65, 84
Wollaston, George Hyde, 65
Wood planing machine, 113
Woodmill, 26, 34, 48, 50, 108, 110
Woodward, 77
Wool, 82
Woollen stockings, 52
Woolsacks, 62
Woolwich, 35, 40, 42, 113
Woolwich Arsenal, 39, 40, 44
Woolwich Church, 113
Woolwich Dockyard, 28, 40, 43, 44
'Woolwich Powder Boy', 113
Workface, 64, 71, 72, 73, 75, 78, 79, 82, 88
World War One, 35
World War Two, 35, 51, 100
Wulewich, 40
Wyatt Parker & Company, 67

Y
York, 58, 59
York Hotel, 51
York Street, Westminster, 110